Understanding China

The series will provide you with in-depth information on China's social, cultural and economic aspects. It covers a broad variety of topics, from economics and history to law, philosophy, cultural geography and regional politics, and offers a wealth of materials for researchers, doctoral students, and experienced practitioners.

Xueyong ZHANG

China's Railway Expansion in 40 Years of Economic Reform and Opening Up

Translated by
Senlin Yu
Ruoyu Dai
Jiawen HUANG

Springer

Xueyong ZHANG
Southwest Jiaotong University
Chengdu, Sichuan, China

Translated by
Senlin YU
Southwest Jiaotong University
Chengdu, Sichuan, China

Ruoyu DAI
Southwest Jiaotong University
Chengdu, Sichuan, China

Jiawen HUANG
Southwest Jiaotong University
Chengdu, Sichuan, China

ISSN 2196-3134 ISSN 2196-3142 (electronic)
Understanding China
ISBN 978-981-95-0332-2 ISBN 978-981-95-0333-9 (eBook)
https://doi.org/10.1007/978-981-95-0333-9

Previously published by Social Sciences Academic Press (China)

Translation from the Chinese Simplified language edition: "《扩张的动脉 —— 改革开放40年的中国铁路》" by Xueyong Zhang, © The author 2021. Published by Social Sciences Academic Press (China).

The National Social Science Fund of China18FZS020This work was supported by The National Social Science Fund of China (18FZS020).

Understanding China

This Springer imprint is published by the registered company Springer Nature Singapore Pte Ltd.
The registered company address is: 152 Beach Road, #21-01/04 Gateway East, Singapore 189721, Singapore

Contents

Chapter 1
Introduction: Reform and Opening Up and the New Starting Point of Chinese Railways' Modernization

Focusing on the author's research objectives and the data collected, this book covers significant events in the construction and development of Chinese railways since the reform and opening-up period, rather than providing a chronological account of railway progress in China over the past four decades. The research examines various aspects of Chinese railways, including railway network planning, line construction, equipment manufacturing, operation and management, and even educational and health work, all in the context of reform and opening up, and delves into the milestones and transformations that have shaped the railway system in China.

Undoubtedly, the past 40 years since 1978 have witnessed the fastest progress of Chinese railways, as in many other fields. It is equally undeniable that such remarkable achievements are primarily attributable to the reform and opening up of Chinese railways. Through such a process, Chinese railways have established a new framework, implemented entirely new management and operational modes, and acquired world-class manufacturing technologies, leading to a revitalized system with a brand-new outlook. Therefore, this book centers on the genesis of Chinese railways and its twists and turns along the way.

We wish to convey an important view in this Introduction that the history of the railways over the past 40 years is not disconnected from the past. Among the various factors that have contributed to the current status of the railway system, we can clearly trace the influence from the policies of 1949, and even from the decisions of the Qing government to build railways in 1881. Over the nearly century-long period since 1881, diverse forces converged to forge the unique identity of Chinese railways, and their influence has persisted, albeit in varying intensities and manifestations, well into the post-1978 era. In other words, there are some "fundamental" elements that underpin the entire trajectory of Chinese railway development, necessitating our primary focus.

The Introduction will discuss three fundamental questions regarding the basic understanding of Chinese railway history: What constitutes "railway history"?

X. Zhang, *China's Railway Expansion in 40 Years of Economic Reform and Opening Up*, Understanding China,
https://doi.org/10.1007/978-981-95-0333-9_1

What factors have shaped the history of Chinese railways? And how should the history of China's railways be narrated against the backdrop of reform and opening-up? For the first question, we will present our understanding through a review and evaluation of relevant literature. As for the second, we will analyze several major issues in railway history from specific perspectives and discuss how these issues continued to influence railway development during the reform and opening-up period. Finally, in response to the third question, we will provide the answer by outlining the fundamental structure of this book.

1.1 What Is "Railway History"?

Research on the history of Chinese railways has spanned over a century. As early as 1906, thirty years after railways were first introduced to China, three seminal works were successively published: *Essentials of China's Railways* by Wang Shengchun, *Compendium of China's Railways* by Liu Fu and Yi Zhenqian, and *A Comprehensive Analysis of the Current State of China's Railways* by Zeng Kunhua, marking the beginning of the studies on Chinese railway history. The following year, P.H. Kent, a British lawyer who worked in Tianjin for four decades, published *Railway Enterprise in China: An Account of Its Origin and Development*,[1] which remains a key reference for understanding the early history of Chinese railways. Over the century that followed, numerous works have been published focusing on different aspects and periods of Chinese railways. Surprisingly, however, none of these works discusses the methodology of railway history writing. While the concept of a railway might seem self-evident,[2] its history should not merely be a collective description of individual rail lines. Unfortunately, most railway history works approach the subject in such a manner, presenting railway history as a series of decision-making processes, construction timelines, and operational accounts for various railways. While this may seem sufficient to meet the general needs of readers, as it mirrors the public perception of railways, a considerable body of research goes beyond this, exploring the broader impacts of railways on economics, politics, society, the military, and even social mentality. In this context, railways become a significant yet vague background to social change, with their own development taking a back seat.

The author is of the opinion that railway history should neither be simplified as a history of railway lines' expansion nor generalized into the social history of railways. Rather, the true focus of railway history should be the railway system and its operations, which includes major subsystems such as planning and management,

[1] Kent, P. H. (1958). *Railway enterprise in China: An account of its origin and development* (Li Baohong et al., Trans.). Beijing: SDX Joint Publishing Company.

[2] Strictly speaking, the definition of the railway also requires careful identification and analysis. The definition of railway presented by Zhu Congbing in *Li Hongzhang and the Development of China's Railways* (Qunyan Press, 2006) is the most detailed exploration of this issue that the author has read.

technology research and development, track construction, industrial manufacturing, and dispatching and command, taking into account the unique characteristics of Chinese railways. Such a history also encompasses related subsystems like basic and professional education, railway health, and even public security and justice. As one of the most integrated and complex systems in modern transportation, our understanding of railways—how they operate and interact with other social systems, exhibit operational characteristics and patterns, and have evolved through different historical periods to their current state—remains limited. These questions can be broadly summarized into two categories: the internal operations of the railway system, and the bidirectional interaction between the railway system and the national and social needs.

Each mode of transportation has its own system and relies on its supporting infrastructure for operation, as automobiles depend on highways, ships need docks, and airplanes require airports. However, when it comes to the intricate connections within their respective systems, the railway undoubtedly differs from other modes of transportation. From a technical perspective, the wheel-rail (train and railway) interaction constitutes one of the three most fundamental issues[3] in railway technology. Moreover, the railway's system of dispatching and command is far more complex and specialized than those of other transportation modes. For example, traffic jams on roads are certainly annoying, but a transfer of similar situation on railways would be catastrophic. Though authorities in different regimes strived to maintain centralized and coordinated operations throughout the history of Chinese railways, the question of what consists of the ideal model has always been the subject of ongoing debates, usually on whether railways should prioritize public welfare or enterprise interests, whether operations should be centralized or open to competition, and whether the railway network should be separated from transportation services. Perhaps these debates will never have universally agreed-upon answers, but they drive the railway towards continuous improvement and progress.

From a different perspective, regardless of how obscure the railway system may be in the eyes of outsiders, the public has its own standards for evaluation: whether the railways can meet social demands. Simply put, what the railways contribute to society is their transport capacity. If railways can transport passengers and cargo to their destinations efficiently, the public will not be concerned about their internal operations. Nonetheless, railways in China often fail to meet such demands. During some periods, delays and cargo backlogs have almost become synonymous with railways. Public pressures have continuously driven railway progress, necessitating constant technological advances, management reforms, and service optimizations to meet public expectations.

In 1949, when the People's Republic of China (PRC) was founded, the government took over from its predecessor a railway network spanning nearly 30,000 kilometers (with an operating mileage of 21,810 kilometers), and this inheritance was

[3] The other two are the pantograph-contact line interaction and the fluid-solid interaction (aerodynamics), but only emerged as necessary and important with the development of electric railways and high-speed rail systems.

by no means terrible in the context of constant wars. Over the next 30 years, Chinese railways demonstrated remarkable growth and efficiency: the length of the main lines expanded from 29,119 kilometers in 1957 (with an operating mileage of 26,708 kilometers) to 58,686 kilometers in 1980 (with an operating mileage of 49,940 kilometers), marking an approximately 100% increase (with the operating mileage up by 87%); the number of locomotives grew from 4251 to 10,278; the number of passenger trains rose from 8566 to 16,157; the number of freight trains increased from 90,249 to 266,376; the volume of passenger traffic surged from 312.62 million to 912.46 million; and freight jumped from 274.21 million tons to 1085.84 million tons.[4] The railways managed to play a "pivotal role" even during the tumultuous decade of the "Cultural Revolution" (1966–1976). Among the achievements from 1965 to 1976, China "completed 5,669 kilometers of main lines, 3,577 kilometers of branch lines, and 1,697 kilometers of double track lines. By the end of 1976, the operating mileage reached 46,262 kilometers, including 7,285 kilometers of double-track sections; the Nanjing Yangtze River Bridge was completed, and the electrification of the Baocheng Railway (Baoji-Chengdu) was achieved; the SS1 Electric Main Line Locomotive and the Dongfeng 4 Diesel Locomotive were successfully developed".[5]

Problems arise when we compare the growth of railways with that of the national economy. From 1965 to 1976, national gross industrial and agricultural output value increased by 133.5%, while railway freight only grew by 69.8%.[6] According to relevant statistics, "Our national economy, from the early days after the founding of PRC to the early 1980s, as reflected in railway passenger traffic and freight, saw a 9.9-fold increase in railway freight from 99.83 million tons in 1950 to 1,085.84 million tons in 1980; and a 4.8-fold increase in passenger traffic from 156.81 million in 1950 to 912.46 million in 1980. During the same period, the railway mileage only increased by 1.25 times, the number of locomotives by 1.4 times, passenger trains by 2.2 times, and freight trains by 3.8 times. In the 1980s, compared with data in the early 1950s, the freight railway counts per passenger-kilometer decreased by 50%, and the number of freights per million ton-kilometer decreased by 66.4%".[7] In other words, the core issue of railway reform in China does not lie in the achievement of railway construction but in the extent to which they can meet the national social demands, which is essentially a relationship between transportation capacity and traffic volume, and the evaluation shall be correlative and dynamic.

The focus of this book is on the internal operation of the railway system and its interplay with national and social demands against the historical backdrop of China's reform and opening up. Specifically, it delves into why reforms in China's railway industry were necessary, how these reforms were carried out, what

---

[4] Xu Zenglin. (1999). *Fifty years of the railroads of the People's Republic of China (1949–1999)* (p. 374). Beijing: China Railway Publishing House.

[5] Miao Qiulin. (1994). *China railway* (p. 59). Beijing: China Railway Publishing House.

[6] Ibid.: 60.

[7] Wang Zhizhong, & Wei Liyin. (2001). *Research on China's railway reform and development (1978–1998)* (p. 19). Beijing: Contemporary China Publishing House.

difficulties occurred during the process, and how they were tackled. Furthermore, the book explores the formulation and development of a Chinese railway modernization model that integrates government leadership with market mechanisms.

From a macro to micro perspective, the issues can be categorized into three hierarchical levels:

Firstly, we need to consider the relationship between the railway industry and the society in China, which is exemplified in the relationship between rail transportation technology and social demand for this technology. To understand it within the context of the reform involves clarifying the precise positioning of railway reform within the historical context of the comprehensive reform. Similar to reforms in other industries, railway reform embodies the traits of comprehensive reform, yet also possesses distinct characteristics that set it apart. In certain aspects, it has been pioneering and leading in the comprehensive reform efforts; in others, it may be trailing behind. As history researchers, we need to identify which aspects are leading the way and which are lagging behind. More importantly, we should delve into how such particularities arose and what implications and values they offer for us to correctly understand China's reform, the monumental historical event.

Secondly, we need to consider the relationship between "the Railway Man"[8] and railway reform. As is the case in all industries in China, reform is progressively implemented through the active engagement and initiative of practitioners across various sections, all under the leadership of the Communist Party of China (CPC). Each individual is both a practitioner of reform within their industry and a beneficiary of reforms in others, acting as both an agent and, in a sense, a recipient. Thus, there exists a set of apparent internal contradictions. On the one hand, over the past 40 years, all Chinese people firmly believed that "reform and opening up are the source of vitality for the development of contemporary China, the important magic weapon for our Party and the people to catch up with the pace of the times, and the only way to uphold and develop socialism with Chinese characteristics".[9]

On the other hand, in different periods and under diversified conditions, different industries have varying understandings of reform, which may even be detrimental to the short-term interests of some industries. Therefore, "it is crucial to have new plans and measures. We should have a strong problem-oriented consciousness, focus on major issues, identify crucial points to find answers, and strive to address a series of prominent contradictions and problems in the development of China".[10] The railway industry is no exception. The process of railway reform reflects how the Railway Man understands and perceives railway reform, and how they comprehend railways and reform, thereby influencing or changing the way the public perceive

[8] The "Railway Man" reflects the self-identity of railway practitioners, and to some extent, it reflects the high degree of systematicity of the railway industry. Although there are also expressions like "Oilman", they seem to be less used in transportation industries such as highway and shipping sectors, as a matter of experience.

[9] Xi Jinping. (2012, December 13). Speech on inspection work in Guangdong. *People's Daily*.

[10] Xi Jinping. (2013, November 14). Speech at the symposium with non-CPC personages held by CPC Central Committee. *People's Daily*.

railways. Ultimately, the Railway Man is a typical representative of industries within the overall reform. The successful experience of the past 40 years demonstrates that this is an ideological issue that should be addressed and resolved during the reform.

As President Xi Jinping has pointed out, "The process of emancipating minds is also one of unifying thoughts. Only by doing so can we maximize the consensus on the reform and develop a cohesive force to move the reform forward".[11] In fact, as Chinese society develops and the comprehensive reform proceeds, new situations and problems will surely arise during reform. Ultimately, these issues boil down to the relationship between "emancipating minds" and "unifying thoughts". Deep reflection on the reform process of the railway industry, a typical industry, also allows us to explore the lessons and experiences needed for future reforms.

Thirdly, we need to consider the fundamental reason and method for the railway reform. China's remarkable achievements in reform are fundamentally attributed to the right leadership of CPC. What lessons can the railway industry, as a specific sector, learn from its own reform process? The railway system is gigantic and complex. In terms of the result, the current achievements are comprehensive and holistic, but when we go back to specific historical periods, they vary. As researchers in history, we should clarify the detailed process of railway reform, which means exploring how breakthroughs were made at different stages and in the face of different problems, and what effects these measures have ultimately generated. More specifically, as for the basic sectors of the railway system, which sector played the vital role in driving the reform forward in different periods? Was it technology that led the way, management that inspired innovation, or operations that radically changed the overall picture? Which direction should China's railway reform proceed?

1.2 Various Factors Shaping the Form of China's Railway

When China's railway system embarked on the path of reform and opening up, both insiders and outsiders enumerated various shortcomings of the railway system and primarily attributed them on the planned economic system, the target of reform. Like other fields, this judgment contains some truth. However, many difficulties encountered during the reform indicated that merely blaming the planned economic system was not tenable. This does not mean that the planned economic system is suitable for the railway industry, rather, it suggests that many problems in the railway industry appear to stem from the planned economy but are attributable to other factors. The reality is that some problems need to be traced back even further, to the very inception of China's railway. Identifying these factors will provide a deeper understanding of China's railway during the period of reform and opening up. At least three aspects need to be highlighted: the emotional significance of the railway

[11] Ibid.

to the nation, the perception of the railway's "closedness", and the understanding of the fundamental attributes of China's railway.

Railways are generally viewed as a symbol of the Industrial Revolution worldwide, and so it is in China. However, according to the Chinese government and people, railways have a special significance beyond their general role in economic and social development. They are tied to the sentiment of national self-strengthening. The existence of railways is one thing, but who controls them is another more important matter. Since their inception in Britain in 1825, railways spread rapidly in mainland Europe and other parts of the world. Only in China did the debate over whether to build railways go through three rounds of discussions, with participation from both the government and the public. One of the focal points was the perceived infringement on Chinese sovereignty by foreign-built railways. The nationalist sentiment surrounding sovereignty even influenced the development of railways throughout the late Qing Dynasty, fermenting into the violent Railway Protection Movement in 1911, which contributed to the fall of the Qing Dynasty. On the other hand, railway construction can also become a symbol of national pride: Zhan Tian you gained great fame for his role in the construction of the Beijing-Zhangjiakou Railway,cwith many overlooking the difficulties and setbacks he faced as a young student sent by the Qing Government to study in the United States. The historical echoes of national sentiment were so profound that even in the 1980s, there were advocates who argued that the Beijing-Zhangjiakou Railway should be considered the first railway in China.[12] Similarly in 1949, while the PRC government was stabilizing its control, the first immediately went to the construction of the Chengdu-Chongqing Railway, largely to showcase the strength of the new regime and its focus on the people's interests. The successful operation of the railway in 1952 indeed produced the intended effect, greatly enhancing public trust and support for the new government.[13]

Excavating and highlighting the significance of railways used to be the main storyline of railway history. The renowned work of Mi Rucheng, a great historian of railways, is entitled *Imperialism and China's Railway*, and his *Materials on the History of Modern Railways in China* and *Materials on the History of Railways in the Republic of China*, which have benefited a lot of researchers, are also based on the right of way as the main narrative line in the collection and editing of materials. As the focus of modern Chinese historical research has shifted, attention to railway rights has gradually waned. It is normal for academics to follow the trends of its time, but the focus of earlier research persists in the study of the subject. This sentiment can even be seen in the high-speed railway technology innovation route of the 1990s. At that time, it was a consensus within the Ministry of Railways (hereinafter

[12] Fang Ju. (2006). *History of Chinese railway* (pp. 17–24). Beijing: Beijing Jiaotong University Publishing House. This discussion also reflects different views on whether the Tangxu Railway or the Wusong Railway is China's first railway, similarly showing the influence of national sentiment on railways.

[13] Tian Yongxiu. (2016). Chengdu-Chongqing railway opened to traffic and public recognition. *Journal of Southwest Jiaotong University (Social Science)*, (6), 7–13.

referred to as MOR) to build high-speed railways, but opinions diverged on how to acquire high-speed rail technology, especially high-speed railway technology. Initially, the decision was to self-develop the technology, out of concern of the containment on technology from its origin country. In 2004, when the decision shifted to upgrading technology through "introduction, digestion, absorption, and re-innovation", widespread debate was sparked, embodying the influence of early railway history on later decisions.

1.2.1 Closedness in the Railway System

Railways are commonly referred to by the public as the "Iron Big Brother".[14] This term not only affirms the importance of the railways but also contains a hint of irony regarding their closed-loop management and indifference to public demands. Many commentators believe that this closed-loop management originated from the management system after 1949. During the era of the planned economy, the offices of various ministries were often called "courtyards", a term that vividly illustrated the closedness of this economy system. Regarding the "Courtyard of MOR", those who were within once recalled,

> From the south of the MOR Office Building at No. 10 Fuxing Road, Haidian District to the north of Lianhuachi Road lies the Yangfangdian residential area of MOR, which covers the entire area on the east and west sides of the North Beehive Road.

Starting from the northernmost MOR building, a series of brick-and-wood structures were built and gradually extended southward. This expansion ultimately led to a super-large compound with neither walls nor gates. Today, most of the eastern and southern sections have now been allocated to various subordinate organizations of MOR, leaving less than one-tenth of the original area as "the courtyard" in the northwest corner. At that time, Bus Route No. 21 had in total six stops, and three of them were at Huichengmen, Yangfangdian, and North Beehive within the compound. To facilitate the lives of the employees and their families, MOR built a kindergarten, a primary school, and a middle school for railway workers' children. It also established a railway hospital, a cultural center, a boiler room, a barber shop, public baths, and other facilities. In addition, the Ministry provided premises for a variety of services, including public security, banking, postal services, grain and oil, catering, coal, vegetables, sideline products, general merchandise, bookstores, photography, Chinese and Western medicine stores, bicycle repair, and waste material recycling. Together, these enterprises collaborated to develop and support the residential area.[15]

[14] Railway in Chinese is "铁路", which can be literally translated as Iron Road. The nickname of Chinese Railway uses Iron to homophonize the first character of Iron Road, which does not necessarily indicate that the Chinese railway is cold, hard and stiff.

[15] (2013, March 23). The little rascals of MOR courtyard. *Beijing Evening*.

This memory naturally evokes thoughts of the planned economy system. However, in terms of railways, the closed-loop management actually dates back even further, to the very early days of its establishment. Building a railway demands for mature technology, enormous funding, and qualified engineers. These requirements cannot be met merely through strong emotions. Instead, it requires immense patience, strong organizational abilities, and scientific management methods. As railways gradually expanded and became transportation networks, the complexity of the railway system increased. When we talk about the railway system, we mean that building the railway is just one aspect—albeit an important one—besides which it includes locomotive and vehicle manufacturing, power supply systems, and effective scheduling and command. However, this is not sufficient,as early railways primarily connected modern large cities but inevitably passed through relatively underdeveloped areas, including remote and impoverished regions. Therefore, at various stations along the route, MOR had to build schools, hospitals, and other living facilities for its employees to ensure a minimal standard of sustainable living. The often-criticized practice of enterprises running social welfare services actually has its origins in the railway sector. This is not an attempt to defend the railways, but rather to highlight the complexity of the issue. In the first half of the twentieth century, the drawbacks of the railway system were not conspicuous under the backdrop of overall industrial lag. Even for a long time after 1949, such practices remained prevalent. As the national industrial system continued to develop and improve, the railways became a "weak link" in the national economy. When this system was forced to adjust its operational methods, its immense historical momentum made reforms extremely difficult.

1.2.2 Basic Attributes of China's Railway System

The "integration of government and business functions" is a typical feature of China's railways. Its essence is "the high degree of integration between government functions and enterprise functions, where the government is essentially the enterprise. The government exercises unified direct control over the enterprises it owns, including micro-management".[16] This characteristic is both historical and realistic at that time in China, which is closely related to the economic and social development of the country.

For a long time, MOR of the People's Republic of China was the governing body under the State Council responsible to oversee national railway affairs, which unified and managed all operational and management tasks of the national railway. In the early 1980s, its functions mainly included four aspects: (1) Administrative management, which primarily involves the management of long-term planning,

[16] Wang Zhizhong, & Wei Liying. (2001). *Research on China's railway reform and development (1978–1998)* (p. 2). Beijing: Contemporary China Publishing House.

development strategies, industry policies, network layout, regulations, design specifications, and technical standards for national railways; (2) Enterprise operation and management, which primarily involves unified management of transportation, infrastructure, industry, materials, and international trade of the national railways; (3) Social management, mainly on education, healthcare, housing, pension, environmental protection, cultural and sports activities, publishing, and newspapers; and (4) National public power, which encompasses the function of public security and jurisdiction.[17] While enterprise operation and management functions are designated for an enterprise, administrative management, social management and public power functions are beyond the scope of what an enterprise can undertake. By 1980, the total number of railway employees had reached 2.6 million, effectively forming a large, comprehensive, and closed social system. The Ministry had put in place production facilities and supportive organizations, providing jobs for the adult children of the employees, placating the parents, pumping up their income, and thus promoting social order. Until 1989, "a total of over 860,000 adult children of the employees have been placed in jobs",[18] evidencing that the railway system operated as a de facto "society within a society".

This phenomenon is closely related to the operation pattern of Chinese Railways, which is characterized by a highly centralized dispatching command at its core, coordinated actions among transportation departments as its basis, and semi-military approach to order execution as the guarantee of the former two.[19] "For decades, due to the adherence to such a pattern, high efficiency was achieved in railway transportation; national transportation plan was completed and pivotal goods transportation missions were accomplished, despite strained railway capacity; different subordinate administrations and interfaces prioritized the central tasks and missions and ensured that local interests align with the general targets; workers were trained to maintain a high level of discipline to secure a smooth flow of railway".[20]

Chinese railways serve both commercial purposes and public interests, and as a matter of fact, the latter plays a more significant role. Some scholars believe that Chinese railways are entirely a public enterprise.[21] This is true because Chinese

[17] Ibid.

[18] Department of Personnel and Labor Relations. (1989). Recognizing the situation, strengthening reform, and advancing the railway collective economy to a new stage. In the Editorial Committee of *Selected essays on railway reform* (Ed.), *selected essays on railway reform* (p. 66). Beijing: China Railway Publishing House.

[19] This characteristic of the railway was summarized from the written instruction by the Central Committee of CPC on the leading party ground of MOR's *Report on Establishing Political Department and Reforming Management System in Railway System*. The instruction emphasized that "Railway is the main artery of national economy, a highly centralized enterprise with semi-militaristic nature" and "All authorities should be centralized at MOR".

[20] Lv Yinhua. (1994). *China's railways advance in reform* (51). Beijing: China Railway Publishing House.

[21] Tu Yourui. (1994). Common issues confronting railways of the world and features of railways in China. In Zhao Guoqian (Ed.), *Object lesson and assessment* (p. 6). Beijing: China Railway

railways perform massive non-profitable missions related to government, national defense, social services, and economic development, usually not the responsibility of an ordinary state-owned enterprise. Therefore, profit is not the first consideration in the operation of Chinese railways. For example, "railway haulage in China is 0.5 cents per ton-kilometer in comparison with India's 1.66 cents, the second lowest throughout Asia. In Europe and the USA, railway haulage is 5 times, 6 times, 8 times, and 10 times that of China, and even reached 16 times in West Germany".[22] Compared to other transportation industries, railway haulage is 1/11 of highway and 1/60 of civil aviation.[23] Low-cost haulage leads to reduced economic benefits. According to statistics in 1982, the rate of profit and tax (pre-tax rate of return) on funds in Chinese heavy industry is 16.77% while it is only 9.3% in the railway industry, higher than the rate of 2.21% in coal industry but is far lower than that of 59.1% in oil and other industries such as metallurgy, power, chemistry and construction.[24] Up to the early 1980s, salary of railway workers only accounted for 18% of the total income of railway transportation, compared to 50% in Japan and the USA, 65% in India, 113% in Austria, and 127% in the Federal Republic of Germany.[25] Chinese railways neither have the pressure and need for profiting, nor the worry of investment and debt. Within the traditional system, the funds needed by Chinese railways were all allocated by the government, and the materials needed were provided under a mandatory plan, typically in sufficient supply. In this sense, Chinese railways was more like a functional department of the government, rather than an enterprise.

The management of Chinese railways was marked by a centralized system that depended on mandatory administrative orders and planning. Specifically, the authority of pricing haulage, adjusting tax rates and applying for loans and credit in the railway rested with the central government, and so it was with that of planning for railway development, formulating industrial policy and making major investment decision, leaving MOR with limited governmental function. Within the railway industry, all authority involving personnel, funds, materials, production, supply and sales was centralized, including making investment decisions, planning for production and business operations, setting product prices, purchasing materials, disposing of production facilities, establishing organizational structures, appointing and dismissing officers, hiring employees, allocating income, etc. As a result, railway enterprises were feeble in taking on any operational responsibility.

Publishing House.

[22] Ibid.: 3.

[23] Lv Yinhua. (1994). *China's railways advance in reform* (22). Beijing: China Railway Publishing House.

[24] Hu Guangrong. (1985). Countermeasures on increasing economic benefit of railways, In China Academy of Railway Sciences (Ed.), *Study on the strategies of railway development* (internal information, p. 204).

[25] Tu Yourui. (1994). Common issues confronting railways of the world and features of railways in China. In Zhao Guoqian (Ed.), *Object lesson and assessment* (p. 3). Beijing: China Railway Publishing House.

This feature fitted perfectly with the macroeconomic system of this country, especially prior to the reform and opening up. Then, Chinese railways was fully self-sufficient and could independently accomplish missions assigned through administrative orders by the country. The absence of market undermined its economic value.

1.3 Structure of This Book

Now that we are going to discuss what has happened to Chinese railways since the reform and opening up, focusing on what kind of reform and opening Chinese railways have been going through. With the basic understanding of Chinese railway history mentioned previously, we do not intend to regard it as linear — such a history would never exist. At any time, reform is not based on momentary impulse—though it constitutes the initial drive—nor is it like a meticulous craftsman who is always trying to perfect his work. The purpose of reform is to find a way out of difficulties, sometimes even crises. As the path ahead is unclear, a successful reform usually ventures out prudently, rather than following a detailed map. The progressiveness of reform does not indicate that consensus on reforms is easily obtained. While Deng Xiaoping's phrase "crossing the river by feeling the stones" has transformed the prudence into a guiding methodology, the initiation and progression of reform demands great courage. This is further illustrated by some other quotes from Deng, such as "we should dare to experiment and break a new path" and "blaze a new trial".

Just like the twists and turns in the history of railways for almost a hundred years, the forty-year development of Chinese railways has not always been smooth since the reform and opening up, and even encountered unprecedented setbacks. Some of the difficulties stemmed from the general situation of the reform, while others from the path chosen and measures adopted in the reform itself. In consideration of these two aspects, this book has divided the progress of Chinese railways' reform and opening up into four stages: initial stage, proactive change, marketization, and comprehensive promotion.

1.3.1 Initial Stage (1978–1985)

As a preliminary experiment of reform in China, the overhaul of Chinese railways in 1975 marked the historical beginning of the reform. In the early stage of the reform and opening up, the contradiction was intensified between railway capacity and the social and national demands. Subsequently, Chinese railways became a severely vulnerability of national economy. On the one hand, the railway industry passively implemented a series of policies formulated by the country to reform state-owned enterprises. On the other, it had to advance with exploratory efforts under the constraints of the economic system. However, due to misjudgments regarding the characteristics of Chinese railways, the historical development stages and primary operational contradictions, the sector's accelerated development failed to keep pace with even faster-growing demands. This growing disparity ultimately compelled the implementation of the next phase "Contract Responsibility System" reform as an institutional response.

1.3.2 Proactive Change (1986–1991)

In the mid-1980s, Chinese railways became the bottleneck of the national economy instead of a previous role as a vulnerability. Other than the fast-growing economic demand, another important factor contributing to the heavy burden of Chinese railways was some reforms in national infrastructure, the most principal one was the grant-to-loan conversion policy implemented in 1984. In order to find a solution, MOR proposed the Contract Responsibility System, hoping to search for more development opportunities. However, railway policies centered on "tapping the potential and expanding capacity" proved to be improvident. The pattern of "Self-Sustaining Approach to Building New Railways" was unable to raise funds for the quick development of the railway. Under the Contract Responsibility System, the contract could be concluded on time but failed to bring the railway into the fast track of development. The bottleneck effect continued to worsen.

1.3.3 Marketization (1992–2003)

In other fields, railway reform has been market-oriented since the reform and opening up. Only after Deng Xiaoping's South Tour Speeches and the 14th CPC National Congress in 1992 did China formally establish its goal of developing a socialist market economy system. As a result, MOR also determined the market-oriented direction of railway reform, and issued *Opinions on Several Issues Concerning the Implementation of the Decision of the Third Plenary Session of the 14th CPC Central Committee and Strengthening the Reform of Railways* (the "Thirty Items for Railway Reform"), which outlined the fundamental concepts, overall goals and basic principles for the reform.

During this period, the reform was pursued with great intensity. Throughout this process, reform measures were taken step by step, including restructuring the railway management system, optimizing the operating mechanism, establishing a modern enterprise system, revitalizing the railway investment system, and developing diversified railway operations, thus effectively facilitating the construction of China's railways. In addition to the development momentum from the market-oriented reform, the substantial increase in construction investment, coupled with the abandonment of the "tapping the potential and expanding capacity" strategy, was another key factor contributing to the rapid progress of the railways. Of course, all these factors are interrelated. It is through the preceding decade of reforms that China accumulated significant experience and technological expertise in railway construction, enabling the country to effectively utilize these new investments, fulfill the tasks assigned by the government, and lay a solid foundation for overcoming the "bottleneck" in the development of railways.

1.3.4 Comprehensive Promotion (2004–2018)

Following a series of reform measures aimed at addressing, alleviating and overcoming the "bottleneck", and after accumulating technological resources, talent and management expertise over nearly 30 years, Chinese railways had then reached a stage where it was fundamentally capable of resolving the "bottleneck" problem. In 2004, the first railway industry planning document in the history of PRC, the *Medium and Long-term Railway Network Plan* was implemented, marking the beginning of a new era of systematic, large-scale railway construction. In 2008, as the pace of industrialization, urbanization, marketization, and internationalization accelerated, the Plan underwent its first adjustment, ushering in a period of rapid progress for China's railways. In 2016, in response to new challenges, the Plan was revised. Under the guidance of the Plan, the overall capacity of "China's High-Speed Rail Network" and the quality of passenger and freight transportation services had been improved. To meet the new requirements for a modern transportation system with Chinese characteristics and the modernization of railways, the comprehensive reform of railways in various fields such as the management system, technological innovation, investment and financing, and safety production also entered a critical period. In this process, out of reflection on speed and safety, we adopted the way of "slow thinking" on the "high-speed" development of railways, which helped China's railway technology, management and operation mature, enabling the railways to achieve a "two steps forward, one step back" progress through a process of reflection and adjustment.

In 2013, MOR was dissolved. Why wasn't that year the start of a new stage? Undoubtedly, the dissolution was a major event in the development of China's railway sector, and the question of whether and when to dissolve it had also been a long-standing political and social hotspot since 1998. For a while, the dissolution of MOR was viewed as a sign of the "last mile" problem to separate government and business functions. But when the dissolution actually happened, it caused a little stir. In retrospect, a series of reform and opening-up measures led by MOR had long paved the way for its eventual dissolution. Therefore, 2013 was not an end, but rather a turning point. Various reform measures had already been planned and promoted, and the development of China's railways before and after 2013 showed continuity rather than disruption.

The above four stages correspond to Chaps. 2 through 5 of this book. In addition, there are two important issues that cannot be fully covered by any single stage, which we have chosen to describe in two additional chapters, namely the emergence and development of China's high-speed railways (HSR) and the export of China's railway technology.

In the 1990s, the Chinese government put forward a proposal to develop the high-speed railway, and repeatedly demonstrated its necessity and feasibility. The technological renovation of the Guangzhou-Shenzhen Railway, along with the completion of the Qinhuangdao-Shenyang Passenger Dedicated Line and Beijing-Tianjin Intercity Railway, marked China's mastery of the high-speed railway construction technology, and the completion of the Beijing-Shanghai High-speed Railway signified the maturity of the high-speed railway technical system. By the end of 2019, China's high-speed rail operating mileage exceeded 35,000 kilometers, accounting for about 70% of the global high-speed rail network. China's HSR has profoundly transformed Chinese society.

In March 2013, MOR was dissolved and replaced by the National Railway Administration affiliated to the Ministry of Transport, and China Railway Corporation, a state-owned enterprise directly under the State Council. Since 1998, the dissolution of MOR had been a hot topic at every National People's Congress. Although ministries with similar functions, such as those overseeing textiles, coal, metallurgy, and machinery, were abolished, MOR was exempted and avoided the institutional reform in 2003 and 2008. Nevertheless, it could not remain entirely unaffected. In the context of reform, institutions formerly under MOR, including those in higher education, engineering construction, and vehicle manufacturing, were gradually spun off, followed by the railway police, procuratorates, and courts. Finally, the reform crossed a threshold.

Half a year later, then-Premier Li Keqiang embarked on vigorously promoting China's HSR during his visit to Thailand. Premier Li characterized China's HSR as "technologically advanced, safe, reliable, and highly cost-effective". Following this, the promotion of China's HSR became a staple of Premier Li's subsequent overseas visits. Shortly after Premier Li's speech, President Xi Jinping proposed the Belt and Road Initiative, with HSR serving as a fundamental pillar of the initiative. Consequently, the media hailed HSR as a "National Symbol". Domestically, HSR lines were constructed in remote and underdeveloped provinces such as Guizhou, Yunnan, Xinjiang, and Gansu.

"Going Global" emerged as one of the central topics deserving attention in the latest development of China's railway sector. From a broad perspective, the history of China's railway technology spreading overseas dates back a long time, with the earliest chapters written through railway assistance projects in regions like North Korea, Vietnam, and East Africa. During this period, most overseas railway projects undertaken by China were driven by political considerations, and the country's railway sector lagged significantly behind developed nations in terms of construction technology and machinery supply. Nonetheless, these early foreign aid railway projects honed domestic enterprises and accumulated invaluable experience for China's subsequent "going global" endeavors.

Since the reform and opening up, China's railway construction capabilities have undergone a remarkable transformation, equipping the country with the prowess to compete on par with the world's leading railway nations. This transformation has been evident in numerous overseas railway projects, accompanied by a notable surge in the export capabilities of railway equipment. Following the rapid advance of China's HSR technology, Chinese enterprises have ventured into the global high-speed rail market. The Ankara-Istanbul High-Speed Railway in Turkey marks the first overseas high-speed rail project implemented by Chinese enterprises, while the Jakarta-Bandung High-Speed Railway in Indonesia represents the first step for China's high-speed railway industry to "go global" with its entire industrial chain and all essential components.

The proposal of the Belt and Road Initiative has ushered in new development opportunities for Chinese railways. Given the current imbalance in sea and land transportation across Eurasia, railway construction holds paramount importance in the steady progress of the Belt and Road Initiative. While Chinese railways currently face certain challenges in their "going global" endeavors, they are poised to play a pivotal role in the future connectivity of infrastructure projects jointly performed by Belt and Road Initiative partner countries. Furthermore, they will exert a profound influence on fostering shared development and prosperity worldwide.

Lastly, it is always difficult and even risky to describe the events that are currently unfolding from a historical perspective. Yet, considering that these texts will also become usable material for future studies of railway history gives us the courage to engage with this subject.

Chapter 2
Initial Stage: Attempt to Eliminate "Weak Links" (1978–1985)

"The relationship between Chinese railways and Chinese society mirrors the interplay between rail transportation technology and the public's demand for such technology".[1] The primary thread in the development of Chinese railways revolves around the ongoing contradiction and its subsequent resolution between national and social needs and rail transportation capacity. Since the reform and opening up, the fundamental approach to addressing this contradiction has been railway reform—a theme that has dominated Chinese railway history over the past 40 years. Consequently, the issue can be divided into two sub-issues: on the one hand, how national and public needs for railways manifest, and in what forms; on the other, how the railway system continuously responds to these needs while adjusting and developing itself? Thus, the crux of railway reform is not merely the achievements in construction (whose development, from a statistical perspective, is growing at a satisfying speed). Rather, it centers on the extent to which the railways can meet national and public needs, specifically, the relationship between the required transportation volume and the capacity provided by railways. Thus, evaluating railway reforms necessitates a framework that is inherently comparative and adaptable.

Since 1975, over a relatively long period, the prevailing view of the country and the general public on the railway was that "the railway is a weak link in the national economy",[2] meaning that its transportation capacity was far from meeting the demand. Clearly, striking a balance between the two was the central task of railway

[1] Zhang Xueyong. (2018). External influences and internal rationale: The reform and opening up history of China's Railways in the perspective of a century of changes. *Journal of Southwest JiaoTong University (Social Sciences)*, (6), 1–8.

[2] Decisions of the Central Committee of CPC on improving railway work ([1975] ZF. No. 9). In the Research Office of the Party History, Party Building, and Political Work of the National Defense University of the Chinese People's Liberation Army (Ed.), *Cultural Revolution Research Materials* (Volume II) (internal information, p. 248). The expression "prominent weak links" is used in *Document No.9*.

X. Zhang, *China's Railway Expansion in 40 Years of Economic Reform and Opening Up*, Understanding China,
https://doi.org/10.1007/978-981-95-0333-9_2

reform during that period. Nevertheless, we should note that after approximately a decade of exploration, around 1990, the railway failed to resolve its awkward position as the "weak link" and instead, its problem became even more severe, ultimately turning into a bottleneck for national economic progress.[3]

In 1992, Deng Xiaoping's South Tour Speeches marked the real acceleration of railway reform in China. Prior to this, the history of railways was characterized by continuous exploration, adjustment, and experimentation. Distinctly, this period can be divided into two stages based on the extent of reform efforts. Marked by the introduction of the Contract Responsibility System[4] in 1986, the Railway Man started proactive reforms and sought changes. Before this, although MOR issued a series of reform documents and made numerous attempts, these efforts, though integral to the reform process, largely entailed the implementation of macro and general reform measures from the Party Committee and the Central Government within the railway system. In other words, the railway industry's unique characteristics, and the proactive and creative spirit of railway reformers were not yet prominent. Thus, the years from 1978 to 1985 constituted the first historical stage of railway reform in China over the past 40 years, characterized by exploration through the restructure of Chinese Railways.

2.1 Pilot Reforms: Restructure of Chinese Railways in 1975

As is widely recognized, the Third Plenary Session of the 11th Central Committee of the Chinese Communist Party (CPC), held in December 1978, marked China's entering a new era of reform and opening up. Additionally, such landmark events acquire their unique historical significance within the grand historical context. This means that when we shift our focus to the history of reforms within specific industries, the overarching historical context often obscures the richness and distinctiveness of the history itself, making it difficult for us to grasp the details. Indeed, the reforms in various industries in China are by no means progressing uniformly, but rather differ from one another. So, where does the history of railway reform in China begin? This is where our discussion should start.

[3] Zhu Rongji. (1994). Speech at the national conference of leaders in the railway sector. In Zhao Guoqing (Ed.), *Challenges and Solutions* (p. 3). Beijing: China Railway Publishing House.

[4] Detailed discussion is provided in Chap. 2 of this book.

2.1.1 *Identifying the Historical Foundations of Chinese Railway Reform*

In the few existing studies on railway reform, 1978 is routinely regarded as the starting point. For instance, *Fifty Years of the Railroads of People's Republic of China (1949–1999)* states: "In December 1978, the CPC Central Committee convened the Third Plenary Session of the 11th Central Committee, making a strategic decision to shift the focus of the Party's work to socialist modernization and putting forward the requirement to address the imbalance in the national economy. This marked a significant turning point since the founding of PRC and the beginning of railway work moving from stagnation to progress".[5] While this view is not incorrect, it may not be entirely convincing, for if railway reform indeed was in step with the overall reforms, why did the railway industry lag far behind the pace of national and social development for a considerable period of time? In fact, China's reform began with great difficulty. The entire social landscape did not immediately transform after the central government's call. On the contrary, it was only after repeated trials and explorations that the process of reform was truly accelerated.

The railway industry was no exception. Examining the period around 1978, it is difficult to simply conclude that the railway industry exhibited signs of rapid change. Much of the railway operations continued to rely on established practices, although reform steps were gradually taken.

In the field of railway reform, three events are considered as significant starting points: the pilot program in May 1979 at the Shanghai Railway Administration Rolling Stock Depot to expand enterprise autonomy in operational management, the establishment of the Office for Economic System Reform under MOR in November 1980, and the publication of *Opinions of MOR on Railway Reforms* in 1985 in light of the documents issued at the Third Plenary Session of the 12th Central Committee. Notably, all three events occurred after 1978.

Which of these, or which one of them, can be deemed the definitive starting point of railway reform?

Firstly, in terms of chronological order, the pilot program at the Shanghai Railway Administration Rolling Stock Depot in May 1979 was the earliest trial in the railway system.[6] However, neither its scale nor impact was sufficient to be historically significant. Even in the *Chronicle of Shanghai Railways* compiled by the Administration itself, it is briefly mentioned with just a single sentence, "Shanghai Railway took the lead in 1979 by initiating reforms and corporate reorganization efforts such as expanding enterprise management authority and the lump-sum contracting of the progressive increase of the profits (the profit-sharing arrangements)" and it was not included in the *Memorabilia* for that year.

---

[5] Xu Zenglin. (1999). *Fifty years of the railroads of the People's Republic of China (1949–1999)* (p. 135). Beijing: China Railway Publishing House.

[6] Compilation Committee of the *Chronicle of Shanghai railways*. (1999). *Chronicle of Shanghai railways*. Shanghai: Shanghai Academy of Social Sciences Press.

Secondly, from the perspective of organizational management, the establishment of the Office for Economic System Reform under MOR on November 6, 1980, was the earliest body tasked with overseeing and coordinating reform within the railway system.[7] The Office was led by Deng Cunlun, then Deputy Minister of Railways, with Deputy Leaders including Geng Zhenlin and Liao Shiquan, then Deputy Ministers, and Ye Keming, then Director of the Ministry's Finance Bureau. Despite the high ranks of the officials and institutional functions, this early reform, conducted under the decrees of the central government and aligned with the broader state-owned enterprise reforms, lacked the distinctive and proactive features specific to the railway industry. In this regard, it cannot be considered the definitive starting point for railway reform.[8]

Lastly, from the perspective of industry-specific characteristics, the *Opinions of MOR on Railway Reforms*[9] on January 1, 1985, in line with the principle proclaimed at the Third Plenary Session of the 12th Central Committee, represented a genuine reform initiative tailored to the railway industry. The Contract Responsibility System implemented the following year was the Ministry's first comprehensive reform package. While this plan embodied the railway industry's initiative to change, it was unlikely to be the starting point of railway reform given that it lagged behind the national reform efforts by six or seven years.

Thus, none of the three events mentioned can be considered the historical starting point of railway reform. Then, where does the real beginning of railway reform lie?

Expanding the scope of research, we can see that the basic work of the railway industry from 1978 to 1985, regardless of its scale, should be considered part of the 40-year history of railway reform and opening up. It is important to identify the guiding principles that truly influenced the railway industry during this period and the central documents that best reflect such principles. As a result, it becomes evident that the railway industry maintained a certain level of stability and continuity for a long time. In fact, while the rest of the country was still reeling from the upheavals of the "Cultural Revolution", the railway industry was among the first to stabilize and adjust itself. Conversely, despite the nationwide surge of reform experiments, the railway industry struggled to find its path. Therefore, this book argues that the starting point of railway reform should be traced back even earlier, specifically to the issue of the 1975 Central Committee document *Decisions of the Central Committee of CPC on Improving Railway Work*, commonly known as *Document No. 9*.

The reasons for this will be explained as follows.

[7] Notice on the establishment of an office responsible for reform of economic system. Archived in Southwest Jiaotong University, No. 801-0-2(14).

[8] For a detailed discussion, refer to the following sections.

[9] Opinions of MOR on railway reforms. Archived in Southwest Jiaotong University, T-7.

2.1.2 *The Content and Impact of* Document No. 9

In 1975, with support from Mao Zedong and Zhou Enlai, Deng Xiaoping was officially appointed as director of the CPC Central Committee and began to lead the comprehensive rectification of the national economy.[10] On January 28, Deng summoned Wan Li, the newly appointed Minister of Railways, and gave instructions to relevant officers of MOR to draft a plan for railway rectification. After discussions on February 6 and February 11, Deng demanded that the plan be completed by February 25, the day of the National Industrial Secretaries' Conference. He stressed the urgent need for a concentrated overhaul of the railways, advocating for a military-style management approach, and verbally presented the key points of *Document No. 9.*[11]

In his speech on March 5, 1975, Deng stated, "There is a major agenda that the whole Party should focus on. What is this major agenda? The Government Work Reports of the First Session of the Third National People's Congress (NPC) and the First Session of the Fourth NPC both outlined a two-step plan for developing our national economy: the first is to build an independent and relatively complete industrial and economic system by 1980; the second is to build China into a socialist powerhouse with modern agriculture, industry, national defense, and science and technology by the end of the 20th century, which gives us twenty-five years from now. The entire party and nation should strive to achieve this great goal. This is the major agenda".[12] So, how can we boost the national economy? Deng emphasized, "The analysis shows that the current weak link is the railways. If the problems with railway transportation are not resolved, the operation will be disrupted, and the entire plan will fall through. Therefore, the central government is determined to address this issue".[13] Thus, the railways became the breakthrough for comprehensive rectification, marking the beginning of the reform process.

Document No. 9 can be viewed from two perspectives. Firstly, the railway industry served as the "vanguard" in the comprehensive rectification. As early as March 5, 1975, at the National Industrial Secretaries' Conference, Deng Xiaoping pointed out that "the experience gained from resolving railway issues would be helpful to other industrial sectors".[14] Consequently, the experience of rectification in the Xuzhou Railway Sub-administration, led by Wan Li, then Minister of Railways, was quickly promoted across various industrial departments. In late March and

---

[10] Literature Research Office of the CPC Central Committee. (1997). *The Chronology of Zhou Enlai (1949–1976) Volume II* (pp. 693–694). Beijing: Central Literature Publishing House.

[11] For details of the last two meetings, refer to: An Interview with Tie Ying (on May 25, 1994) and Wang Zhen's speech at the National Conference of Leading Officers in the Railway Sector on March 28, 1975. Cheng Zhongyuan et al. (2009). *Prelude: Deng Xiaoping and the rectification move in 1975* (p. 35). Shijiazhuang: Hebei People's Publishing House.

[12] Literature Research Office of the CPC Central Committee. (1994). *Selected works of Deng Xiaoping* (Volume II) (p. 4). Beijing: People's Publishing House.

[13] Ibid.: 5.

[14] Ibid: 7.

April, *People's Daily* repeatedly praised the achievements of the Xuzhou Railway Sub-administration. Notably, on April 22, 1975, *People's Daily* published an article entitled *Link Theory with Reality, Follow the Strategies and Party Spirit, and Enhance Revolutionary Unity and the Holistic Sense,*[15] which introduced and summarized the Xuzhou experience in railways. In May, Deng mentioned "the Xuzhou experience" repeatedly in his *Speech on Several Issues that Should be Addressed in the Current Steel Industry.*[16] In practice, *Document No. 9* became not only a directive for the railway sector but also a guiding document for other production sectors such as coal, metallurgy, electricity, and chemical fertilizers. It wasn't until the Central Work Conference held in April 1979 that the Four-Word Guiding Principle of "Adjustment, Reform, Rectification, and Improvement"[17] for the national economy was officially proposed, in line with the principles of the Third Plenary Session of the 11th Central Committee. The Guiding Principle played a crucial role in the early stage of reform and opening up and was decisive in the recovery of the national economy. Even before nationwide adjustments began, the railway industry had already taken the lead in initiating adjustment and rectification efforts.

Secondly, the railways served as a pilot for comprehensive reforms. The overall rectification, with that of the railways as a breakthrough, was essentially a thorough reflection on and negation of the entire "Cultural Revolution". In May 1975, Deng summarized the Xuzhou experience into the "Four Musts", namely, "a strong leadership team must be established", "a resolute struggle against factionalism must be waged", "policies must be conscientiously implemented", and "necessary rules and regulations must be formulated".[18] In July of the same year, Deng formally proposed the idea of "Taking Three Directives as A Set of Policy Guideline"[19] in his *Speech on Strengthening the Party's Leadership and Rectifying the Party's Conduct.* He also presided over the formulation of three major documents, *The General Outline of the Tasks of the Party and Country* (drafted by Hu Qiaomu and Deng Liqun, with the first draft completed on October 7, 1975), *Some Issues on Accelerating Development of the Industrial Sector* (drafted by Hu Qiaomu, finalized on October 26, 1975), and the *Outline of the Report on the Work of the Academy of*

---

[15] (April 22, 1975). Link theory with reality, follow the strategies and party spirit, and enhance revolutionary unity and the holistic sense, *People's Daily*.

[16] Literature Research Office of the CPC Central Committee. (1994). *Selected works of Deng Xiaoping* (Volume II) (p. 9). Beijing: People's Publishing House.

[17] Literature Research Office of the CPC Central Committee. (1989). *Selected works of Li Xian'nian* (p. 353). Beijing: People's Publishing House.

[18] Literature Research Office of the CPC Central Committee. (1994). *Selected works of Deng Xiaoping* (Volume II) (pp. 8–11). Beijing: People's Publishing House. The "Four Musts" were proposed at a conference on steel work, but Deng Xiaoping repeatedly emphasized the Xuzhou experience in his discussions.

[19] "In a previous period, Mao Zedong issued three important directives: first, to study theory and guard against revisionism; second, to maintain stability and unity; and third, to boost the national economy. These three directives are interconnected and form an integral whole, and none of them should be discarded". Literature Research Office of the CPC Central Committee. (1994). *Selected works of Deng Xiaoping* (Volume II) (p. 12).

Sciences (drafted by Hu Qiaomu, with the sixth draft completed on October 24, 1975). Scholars of the Communist Party of China's history generally view the "Three Directives" and the "Three Major Documents" as a summary of the experiences of socialist construction development in the first 17 years of PRC, a comprehensive negation of the Cultural Revolution, the earliest systematic exposition of Deng Xiaoping Theory, and a prelude to the rectification and reform era.[20] Furthermore, Deng pointed out, "When talking about reform, we actually experimented with it between 1974 and 1975... At that time, the reform was called Rectification Move, stressing the need to boost the economy, with a primary focus on restoring production order".[21] Similarly, Jiang Zemin also noted, "This rectification was essentially an experiment for the later reforms, reflecting the aspirations of the broad masses of officers and the correct leadership of the Party, and achieving remarkable results in a short period".[22] For these reasons, the Rectification Move on all fronts is regarded by scholars as the vanguard to comprehensive reforms.[23] In this sense, Chinese railways played a pioneering role in China's reform process.

In retrospect, no new measures were introduced during the Rectification Move in 1975, which might not be easily classified as a reform. Nevertheless, since "reform is also about unleashing productive forces",[24] the Rectification Move should undoubtedly be recognized as a reform. In fact, during the Rectification Move on all fronts, Deng Xiaoping's summation of the Xuzhou experience already outlined the fundamental principles that should be adhered to in future reforms, particularly foreshadowing the articulation of the "the Four Cardinal Principles, including the principles of upholding the socialist path, the people's democratic dictatorship, the leadership of the CCP and Mao Zedong Thought and Marxism–Leninism", which, as we all know, form the cornerstone of our nation. As a unique industry featured with "highly centralized dispatching command at its core, coordinated actions among transportation departments, semi-military approach to order execution", the

---

[20] In "Answers to Questions from the American Journalist Salisbury" in 1987, Hu Qiaomu stated, "The three directives were framed with a focus on maintaining stability and unity, as well as boosting the national economy. These two slogans continued to serve as guiding principles until the Third Plenary Session of the 11th Central Committee was held, and even after the 13th National Congress. This indicates one central task and one basic point; at that time, reform and opening up, the other basic point, was not yet possible to propose, and what was termed 'rectification' actually included both reform and opening up". Hu Qiaomu. (1994). *The Collected works of Hu Qiaomu* (Volume II) (p. 248). Beijing: People's Publishing House). For related studies, refer to: Cheng Zhongyuan. (1995). Three notable documents in the 1975 rectification. *Contemporary China History Studies,* (1), 72–82; Liu Guijun. (2017). On the "Taking three directives as a set of policy guideline" in the 1975 rectification. *Deng Xiaoping Studies*, (2), 112–120.

[21] Literature Research Office of the CPC Central Committee. (1994). *Selected works of Deng Xiaoping* (Volume II) (p. 255). Beijing: People's Publishing House.

[22] Literature Research Office of the CPC Central Committee. (2006). *Selected works of Jiang Zemin* (p. 630). Beijing: People's Publishing House.

[23] For related studies, refer to: Cheng Zhongyuan et al. (2009). *Prelude: Deng Xiaoping and the rectification move in 1975* (pp. 306–309). Shijiazhuang: Hebei People's Publishing House.

[24] Literature Research Office of the CPC Central Committee. (1994). *Selected works of Deng Xiaoping* (Volume II) (p. 255). Beijing: People's Publishing House.

railway industry not only contributed significantly to the rapid turnaround of the national economy but, more importantly, its socialist attributes epitomizing the Four Cardinal Principles.

"The Rectification Move was essentially a targeted effort to address specific issues within the railway industry, marking the beginning of the repudiation of some specific policies and guidelines of the Cultural Revolution. The emphasis on centralization and unity not only aligned with the semi-military nature of the railway industry itself but also countered the one-sided criticism of over-centralization and the rise of decentralization. Demands for uninterrupted, safe and punctual operations, and boosting transport production were undoubtedly rebuttals to the criticism of the idea that 'only the development of productive forces is the driving force for the development of human society'; The reshuffling of leadership weakened the factional influence of the 'Gang of Four' and shook their ideological foundations. Establishing and improving rules and regulations, strengthening organizational discipline, and restoring public order negated anarchism and countered criticisms against 'over control, restriction, and suppression'".[25] Deng Xiaoping's choice of the Rectification Move as a breakthrough was based not only on economic consideration but also demonstrated his profound political wisdom. The Rectification Move on all fronts in 1975 "centered on whether to develop productive forces or to take class struggle as the guiding principle. The contradiction between these two remained the central issue on the political front at that time".[26]

The history underscores the undeniable significance of the 1975 Rectification Move in swiftly ending the turmoil of the Cultural Revolution and ushering in a new era of comprehensive reform by rectifying past wrongs.

The principle of *Document No. 9* continued to play a pivotal role in railway work for an extended period after the end of the Cultural Revolution.

At the National Railway Work Conference in February 1977, it was emphasized that *Decisions of the Central Committee of CPC on Improving Railway Work*, approved in 1975 by Chairman Mao Zedong, the great leader, are a concrete manifestation of Chairman Mao's concepts on railway work. In response to CPC Central Committee's call to adhere to the outline of governance by working hard and making great strides forward, the railway system put forward the idea that comprehensive rectification should be carried out for the country and the railway industry. We should boost the national economy, and railway work should strive to take the lead. The railway serves as the link between cities and villages, industry and agriculture, and is the lifeblood of the national economy. When well-managed, it will promote the development of the entire national economy. Therefore, railway transportation should be unimpeded, safe, punctual, and efficient, setting the pace for the nation.

---

[25] Cheng Zhongyuan et al. (2009). *Prelude: Deng Xiaoping and the rectification move in 1975* (p. 51). Shijiazhuang: Hebei People's Publishing House.

[26] Hu Qiaomu. (1994). *The Collected works of Hu Qiaomu* (Volume II) (p. 248). Beijing: People's Publishing House.

Specifically, the focus was on "further strengthening the Party's leadership over railway work. The railway should have centralized and unified command while also accepting the leadership of local Party committees, thereby giving full play to the initiative and creativity of both the central and local levels. Under the unified leadership of local Party committees, a set of rules and regulations for managing socialist enterprises should be established. Efforts should be made to vigorously restore the order of railway public security, protect state property, crack down on sabotage activities by class enemies, combat capitalism, and seriously improve the public security work at trains, stations, and along the railway lines".[27] In a short period, the vast majority of railway practitioners burst with high work enthusiasm, giving full play to the existing transportation capacity of the railway, improving efficiency, and maximizing the transportation capacity. By 1981, Wan Li, then Vice Premier of the State Council, reiterated the importance of *Document No. 9* during his inspection of railway work. In his speech, he highlighted, "the current problem is mainly that some railway administrations lack enthusiasm and are poorly managed. Whenever there are 'bottlenecks', some administrations immediately ask the central government for funding or other resources. This operational passivity reflects a problematic mindset. Railway practitioners should recognize that national financial resources are limited, and that securing more investments in the next five years is impossible. Only by perking up, and improving management, can we achieve or exceed historical highs. The slogan put forward in 1975 based on the principle of *Document No.9*, 'to be a pacesetter in ensuring safe and punctual operations, and building unimpeded and well-connected railway lines', is not only applicable to the past practice but also to the present. MOR should vigorously promote and propagate this slogan, which embodies our glorious traditions and vitality, to inspire officers and the masses to revitalize their revolutionary spirit and embrace new campaigns".[28]

This speech roughly represents the central government's basic attitude towards railway construction at that time. According to *the Speech at the National Conference on Traffic Safety* in late November 1984, then Minister of Railways Chen Puru pointed out, "unimpeded transportation and safe punctuality will always be important indicators for evaluating our railway work, as summarized in *Document No.9* of the Central Committee".[29] This demonstrates the profound and lasting influence of *Document No. 9*.

---

[27] (1977, February 22). Strengthening railway construction and improving operation and management. *People's Daily*.

[28] Wan Li. (1997). Strengthen railway construction and improve operation and management, In the Editorial Group of *Wan Li's views on railway reform and construction* (Ed.), *Wan Li's views on railway reform and construction* (p. 75). Beijing: China Democracy and Legal System Publishing House.

[29] Chen Puru. (1985). Speech at the National Conference on Traffic Safety. *Railway Transport and Economy*, (1), 1–9.

2.1.3 *Implementation of* Document No. 9 *and Rectification Move*

The phrases "one prominent issue and two inabilities to adapt" in *Document No. 9* summarize the fundamental assessment of and basic demands for the railway industry at that time. In other words, "railway transportation remains a prominent weak link in the national economy, unable to meet the needs of industrial and agricultural development, and unable to meet the needs of strengthening war preparedness".[30] The conclusion that "railways are a weak link in the national economy" persisted for a considerable period of time and became the core issue of Chinese railway reform from 1975 to 1992. As a result, during the initial stages of reform, the railway industry consistently grappled with the issues raised in *Document No. 9*.

Deng Xiaoping pointed out, "Over the years, there has been an increase in railway practitioners and equipment such as locomotives, vehicles, and tracks, but due to the weakening of centralized and unified leadership, railway transportation has failed to improve. Currently, only around 40,000 cars are loaded daily. Some said that there should be no problem loading 55,000 cars based on our current actual capacity".[31] Wan Li, then Minister of Railways, shared a similar view, "Railway transportation is a prominent weak link in the national economy, mainly manifested in the following aspects: First, production has declined. Over the years, the number of railway workers has increased significantly, and there has also been an increase in equipment such as locomotives, vehicles, and tracks, yet production has instead dropped significantly. Based on our current actual transportation capacity, we should be able to load more than 55,000 cars daily, but currently, we can load around only 40,000, and sometimes even fewer than that. Second, the number of accidents is alarming. Last year, there were over 750 major and serious train accidents, more than eight times the 88 accidents a decade earlier in 1964. Third, rules and regulations are not strictly enforced, and labor discipline is lax. Many accidents caused by negligence (including those resulting from issues with locomotive and vehicle maintenance) are due to non-compliance with rules and regulations. Last year, the national railway locomotive serviceability rate was less than 70%, with most locomotives not being repaired on schedule, and many operating with defects. Fourth, congestion is severe, and most trains cannot operate on schedule".[32] After thoroughly

[30] Decisions of the Central Committee of CPC on improving railway work (ZF [1975] No. 9). In the Research Office of the Party History, Party Building, and Political Work of the National Defense University of the Chinese People's Liberation Army (Ed.), *Cultural revolution research materials* (Volume II) (internal resources, p. 248).

[31] Literature Research Office of CPC Central Committee. (1994). *Selected works of Deng Xiaoping* (Volume II) (p. 255). Beijing: People's Publishing House.

[32] Wan Li. (1997). Oppose to factionalism and resolve Xuzhou issue within the prescribed time. In the Editorial Group of Wan Li's Opinions on Railway Reform and Construction (Ed.), *Wan Li's Opinions on Railway Reform and Construction* (pp. 3–4). Beijing: China Democracy and Legal System Publishing House.

studying the principle of the central government's documents, the railway system at that time aimed "to be a pacesetter in ensuring safe and punctual operations, and building unimpeded and well-connected railway lines".[33]

Document No. 9 also outlined the approaches for the Rectification Move, emphasizing the need for strengthened management. Specifically, these approaches included implementing a management system where MOR primarily takes the lead in managing national railways, and provincial, municipal and autonomous region party committees should continue to strengthen their leadership over railway work. Moreover, they called for establishing and improving necessary rules and regulations, strengthening organizational discipline, and ensuring safe and punctual transportation, as well as "rectifying railway transportation order, combating all forms of sabotage, and strengthening proletarian dictatorship".[34]

Following the issue of *Document No. 9*, Wan Li personally traveled to Xuzhou to lead the Rectification Move. Taking Xuzhou as the hub, he swiftly initiated a comprehensive overhaul of the notorious "problematic" subordinate organizations across the railway system, including China Railway Zhengzhou Group Co., Ltd., China Railway Nanchang Group Co., Ltd., China Railway Kunming Group Co., Ltd., and CRRC Zhuzhou Locomotive Co., Ltd. This approach ensured that the rectification work covered a broader area.

Within just a few weeks of the implementation of *Document No.9*, the nation's railway transportation continued to improve. In mid-March, the average daily number of loaded cars reached 46,741, surpassing 50,004 by the middle of the month and further rising to 52,140 in the first seven days of the latter half of the month. In February, the maximum number of trains retained on the main trunk lines across the network peaked at 217, decreasing to 140 in early March and then to around 80 in recent days. There was a decline in train accidents, and both infrastructure development and railway industrial production also achieved significant results.[35]

Notably, these figures were already close to the central government's expectations for railway carrying capacity, demonstrating the immediate effectiveness of the Rectification Move. By the end of 1975, railway transportation had essentially restored normal operational order, with various statistics reaching new highs during the Cultural Revolution.

---

[33] Wan Li. (1997). *The struggle for the continued improvement of the railroad transport situation.* In the Editorial Group of Wan Li's Opinions on Railway Reform and Construction (Ed.), *Wan Li's Opinions on Railway Reform and Construction* (p. 56). Beijing: China Democracy and Legal System Publishing House.

[34] Decisions of the Central Committee of CPC on improving railway work (ZF [1975] No. 9). In the Research Office of the Party History, Party Building, and Political Work of the National Defense University of the Chinese People's Liberation Army (Ed.), *Cultural Revolution Research Materials* (Volume II) (internal information, pp. 248–249).

[35] Wan Li. (1997). Implementing the Central Government's Document No. 9 earnestly, continuing rectification work and reversing the situation. In the Editorial Group of Wan Li's Opinions on Railway Reform and Construction (Ed.), *Wan Li's Opinions on Railway Reform and Construction* (p. 7). Beijing: China Democracy and Legal System Publishing House.

The essence of *Document No. 9*, "to strengthen management, unify command, and restore normal order of railway operations", constitutes an effective resolution under the specific political and economic situation during the particular period of time. Undoubtedly, the most prominent strengths of Chinese railways lie in their consistent emphasis on political theory studies, ideological and political education, and the reinforcement of management and mobilization. "The long-term effective operation of the traditional railway system is attributable to the dedication and sacrifice of millions of railway practitioners in China; it benefits from the stringent discipline traditionally upheld by Chinese railway practitioners and their unwavering implementation of superior directives; it also stems from their enduring acceptance of a relatively low average wage system".[36] Despite a myriad of issues that exist within China's railway industry, the railway spirit, passed down from generation to generation, has created numerous miracles in China's railway operations.

Concurrently, the positive outcomes of the Rectification Move directly influenced the central government's subsequent overall judgments and decisions regarding the railway industry. It is evident that the railway industry is highly integrated with modern technology. While the high level of working enthusiasm among railway practitioners is undoubtedly a necessary condition for success in all endeavors, technological advancement serves as the core competitiveness. In fact, many pressing issues that the railway reform should address were temporarily shelved, particularly after 1978, as the development strategy centered on "tapping the potential and expanding capacity".

2.2 Tapping the Potential: Countermeasures Against the "Weak Links"

For a considerable period after 1975, the railway issues were consistently perceived as "solvable yet unsolved".[37] In other words, for the central government and relevant management departments, the primary and most prominent issue with the railways was not technical but managerial. Consequently, there was a call to "tap the potential vigorously"[38]—a strategy to remove "weak links".

"Tapping the potential and expanding capacity" refers to "enhancing transportation capabilities by unlocking hidden potential". In 1979, Hua Guofeng, then Premier, stated in the *Government Work Report* at the Second Session of the Fifth

[36] Wang Zhizhong, & Wei Liying. (2001). *Research on China's railway reform and development (1978–1998)* (p. 10). Beijing: Contemporary China Publishing House.

[37] This judgment directly influenced and even determined the thinking behind railway reform in the early stages of Reform and Opening Up. Detailed discussions will follow in subsequent sections.

[38] Wan Li. (1997). Implementing the Central Government's Document No. 9 earnestly, continuing rectification work and reversing the situation. In the Editorial Group of Wan Li's Opinions on Railway Reform and Construction (Ed.), *Wan Li's Opinions on Railway Reform and Construction* (p. 14). Beijing: China Democracy and Legal System Publishing House.

National People's Congress, "To achieve the Four Modernizations, we must certainly establish a number of new enterprises where necessary, but primarily, we must rely on tapping the potential, innovating, and reforming a large number of existing enterprises to gradually bring them closer to or to achieve modernization".[39] This marked the central government's official introduction of the economic development strategy of "tapping the potential", which was part of the policy for the Rectification Move and colloquially known as "tapping, reforming, and transforming". Consequently, more in-depth discussions within academic circles explored relevant ideas in Marxist political economy, and a consensus has gradually emerged: industrial scaling production can be categorized into two types—intensive and extensive. The primary theoretical foundation derives from Marx's *Capital,* which states, "The annual expansion of production is attributed to two reasons: the continuous growth of capital invested in production and the continuous improvement of capital efficiency".[40] This is also one of the essential theoretical principles guiding the socialist economic development in China.

In the railway industry, the strategy of "tapping the potential and expanding capacity" represents an industry-specific interpretation of the central government's "tapping, reforming, and transforming" policy and the theory of "scaling production". This strategy has also been referred to as "tapping the potential and enhancing efficiency" or "expanding capacity and enhancing efficiency", which implies the goal of "unlocking the transport potential and improving operational efficiency", as can be illustrated by the requirements in *Document No. 125* of MOR in 1982.

"To properly reorganize enterprises, strengthen technical management, and reorganize the fundamental work of stations are also effective measures to tap transport potential, enhance operational efficiency, and adapt to transportation trends".[41] At the National Passenger Transport Conference held by MOR in 1983, "developing socialist culture and ideology" and "tapping the potential and expanding capacity" were summarized as the two core tasks of railway passenger transport in 1982. Ever since then, "tapping the potential and expanding capacity" became a "proprietary term"[42] in the railway industry and a significant guiding principle during the reform and opening up, representing one of the deepest industry memories for the Railway Man.

---

[39] Hua Guofeng. (1979, June 26). Government Work Report to the Second Session of the Fifth National People's Congress. *People's Daily.*

[40] The Compilation and Translation Bureau of Works of Marx, Engels, Lenin and Stalin of CPC Central Committee. (2008). *Collected Works of Marx and Engels* (Volume 34) (p. 594). Beijing: People's Publishing House.

[41] Li Chunbo. (1983). Improving the inspection and compilation to tap transport potential. *Railway Transport and Economy*, (1), 12–15.

[42] Taking CNKI as an example, since 1983, there have been a total of 1225 academic papers related to "tapping the potential and expanding capacity", with over 1200 of them originating from the railway industry.

2.2.1 Background behind the Introduction of the "Tapping the Potential and Expanding Capacity" Strategy

After the Cultural Revolution, the entire nation, from the central to the local levels, was shocked at the booming global economy and lamented over the detours taken by the country during its progress. This prompted a new craze for economic development. On April 11, 1977, an editorial in the *People's Daily* indicated, "The people's war to expose and criticize the 'Gang of Four' is advancing triumphantly, driving all aspects of work to progress rapidly and strive for excellence. The national economy is gradually returning to a rosy development track, and a new leap forward is taking shape".[43] A few days later, another editorial entitled *Adherence to the Outline of Governance to Promote a New Leap Forward in the National Economy* stated that "A new leap forward is taking shape. By overthrowing the 'Gang of Four', our thoughts have been greatly liberated, and so have our productive forces. The railway transportation, severely disrupted and damaged by the 'Gang of Four', has progressed rapidly. In March, the daily loading volume exceeded the best level of the same period in history. In early April, it continued to make steady progress, with a new increase in daily loading. The significant improvement in railway transportation has promoted a comprehensive increase in industrial production. It is particularly gratifying to see the tremendous transformation in industrial and mining enterprises in regions severely disrupted and damaged by the 'Gang of Four', where both revolution and production are thriving. The workers in places like Shanghai vowed to reach and surpass 'three levels', namely, first reaching and exceeding the highest historical production level of their own institutions or companies, then surpassing the highest of the same industry nationwide, and finally catching up with and surpassing the world's advanced level".[44]

Encouraged by the enthusiasm of the entire nation to recover wasted time, the central government revised the *Outline of the Ten-Year Plan* (Draft Version) in the context of the generally favorable trend of national economic progress in 1977. In the *Government Work Report* presented at the First Session of the 5th NPC in the following year, it was proposed that "By the end of the 20th century, China will become the highest-yielding country in the agriculture sector around the world; the industrial levels of many provinces should catch up with and surpass those in some industrially developed countries in Europe; agricultural sector should achieve the highest level of mechanization and electrification, major parts of industrial production should be automated, and transportation should be significantly sped up; and various economic and technical indicators should approach, catch up with, and

[43] (1977, November 11). The strategic decision to implement the Outline of Governance. *People's Daily*.

[44] (1977, April 19). Stick to the Outline of Governance to promote a New Leap Forward in the national economy. *People's Daily*.

exceed those of advanced countries in the world".[45] To achieve these goals, the *Government Work Report* emphasized that a series of economic targets needed to be met between 1976 and 1985.

Regarding railways, it stipulated, "To establish transportation and postal networks meeting industrial-agricultural needs, essentially realize electrification and dieselization of locomotives, … and complete six new trunk rail lines within the next three years as part of 120 large-scale national projects".[46] At the National Capital Construction Work Conference held in May 1978, mobilization slogans included "total state capital construction investment over the next eight years should exceed the cumulative sum of the previous 28 year" and "concentrate forces to achieve decisive victories".[47]

As a consequence, "the number of completed large and medium-sized projects decreased by 18.2% although infrastructure investment in 1978 increased by 31.6% compared with 1977, and the number of individual projects put into production fell by 29.1%… Last year, the fiscal revenue, calculated on a comparable basis, increased by 20 billion yuan compared with that of the previous year, of which 15 billion yuan was allocated to the development of infrastructure".[48] This economic progress, criticized by Chen Yun as the "New Leap Forward", far exceeded the capacity of the then fiscal revenue. By 1978, the national finances were already quite strained. On August 5, 1978, the *People's Daily* published an editorial entitled *Ongoing Capital Construction Projects Should Be Thoroughly Streamlined*, which pointed out that "infrastructure development suffers from haste expansion of work fronts, dispersed material use, and disorganized management. As a result, the construction period has been greatly extended, project costs have generally risen, construction quality has significantly declined, and investment efficiency has become increasingly poor. Large amounts of state funds have been tied up in ongoing projects, leading to annual calls for concentrating efforts to settle key problems, while in reality, some construction projects are consuming resources in a protracted way".[49] This indicates that there were already differing voices regarding the economic leap forward in 1977. Given the limited national fiscal revenue at that time, infrastructure development had become a huge burden and severely affected the normal progress of the entire economy.

Thus, the proposition that economic development should be "comprehensively balanced" gradually took shape. It advocated that "the quality of comprehensive

---

[45] (1978, March 7). *Uniting for the construction of a modern and strong socialist country: Report on the work of the Government at the First Session of the Fifth National People's Congress on February 26*, 1978. *People's Daily.*

[46] Ibid.

[47] Wang Zhenchuan. (2015). *Yearbook of the new period of China's Reform and Opening up (1978)* (p. 327). Beijing: China Democracy and Law Publishing House.

[48] Literature Research Office of CPC Central Committee. (1989). *Selected works of Li Xian'nian* (p. 348). Beijing: People's Publishing House.

[49] (1978, August 5). *Ongoing capital construction projects should be thoroughly streamlined, People's Daily.*

balance is related to whether the socialist economy can develop in a planned, proportional, and expedient way, and whether the merits of the socialist system can be given full play. If the scale of infrastructure development is too large and exceeds the limits of material and financial resources, the construction period would become longer, and the investment returns increasingly worse. Many projects were rushed into construction without adequate preliminary preparation or consideration of actual conditions and economic efficiency, which compromised the development of optimal construction plans. Instead, the decision-makers then blindly pursued scale and speed, ignored costs, and employed a war of attrition approach, leading to astonishing waste in resources".[50] The term "comprehensive balance" represents Chairman Mao Zedong's reflective viewpoint on the "Great Leap Forward" before the Lushan Conference in 1959. The central government's re-emphasis on "comprehensive balance" at this time also clearly reflects a re-evaluation of the economic "leap forward" after the Cultural Revolution.

It is also easy to see that there was a shift in the national economic development strategy during the first half of 1979. From the passionate New Year's message of the Xinhua News Agency on January 1, 1979, to the more cautious Four-Word Guiding Principle of "Adjustment, Reform, Rectification, and Improvement"[51] announced at the Central Work Conference in April, the central government's guidelines were transitioned from a radical to a prudent approach on economic development.[52]

Railways have always been a major part of infrastructure construction. During the Fifth Five-Year Plan period, especially when the railway modernization plan was formulated at the beginning of 1978, it was proposed to build 13,000 kilometers of new railways and 16,000 kilometers of electrified railways within the next 8 years. The investment for the first two years was entirely allocated to new line construction in the west, which was obviously unrealistic. In the *Railroad Construction Plan in 1979* (Draft Version) released by MOR in early 1979, there were 63 total investment projects, including 23 new railways and 7 projects for the transformation and electrification of existing lines.[53] The complex terrain and landforms in the west, compounded with hasty infrastructure decisions, underdeveloped technology, chaotic management, and rising project costs, have contributed to significant delays in many railway infrastructure projects. As a result, these projects have become a heavy burden on national finances. In line with the principles outlined at the 1979 Central Committee Work Conference, "priority should be given to the development of urgently needed and feasible 'short lines'. Meanwhile, 'long lines' that are challenging to construct should be set aside temporarily or have their

[50] (1979, February 24). Doing well in ideological liberation and comprehensive balance. *People's Daily*.

[51] Literature Research Office of CPC Central Committee. (1989). *Selected works of Li Xian'nian* (p. 353). Beijing: People's Publishing House.

[52] For related discussions, refer to: Li Zhenghua. (2002). *The conception and beginning of China's Reform and Opening up*. Beijing: Contemporary China Publishing House.

[53] Railroad Construction Plan in 1979 (Draft Version). Archived in Southwest Jiaotong University.

development pace appropriately slowed. This approach will free up manpower, materials, and financial resources to support the development of 'short lines', to invigorate the national economy and promote progress on a new foundation".[54]

After the Central Committee Work Conference, MOR also adjusted its new railway development strategy to "focus on key points and steadily advance during adjustments".[55] Issues with new line construction and existing line renovation were formally raised. In August of the same year, then Minister of Railways Lu Zhengcao published an article stating that "building a new line involves high investment, a long construction period, and slow returns. Currently, due to design, planning, materials, and other reasons, there is significant waste. Although the Beijing-Tongliao Railway is one of the lower-cost lines in recent years, it resulted in a waste of over 300 million yuan. In the long run, the waste of operating costs will be even greater. Building a new line takes at least three to five years, and the Beijing-Tongliao Railway, by the end of this year, will have cost seven years, and cannot be put into full operation in a short time. Therefore, building a new line must be carefully considered".[56] Investment in most new railway projects was reduced or halted. The railway construction saw a sudden shift from "fast-track development" to "renovation of existing lines".

Although the term "tapping the potential and expanding capacity" was not officially used at that time, it had become the basic strategy for all industrial sectors and even the entire economic development. Li Xiannian raised the questions, "Do our enterprises still have potential? Why can't we tap it out?"[57] Chen Yun also pointed out that "we should focus on the potential, innovation, and transformation of existing domestic enterprises. The foundation of our domestic enterprises is substantial. We need to introduce new technology (software) on this basis, either to fill gaps or to support the development of related industries, using these methods to enhance our production capacity".[58] A year later, Chen Yun reiterated the central government's determination that "I have an idea that infrastructure should be constructed at the lowest cost possible, with every penny counted, 'like an iron rooster with no

---

[54] Literature Research Office of CPC Central Committee. (1989). *Selected works of Li Xian'nian* (p. 354). Beijing: People's Publishing House. In the following years, the central government repeatedly emphasized that infrastructure development should follow the principle of "spending within one's means".

[55] (1979, June 21). Performing well in shifting priorities and making steady progress in the adjustment: The principle of "Adjustment, Reform, Rectification and Improvement" should be put into place in railway-related work, *People's Railway Daily.*

[56] (1979, August 3). Lv Zhengcao, Zhao Wenpu, & Bu Ke. Several issues in railway construction: Taking theconstruction of the Beijing-Tongliao Railway as an example. *People's Daily.*

[57] Literature Research Office of CPC Central Committee. (1989). *Selected works of Li Xian'nian* (p. 367). Beijing: People's Publishing House.

[58] Chen Yun. (1986). *Speech at the Debriefing Meeting held by the Central Financial and Economic Affairs Commission (*1979, *September 18). In the Institute of Industrial Economics of* Chinese Academy of Social Sciences (Ed.), *Selected literature on economic policy since the Third Plenary Session of the Eleventh CPC Central Committee (Revised Edition)* (p. 36). Beijing: China Economic Publishing House.

feather pluckable'. My plan is even stingy, to the point of doing nothing at all. If we need to proceed, there are reasons. Not doing anything for three years, the People's Republic of China is still there, and a delay of only three years does not matter that much".[59]

In summary, with the contraction of infrastructure development in all respects and a significant suspension of railway projects, especially new line construction projects, the railway industry, along with the entire industrial sector, has entered a phase of "adjustment and rectification" to tap the potential.

2.2.2 *Implementation of "Tapping the Potential and Expanding Capacity" Strategy in the Railway Industry*

According to the central directives, MOR summarized the issues with railway transportation at the time as "three imbalances": "First, the imbalance between the development of industrial and agricultural production and the growth in railway transportation capacity; Second, the imbalance between the construction of new lines and the renovation of existing lines; Third, the imbalance between the increase and upgrading of railway technology and equipment and the continuously growing passenger and freight volumes".[60]

Upon closer analysis, the three imbalances are not parallel. If the first one is considered a symptom, the other two are the causes. This is because that "In general, for every ton increase in industrial production, the freight volume of raw materials and products will increase by two to three tons correspondingly. However, in recent years, for every ton increase in industrial production, transportation capacity has only increased by 0.9 tons, resulting in a severe imbalance. Transportation capacity serving coal, steel and mining bases, as well as coastal ports, is insufficient with some weak sections of major railway lines meeting only 40% to 70% of the demand".[61]

"In 1980, according to the transportation demands at that time, China faced a shortfall of approximately 10,000 freight cars, 1,800 passenger coaches, and 600 locomotives. Such shortage in equipment capacity led to restriction or even suspension of operations at 70% of the large freight stations across the entire railway network. Within the entire railway network, 30% of freight stations had no cargo warehouses, 41% lacked cargo platforms and loading/unloading lines, and 73% had

[59] Literature Research Office of CPC Central Committee. (2005). *The collected works of Chen Yun* (Volume 3) (P. 471). Beijing: Central Literature Publishing House.

[60] (1979, June 21). Performing well in shifting priorities and making steady progress in the adjustment: The principle of "Adjustment, Reform, Rectification and Improvement" should be put into place in railway-related work, *People's Railway Daily.*

[61] Ibid.

no cargo sheds. At that time, the transport capacity of 25% of railway sections was already saturated, over 20 main lines and 1/3 of the marshalling stations were under severe strain. The number of capacity-restricted railway sections increased rapidly from 5 in 1978 to 20 in the early 1980s, and the entire railway could only meet 50% to 70% of the national transportation demands".[62] All these problems ultimately boil down to two issues: insufficient lines and inadequate equipment.

The *Government Work Report* of the second session of the Fifth National People's Congress proposed to accelerate the technological transformation of major railway lines to enhance transportation capacity. However, in the context of "tapping the potential", MOR summarized the imbalances in the second and third proportions as a response to this directive. Therefore, the *Direction of Adjustment of Railways in the Next Three Years and the Corresponding Plan This Year* issued in May 1979 stipulated that "railway adjustments should be centered on resolving the imbalance issues".

Specifically, it outlined that "based on the national economic demands for railway transportation and resolution of the problems in railway work, the direction and tasks for railway adjustment are summarized as follows: First, we should properly manage the relations between the renovation of existing lines and the construction of new lines, with the immediate priority given to the technological transformation of existing railways; second, we should resolutely improve traction power; third, we should resolutely enhance equipment quality; and fourth, we should emphasize extensive personnel training to improve technology and management levels".[63] The specific approach to railway development was thus determined.

The measures for tapping the potential of the railway industry are comprehensive. For example, there have been extensive and enduring "Railways for the People" education campaigns, widespread efforts to build road ethics, continuous improvements in management and scheduling, and efforts to enhance railway safety. However, these initiatives are themselves prominent advantages and characteristics of Chinese railways and have been ongoing for a long time, so they should not be regarded as the characteristics of this phase. During the years starting from 1978, the railway's response to the issues of the time primarily focused on the renovation of existing lines and the reform of traction power.

MOR had proposed four main plans for the renovation of existing railway lines, which include significantly increasing the double-track rate; extending station lines, improving the vertical and horizontal profiles of the tracks to meet the increased requirements for train weight and speed, and enhancing the strength of the tracks; modernizing key hubs and marshalling yards; and electrifying major trunk lines with high traffic volumes, as electrification directly determines the traction power.

---

[62] Wang Zhizhong, & Wei Liying. (2001). *Research on China's railway reform and development (1978–1998)* (p.20). Beijing: Contemporary China Publishing House.

[63] Miao Qiulin. (1994). *China's railway transport* (p. 63). Beijing: China Railway Publishing House.

At the beginning of the new era of reform and opening up, the imbalance between new and old railway lines in China mainly included the following two aspects, which were largely due to historical reasons.

Firstly, a serious issue with the investment allocation ratio arose in the construction of new and old railway lines.

In the first 30 years of railway infrastructure investment in China, the allocation ratios of funds were generally as follows: 55.8% for new line construction, 19.3% for strengthening existing lines, and 24.9% for locomotive and vehicle purchases, factory construction, and other expenses. However, the ratios varied over different periods. From the founding of PRC until the end of the Second Five-Year Plan, the proportion of funds used for new line construction was 39.2%, while that for strengthening existing lines was 30.3%, with a new-to-existing line investment ratio of 1.29:1. After the end of this plan, starting from 1962, the proportion of funds for new line construction increased to 60.6%, while that for strengthening existing lines dropped to 16.8%, resulting in a new-to-existing line ratio of 3.6:1. During the Third and Fourth Five-Year Plans, the new-to-existing line ratio reached an astonishing 6.5:1.[64] This indicates a severe imbalance in the funds allocated to existing lines, which exacerbated the second imbalance.

Secondly, there is a serious imbalance in the proportion of railway construction between the eastern and western regions in the national railway construction planning. According to relevant data, "from 1950 to 1977, among the 96 railways constructed in China, 75.4% by mileage and 86% by investment were located in the area west of the Beijing-Guangzhou Railway. Basically, no new main railways were constructed east of the Beijing-Guangzhou Railway, except for the construction of some port railways, forest railways, and railways for transporting oil and coal. Provinces such as Jiangsu, Guangdong, Jiangxi, and Zhejiang did not build new railways after the founding of PRC".[65] As a result, the length of the railway network in the area west of the Beijing-Guangzhou Railway had increased from 4000 kilometers to 20,000 kilometers since the founding of PRC, rising from about 17.7% of the national total railway mileage before 1949 to 43% at that time.

We certainly cannot view the above two imbalances as merely a historical legacy issues. On the one hand, this is related to the starting point of railway construction in PRC; on the other, it is fundamentally determined by the socialist nature of Chinese railways. Railway construction in the first 30 years of PRC, particularly in the western regions, has undoubtedly improved the regional distribution of railways to a great extent. It provided favorable conditions for socialist development in the western regions, and contributed to national unity and the identification with the socialist New China in remote areas, embodying the merits of the socialist system. This situation also fully demonstrates the dual characteristics of Chinese railways,

---

64 The relevant data is calculated based on: Statistics Bureau of MOR. (2008). *National compilation of railway statistics* (internal resources). Beijing. China Railway Publishing House.

65 Hu Guangrong. (1979). Discussion on the investment proportion of railway construction in China. *Railway Transport and Economy*, (1), 2–11.

which are both political and economic. During these 30 years of development, the political function of Chinese railways far exceeded their economic value.

Specifically, MOR adjusted the investment ratio: focus was put on the construction of new railways transporting coal from Shanxi to other provinces and on the upgrade of existing railways in East China. At the end of the Sixth Five-Year Plan, the total length of operational railways reached 52,119 kilometers, an increase of 4.4% compared to 1980.[66] Railways with newly constructed second track included the Shijiazhuang-Dezhou Railway, the Shuangqiao-Huairou Railway, the Shenyang-Dandong Railway (the Wujiatun-Nanfen section), the Taiyuan-Jiaozuo Railway (northern Changzhi-Yueshan section), 345.7 kilometers in total. Railways that had been renovated and electrified included the Shijiazhuang-Taiyuan Railway, the Xiangyang-Chongqing Railway (the Xiangfan-Da County section), the Beijing-Baotou Railway (the Fengtai-Datong section), the Lanzhou-Lianyungang Railway (the Baoji-western Lanzhou section) and the Chengdu–Chongqing Railway, with a total length of 2,275 kilometers. 6,142.2 kilometers of newly constructed national railways were put into operation, and 789.2 kilometers of regional railways were completed. However, construction of all these railways had started a long time back except for the Yanzhou-Shijiusuo Railway (307 kilometers in total). The construction of the Anhui-Jiangxi Railway (the Huolonggang-Guixi section) began in 1958 and the Xiangyang-Chongqing Railway (the northern Xiangfan-western Chongqing section) in 1968. Newly launched railway construction projects included the Xinxiang-Heze Railway, the Houma-Xi'an Railway (the Houma-Yumenkou section), the Beijing-Qinhuangdao Railway (the Shuangqiao-Qinhuangdao section), Northern Xinjiang Railway (Urumqi-Alashankou) and the Datong-Qinhuangdao Railway (the Hanjialing-southern Liucun section), with a total length of 1,638.5 kilometers. All the railways were dedicated to transporting coal from Shanxi to other provinces, except for the Northern Xinjiang Railway.[67] Consequently, three related marshalling stations including the Shenyangxi Railway Station were built, and junctions and marshalling stations in Harbin, Xuzhou, Nanjing and Shijiazhuang were renovated and expanded.

Regarding traction power, there were also two imbalances.

The first imbalance manifested itself in the investment ratio between railway construction and rolling stock procurement. Following the founding of PRC, approximately 19.3% of national railway infrastructure investment was allocated to locomotive and rolling stock procurement—a proportion that cannot be considered low. However, "after 1965 with the launch of the Third Front Construction campaign, railway investments became disproportionately concentrated on new line construction, leading to a continuous decline in rolling stock funding. Therefore, the increase of rolling stock lagged behind that of traffic volume. This problem was even more evident in consideration of the density of passenger and freight traffic

[66] Miao Qiulin. (1994). *China's railway transport* (p. 71). Beijing: China Railway Publishing House.

[67] Xu Zenglin. 1999. *Fifty Years of the Railroads of People's Republic of China (1949–1999)*. Beijing: China Railway Publishing House. Counted by the author, the mileage only covers statics from 1979 to 1985.

and the increased speed of rolling stock per hundred kilometers in operation".[68] According to statistics, from the year PRC was founded to 2007, the number of locomotives had increased by 3.51 times and their output 2.35 times. Together they had an increase of 8.25 times, slightly lower than the increase in overall density of passenger and freight traffic which was 8.44 times. The supply and demand of railway capacity could basically match, but backup capacity is missing. The number of freight cars increased by 4.98 times and the daily output 2.72 times. The total increase was 13.6 times whereas the volume of freight traffic rose by 16.6 times. Investment in passenger coaches was the lowest in comparison to other sectors, accounting for 1.9% of the total on railway at the time. As a result of the severe shortage of passenger coaches, boxcars were deployed as substitutes during the Spring Festival transportation.[69]

It should be emphasized that these statistics can only indicate the overall situation and do not reflect the disparity in different areas. The relationship between passenger and freight transport demand and the number of locomotives on the eastern and western sides of the Beijing-Guangzhou Line is quite different. Considering the investment ratios and transport demands in the eastern and western regions, the number of locomotives in the east is far from meeting the actual demand.

The second imbalance was the disproportion in the numbers of steam locomotive, internal combustion locomotive and electric locomotive. Strictly speaking, the development of a certain kind of locomotive should not simply be assessed by the correlation between passenger and freight transport demand and the number of locomotives. Instead, it should be evaluated based on the transportation capacity of different traction types, factoring in operational costs per ten thousand kilometers and fuel consumption. The transportation capacity of a railway is also determined by the weight of the trains, its traffic density and operation speed.[70] To put it simply, based on these three criteria, the electric locomotive has the best transportation capability, with the internal combustion locomotive ranking second and the steam locomotive third.[71] However, data shows that in 1980, the total number of locomotives was 10,278, including 7,801 steam locomotives, 2,190 internal combustion

[68] Hu Guangrong. (1979). Discussion on the investment proportion of railway construction in China. *Railway Transport and Economy*, (1), 2–11.

[69] Planning and Statistics Bureau of MOR. (2008). *National compilation of railway statistics* (internal resources). Beijing. China Railway Publishing House.

[70] The Planning Institution of MOR. (1980). Views on the development and application of traction power in China. *Railway Transport and Economics,* (2), 21–25.

[71] In terms of transport capacity on single-line semi-automatic block lines, electric traction improves by 37.5% to 118% compared to steam traction, and by 29% to 41% compared to internal combustion traction, with greater increases as the gradient limits rise. Internal combustion traction improves by 0% to 54% compared to steam traction, also with greater increases as gradient limits rise. On double-line automatic block lines, electric traction improves by 13% to 72% compared to steam traction, and by 13% to 14% compared to internal combustion traction, with increases also growing with gradient limits. Internal combustion traction improves by 0% to 53% compared to steam traction, with similar gradient-related increases. Additionally, the operating cost per ten thousand ton-kilometers for electric traction decreases with increasing transport volume on the

locomotives, and 287 electric locomotives, which accounted for 75.9%, 21.3%, and only 2.8% of the total respectively.[72] From the perspective of locomotive gross ton-kilometers i.e., the total workload transported by trains that are pulled by locomotives, the total locomotive gross ton-kilometers was 1.0718 trillion in 1980, of which steam locomotive accounted for 79.9%, internal combustion locomotive 18.0% and electric locomotive 2.1%.[73] The number of electric locomotives was associated with the mileage of electrified railways. The 287 electric locomotives were mainly distributed among the Baoji-Chengdu Railway, the Ankang-Yangping Railway, the Xiangyang-Chongqing Railway (the Xiangyang-Ankang section) and the Shijiazhuang-Taiyuan Railway (the Shijiazhuang-Yangquan section). However, steam locomotives primarily operated on the busiest routes, including the Beijing-Guangzhou Railway, the Guangzhou-Shenzhen Railway, and the Tianjin-Pukou Railway and Harbin-Dalian Railway, with a limited number of internal combustion locomotives complementing their use.[74]

In fact, the upgrade of traction power in Chinese railways started quite early. Marked by electrification, traction power upgrade in China began in 1956. The USA, the earliest among developed countries, initiated the upgrade in 1945. The UK, Japan, France and the Federal Republic of Germany took similar steps at about the same time as China. The USSR, however, started even later in 1959. Different from China, these countries took approximately 20 years to complete the upgrade, transitioning to electric traction and phasing out steam locomotives by the mid-1970s. Thus, China's experience with traction power reform exemplifies the paradox of "an early start culminating in delayed completion". The reasons for this are multifaceted. The most crucial reason is the "lack of consensus on the technical policies for traction power upgrade".[75] Due to ambiguities in policies, various departments held differing opinions, and locomotive factories persisted in producing locomotives independently. Under the "monopoly of the purchase and marketing" management system, the operating departments had no choice, which consequently delayed the progress of traction power upgrade.

In 1977, MOR adopted the approach of "concurrent upgrades to both internal combustion and electric traction power, with an emphasis on the latter". As explicitly stated in the *Main Railway Technological Policies* in 1983, the targets of traction power upgrade included "gradually replacing steam locomotives with electric and internal combustion locomotives that have better traction performance and

same gradient, with only a slight increase in cost as the gradient rises, a benefit not shared by internal combustion or steam traction. Ibid.

[72] Planning and Statistics Bureau of MOR. (2008). *National compilation of railway statistics* (internal resources). Beijing. China Railway Publishing House.

[73] Planning and Statistics Bureau of MOR. (2008). *National compilation of railway statistics* (internal resources). Beijing. China Railway Publishing House.

[74] Miao Qiulin. (1994). *China's railway transport* (p. 271). Beijing: China Railway Publishing House.

[75] The main arguments were which should be the priority in the combining development of internal combustion and electricity, whether locomotive selection should be standardized or diversified and which had superior transmission method, among other issues. Ibid.: 266.

higher thermal efficiency… Railway traction power should be primarily electric, with the proportion of traffic handled by electric locomotives gradually increasing".[76] Following the issue of the policies, the upgrade of traction power accelerated. Especially after 1983, the number of internal combustion locomotives and electric locomotives increased significantly, with improvements in production layout, capacity, technology, and product quality.[77] By 1985, the total number of locomotives reached 11,772, a 7.3% increase from 1980, with an average power of 1,934.5 kW per locomotive, a rise of 11.0%.

"Notable developments included the introduction of 421 units of the 2,550 kW GE C36-7 electric transmission internal combustion locomotives from the United States, as well as the import of production technology for 10 key components of internal combustion locomotives. Additionally, 150 units of 6,400 kW eight-axle high-power electric locomotives were ordered from the European company 50Hertz. The domestic 4,800 kW Chinese Railway SS3 electric locomotive was manufactured, and the 6,400 kW eight-axle Chinese Railway SS4 electric locomotive was developed as a prototype. Domestic internal combustion locomotives put into operation included the 3,018 kW DF8, the 1,460 kW DF7, and the 1,213 kW DF5 shunting locomotives".[78] Despite these advances, during the Sixth Five-Year Plan period, the number of electric locomotives only increased by 300, while internal combustion locomotives increased by about 1,400, and 65.2% of all locomotives were traditional steam locomotives.[79]

In addition to the two major tasks, the railway system also carried out reforms in institutional structure, management systems, multi-channel financing, rectification of conduct, and transportation organization. From the perspective of absolute quantity, the achievements during this period were quite optimistic, especially considering the decreasing investment from the State in railways, which highlighted the enormous efforts made by the Railway Man.

However, in terms of the imbalance between transport volume and capacity, the problem not only remained unresolved but became more severe. In other words, the development of railway capacity fell far behind the national and societal demand for railway transport volume.

[76] MOR. (1983, October). Main railway technological policies in China.

[77] Xu Zenglin. 1999. *Fifty Years of the Railroads of People's Republic of China (1949–1999)* (p. 156). Beijing: China Railway Publishing House.

[78] Miao Qiulin. (1994). *China's railway transport* (pp. 72 + 271). Beijing: China Railway Publishing House.

[79] Planning and Statistics Bureau of MOR. (2008). *National compilation of railway statistics* (internal resources). Beijing. China Railway Publishing House.

2.2.3 *Reflection on and Impact of the "Tapping the Potential and Expanding Capacity" Strategy*

First, it should be noted that the "tapping the potential and expanding capacity" strategy was not only a requirement of the Party and the State for the entire industrial sector during the early years of reform and opening up but also a proposal by MOR based on the central government's guiding principles for economic development. It represents the implementation and enforcement of the Central Committee's policies and directives. We cannot simply categorize it as correct or incorrect, or scientific or unscientific. It was a necessary choice based on the comprehensive national strength and fiscal situation at that time, reflecting a certain understanding of the developmental stage of the railway industry.

More specifically, after 40 years, we need to find reasons why the achievements in railway development during the early reform period were less than satisfactory. While it is undeniable that the disproportionate imbalances summarized by MOR were a fact, is this the fundamental reason for the severe shortage in railway transport capacity? Further, while historical facts show that the central government's decisions were simply true on a macro level, did they align with the specific circumstances of the railways on a micro level? We shall acknowledge that the overall inadequacy of national fiscal revenue determined the railway development strategy of that period, but this does not preclude further analysis and exploration of the railway reforms themselves.

First, the primary challenge of the railway system during the early reform and opening up period involved routes that operated independently rather than as part of a cohesive network.

As mentioned earlier, the core issue of railway reform was the relationship between capacity and volume. During the early reform and opening up period, the western region had a significant excess of capacity, while the eastern region suffered from a severe shortage. This was mainly because the western regions being economically underdeveloped compared to the eastern regions, had far less demand for railway transportation. Specifically, the six major north-south lines (the Harbin-Dalian, Beijing-Shenyang, Tianjin-Pukou, Shanghai-Nanjing, Shanghai-Hangzhou, and Beijing-Guangzhou Railways) and six major east-west lines (the Beijing-Baotou, Shijiazhuang-Taiyuan, Shijiazhuang-Dezhou, Jiaozhou-Jinan, Lanzhou-Lianyungang, and Zhejiang-Jiangxi Railways) totaled about 10,000 kilometers, accounting for 20% of the national railway network. These lines had been in operation for 70 to 80 years, traversing densely populated and prosperous areas. In the 30 years between 1949 and 1978, China's railway freight volume grew by 18.2 times and passenger volume increased by 6.8 times. Notably, three-quarters of the cargo was transported via these lines.[80] This means that 20% of the railway lines handled 80% of the traffic, which was a significant factor contributing to the

[80] Liu Ruilin. (1980). Several problems in the development of Chinese railway transportation. *Railway Transport and Economy*, (1), 1–8.

"imbalance between industrial and agricultural production growth and railway capacity expansion". As a result, "existing lines were insufficient in capacity, while new lines had high capacity but suffered large losses. Technology and equipment were outdated, with low proportions of double tracks and electrified lines, and locomotives were insufficient and primarily composed of steam-powered ones".[81]

Data shows that from 1965 onwards, the 33 new lines had a transport density of only 8.5 million converted ton-kilometers per kilometer, 46.5% of that of the existing lines, and incurred annual losses of 850 million yuan, averaging a loss of 86,000 yuan per kilometer each year, with a profit rate of −4.8%. From the perspective of economic return, the eastern regions needed railway transport capacity and the operation of railways could be profitable, but were underserved, while the western regions had low demand and excessive capacity, leading to inefficiencies such as "in 1978, where the fixed assets of the Chengdu, Kunming, and Xi'an Railway Administrations accounted for 30.4% of the total railway assets but only completed 9% of the railway's transportation workload".[82]

In this sense, upgrading the existing lines may seem inevitable, but it was not necessarily so. In fact, the idea of "tapping the potential and expanding capacity" harbored a logical problem. The strategy of focusing on "upgrading existing lines" was based on two facts: Firstly, there was a severe shortage of capacity in the eastern region while excess capacity existed in the western region; secondly, the eastern region was mainly served by existing lines, while the western region predominantly new lines. However, these two facts do not necessarily mean that simply upgrading the existing lines in the eastern regions would resolve the issue of insufficient capacity. The underlying assumption held that the operational mileage of eastern railways was sufficient and that the railway network was well-established, thus making the construction of new lines a lower priority.

Although China had constructed approximately 30,000 kilometers of railways by 1980, it had only established a skeletal railway framework, which was far from complete. This was mainly reflected in several ways: "In coastal transport busy areas, there was a lack of radiating lines, diversion lines, and connecting lines between trunk lines, leading to increasingly concentrated traffic on a few main lines with minimal maneuverability. Many economic regions were not connected by rail. For example, Fujian and Guangdong were not directly connected and transportation between them had to take a detour through the Zhejiang-Jiangxi Railway. There were disruptions in railway access to coastal ports, and some of them even did not have railway connections. Between the major regions, especially in remote and coastal areas, there were few connecting routes with insufficient transport capacity. For instance, only the Baoji-Lanzhou Railway linked the Central Plains to the Northwest. Urgent demand arose for the construction of railways to transport goods

[81] Hu Guangrong. (1985). Countermeasures on increasing economic benefit of railways, In China Academy of Railway Sciences (Ed.), *Study on the strategies of railway development* (internal information, p. 205).

[82] Hu Guangrong. (1979). Discussion on the investment proportion of railway construction in China. *Railway Transport and Economy*, (1), 2–11.

in many resource-abundant areas, such as coal in northern Shaanxi and Inner Mongolia, and oil in Xinjiang".[83]

Therefore, people in the railway industry had doubts about the concept of "upgrading existing lines". At the MOR's leadership conference in 1979, a representative from China Railway Siyuan Survey and Design Group proposed, "MOR should develop a railway network plan based on the requirements of national economic development and the actual conditions of the railways. In this regard, an annual plan should be formulated to coordinate the construction of new lines and the upgrading of existing lines. Without a network plan, a lack of direction is inevitable. Currently, our colleagues tend to emphasize the importance of upgrading existing lines without question. In my opinion, we need to evaluate the situation. Does the existing line necessarily have insufficient capacity? What changes will occur in passenger and freight traffic after the new lines are built? A detailed analysis should be conducted. For example, if the Anhui-Jiangxi Railway can be completed earlier, it will divert three million tons of cargo from the old Nanjing-Shanghai-Hangzhou railway, reducing detours by 190 kilometers and meeting wartime needs. It could avoid major cities like Shanghai, add a route to South China, and benefit the construction of many factories in small third-front cities of East China. Why not to expedite its completion?"[84]

In summary, China's railways suffered from "having independent routes not the cohesive network". This was a problem that could never be resolved by upgrading existing lines.

China's High-Speed Rail Network is a product of the complex system. Broadly speaking, the formation of the Network represents a dynamic and relative indicator. "It should be based on the requirements of the national economic development, in conjunction with the appropriate arrangement of various transportation modes. When railways can meet current and foreseeable transportation needs, the High-Speed Rail Network can be considered complete".[85] Several key indicators are involved, such as mileage of operational railways (including double-track mileage), comprehensive transportation capacity, passenger and freight traffic density, distribution of hubs and marshaling yards, locomotive power, etc. In any case, around the early 1980s, China's railway indicators were far from meeting these standards.

Despite three decades of development following the founding of PRC, the country's total railway operating mileage had reached nearly 50,000 kilometers, ranking third in the world at that time behind only the United States and the Soviet Union. However, "in terms of land area, the statistics showed that for every thousand square kilometers, there were 36.2 km of railway in the United States, 78.5 km in the United Kingdom, 60 km in France, 119.5 km in West Germany, 57.5 km in Japan,

[83] Lv Yinghua. (1994). *China's railway advancing in the reform*. Beijing: China Railway Publishing House.

[84] Bulletin of the National Railway Leading Officers Conference (No. 21). Archived in Southwest Jiaotong University, ([1979] No. 1-0-2 (1)).

[85] Liu Jitang. (1979). Past, present and future of railway: Discussion on several problems in development. *Harbin Railway Science and Technology*, (6), 1–17.

19.8 km in India, and 6 km in the Soviet Union, whereas China had just 5 km. On a per capita basis, the railway length was 1.66 km per thousand people in the United States, 0.21 km in Japan, 0.56 km in the Soviet Union, 0.1 km in India, and only 0.05 km in China".[86]

In contrast, during this period and for a considerable time afterward, other forms of transportation in China, such as maritime shipping, highways, aviation, and pipelines, were still far from capable of handling large transportation volumes. Consequently, railways were responsible for carrying 80% of the nation's total freight volume and 70% of its total passenger traffic.

In this sense, at the dawn of the new era of reform and opening up, the core issue of China's railway development was not merely the renovation of existing lines, but rather the construction of new ones.

Secondly, adjustment to the macroeconomic development strategy effectively pushed railway construction into a subordinate role.

From a macro perspective, the Third Plenary Session of the Eleventh Central Committee of CPC proposed that "the living standards of urban and rural residents should be gradually improved based on the development of production".[87] The core guiding principle was to strive for a rapid improvement on the quality of life for the general public. Given the limited national fiscal revenue, the focus was put on efforts to develop industries that the people urgently needed. More specifically, this involved addressing the imbalance in the economy with an excessive proportion of heavy industry by prioritizing the development of agriculture and light industry. This was both an economic and political decision.

The *Report of the 12th National Congress of CPC* emphasized, "Over the next twenty years, we should concentrate our efforts on the fundamental areas of agriculture, energy and transportation, as well as education and science, regarding them as strategic priorities for economic development. By addressing these issues in a balanced way, we can promote faster growth in the production of consumer goods, drive the development of the entire industrial sector and other production and construction endeavors, and secure the improvement of people's livelihood".[88]

Regarding transportation issues, the Central Working Conference of CPC in 1979 noted that the tension in transportation remained unresolved and the capacity of some weak sections of major railway lines could only meet fifty to seventy percent of the demand. This reasserted the judgment made in *Document No. 9*. The

---

[86] Fu Zhong. (1982). Discussion on the speed of railway construction in China. *Journal of Northern Jiaotong University*, (2), 69–78 + 182.

[87] Literature Research Office of CPC Central Committee. (2008). *Anthology of important documents in the past 30 years of Reform and Opening up* (Volume I) (p. 17). Beijing: People's Publishing House.

[88] Hu Yaobang. (2008). Comprehensively creating a new situation for socialist modernization: Report to the 12th National Congress of the Communist Party of China. In the Editorial Group of the following book (Ed.), *Reports, communiques, resolutions, and decisions of All CPC National Congresses and Central Committee Plenary Sessions since the Third Plenary Session of the 11th CPC Central Committee* (p. 137). Beijing: China Fangzheng Press.

term "tense transportation situation" would continue to recur frequently in many important documents of the state and the Party for a considerable period of time. However, transportation was viewed as the solution to energy problems. "Currently, the most pressing issue is the shortage of energy and insufficient transportation capacity, which severely constrains the development of the national economy. According to estimates by relevant departments, approximately twenty percent of the nation's industrial production capacity is idle due to a lack of coal and electricity. …The problem of insufficient transportation capacity is also severe. While coal is in short supply, some mine operators are unable to transport the coal they produce. It is clear that without progress in infrastructure such as energy and transportation, any effort to expand processing industries would be futile. In other words, unless the issues of energy and transportation are resolved, the overall economic vitality cannot be revitalized".[89] The central government held the view that transportation construction should be used to resolve energy issues, which in turn would serve the needs of industry and agriculture. From this macroeconomic perspective, railway construction was, in fact, assigned a relatively lower priority.

In 1982, the World Bank Mission provided a pessimistic forecast of the future impact of China's transportation on the national economy in its report, *China: The Development of a Socialist Economy Annex VI: Transportation Sector*. While the report acknowledged the sound rationale behind Chinese railway construction strategies at that time, it also pointed out numerous unresolved issues within the railway sector. These challenges were expected to have far-reaching consequences, particularly in the areas of energy, transportation and other related sectors. According to the report, Chinese railways prioritized the development of heavy industry in the past, resulting in the transportation sector centering on freight for industrial goods rather than passenger services. In addition, rail was used to transport workers and facilitate population movements to remote areas of China, aiding in the regional administration. However, with the adjustment of economic priorities in the new era, there would be a greater demand for the transportation of light industrial products, bulk goods, and passengers. According to data provided by MOR to the World Bank, the annual growth rate for freight volume was projected to be 5%, while for passenger volume, it was 6%. This suggested that by 1990, transportation volumes would at least double. However, given the overall state of China's transportation sector, particularly the fact that railways were expected to handle 75% of this increased volume, the pressure on the railway system was anticipated to be immense.[90]

Additionally, during this period, there was a widespread trend in developed countries of dismantling their railway networks, resulting in the perception of

---

[89] (1986). Centralized Financial and Material Resources to Ensure Key Construction. In the Institute of Industrial Economics of Chinese Academy of Social Sciences (Ed.), *Selected literature on economic policy since the Third Plenary Session of the Eleventh CPC Central Committee* (Revised Edition) (p. 169). Beijing: China Economic Publishing House.

[90] World Bank Mission. (1982). *China: The development of a socialist economy—Annex VI: Transportation Sector*. Beijing: *China Financial & Economic Publishing House.*

railways as a sunset industry. For example, between 1970 and 1980, the United States saw its railway network decrease from 314,500 kilometers to 294,700 kilometers, while the United Kingdom experienced a reduction from 19,000 kilometers to 17,900 kilometers.[91] Coupled with the tendency to understand the issue in terms of absolute operational mileage, there was a misperception of the historical stage at which Chinese railways found themselves. In short, we overlooked that developed countries not only went through a period of intense railway construction and development, completing their railway networks, but also diversified their transportation modes that can absorb large volumes of traffic, leading to a significant decrease in railway traffic. In contrast, our country's railway network was under development, similar to that of developed countries in the early twentieth century. In fact, even around the 1980s, many railway experts were calling for the construction of a 100,000-kilometer railway network.

In short, the strategy of "tapping the potential and expanding capacity" ultimately stems from the nation's insufficient capacity to allocate the necessary resources for construction. Over the next few years, the national financial allocations for railways steadily declined. "In 1979, the investment in MOR was slashed from the initially planned 3.24 billion yuan to 2.494 billion yuan, a reduction of almost one-fourth. To align with the policy changes and the national plan, MOR firmly canceled 12 out of 63 major infrastructure projects. Priority was given to upgrading critical sections of the six main coal transportation routes, including those for transporting coal from Shanxi. The new track-laying plans were reduced by half. In 1980, the infrastructure investment for MOR was only 1.473 billion yuan, barely enough to support projects related to coal transportation from Shanxi and Henan provinces. In 1981, the investment was further reduced to just 1.003 billion yuan".[92] For example, in 1979, despite dealing with existing lines, the Shanghai Railway Administration "faced a 25 percent reduction in investment. Among these cuts, the Shanghai-Hangzhou line, a critical section, saw a 50 percent reduction, while the Shanghai-Nanjing line had a 10 percent cut".[93] Consequently, key projects were impacted, and the transportation capacity of the main lines did not see much improvement. In 1982, 1983, and 1984, the infrastructure investment of MOR saw some increases, mainly directed towards the renovation and upgrading of existing railway lines. Overall, throughout the Sixth Five-Year Plan period, the total new railway mileage added in our country was just over 1,500 kilometers, averaging only about 300 kilometers per year. The share of railway investment in the national infrastructure investment dropped to 7.0%, only marginally higher than that during the Fifth Five-Year Plan period.

---

[91] Gu Yonglin. (1983). Reduction of railways in major capitalist countries from 1970 to 1980. *Foreign Railway Locomotive and Motor Car*, (8), 49.

[92] Xu Zenglin. 1999. *Fifty Years of the Railroads of People's Republic of China (1949–1999)* (p. 150). Beijing: China Railway Publishing House.

[93] Bulletin of the National Railway Leading Officers Conference (No. 21). Archived in Southwest Jiaotong University, (No. 1–0-2 (1) [1979]).

2.2.4 Continuation of the Strategy of "Tapping the Potential and Expanding Capacity"

After several years of efforts aimed at "tapping the potential and expanding capacity", the railway industry not only failed to alleviate the issues in its "weak links", but these problems have, in fact, become more pronounced.

The tense situation in railway transportation has put significant pressure on the Railway Man. At the closing ceremony of the first session of the Railway Transportation Modernization Management Seminar held by MOR in August 1985, Ding Guangen, then Minister of Railways, stated, "Currently, both our railway passenger and freight services cannot meet the demand. Although we fulfill the tasks assigned by the state each year, there is still a significant gap between our performance and the transportation requirements of the people and the state. Our freight capacity can only meet 70% of the demand, and passenger services on some major lines are experiencing severe overcrowding".[94] In December 1985, in *A Blueprint for the Development of Chinese Railways* (the Third Draft) compiled by MOR, it was clearly stated that "Chinese railways have not experienced a phase of development that was either ahead of or in parallel with the nation's industrialization. The growth in railway transport capacity has not kept pace with the needs of the national economy. In recent years, this disparity has become increasingly evident, leading to backlogs in goods transportation and severe overcrowding on some main passenger lines. The railways have thus become a prominent bottleneck in the national economy, limiting economic and social progress. In comparison to the needs of the state, Chinese railways are considered a constrained mode of transportation, facing severe challenges".[95]

Progress of railways had always been falling behind that of the national economy, and the gap was continuously widening. One year later, in February, at the National Railway Work Conference, Ding emphasized once again, "The main contradiction in current railway work lies in the critical issue of severely insufficient transport capacity, which is constraining the development of the national economy".[96] In October 1987, a senior official from MOR responded to a journalist's question, stating that "The backlogged goods and overcrowded passenger dedicated lines are indeed social problems that constrain economic development and cause public

---

[94] Ding Guangen. (1985, November 15). Prioritizing reform and striving to improve railway transport capacity: Speech of Minister Ding Guangen at the Closing Ceremony of the First Railway Transportation Modernization Management Seminar on August 28, 1985 (Excerpt), *People's Railway Daily*.

[95] (1985). *A Blueprint for the Development of China's Railways (the Third Draft)*. Archived in Southwest Jiaotong University, T-272.

[96] 1987, September 19). Proof from Guangzhou Railway Administration's pilot program: Contract Responsibility System outperforming Partial Contract System—MOR promotes Contract Responsibility System covering operations, infrastructure development, and technological transformation. *People's Daily*.

concern. These issues also impose a heavy burden on our 3.2 million railway employees, causing them great anxiety".[97]

However, after 1985, the railway industry continued with the basic strategy of "tapping the potential and expanding capacity". There are roughly two reasons for this continuation.

Firstly, the central government adhered to the principle of "pursuing scaling production primarily through intensive development and improvement". In 1983, the Central Committee of CPC began drafting the *Seventh Five-Year Plan*. At the National Congress of CPC in 1985, the *Proposals of CPC Central Committee on the Seventh Five-Year Plan for National Economic and Social Development* was formulated. The document highlighted "the need to shift the focus of construction to the technical transformation and expansion of existing enterprises, and take a path of scaling production primarily through internal development and improvement".[98] Released in the same year, *Questions and Answers on the Seventh Five-Year Plan Proposals* explained, "We should prioritize leveraging the existing capabilities of enterprises rather than primarily relying on increasing investment in infrastructure and the labor force. Through technological advancement, equipment renovation, enhanced management, and the tapping of internal potential, we can achieve expanded reproduction. Once a certain foundation is established and enterprises reach a certain scale and level, pursuing expanded reproduction primarily through intensive development and improvement can reduce fixed asset investment, make more effective use of construction funds, improve economic efficiency, and accelerate technological progress".[99] In this context, the railway sector clearly meets the criteria of "having an established foundation and enterprises reaching a specific scale and level". This guiding principle is undoubtedly correct.

Secondly, the state made adjustments to its development strategies for the transportation sector. In the transportation sector, according to the *Proposals of CPC Central Committee on the Seventh Five-Year Plan for National Economic and Social Development*, we should concentrate the necessary financial, material, and technical resources on construct key projects in energy, transportation, communications, and raw materials industries with high quality and efficiency. ... We would continue to focus on railway construction while also stepping up efforts on the construction of roads, waterways, and air transportation, improving the transportation structure, promoting the modernization of transportation and the rational organization of various transport modes, thereby enhancing the efficiency and quality of

---

[97] Zhang Guorong. (1987, October 24). Railway development lies in reform: An interview with a responsible from MOR. *People's Daily.*

[98] (2008). Proposals of CPC Central Committee on the Seventh Five-Year Plan for national economic and social development. In the Editorial Group of the following book (Ed.), *Reports, communiques, resolutions, and decisions of All CPC National Congresses and Central Committee Plenary Sessions since the Third Plenary Session of the 11th CPC Central Committee* (p. 137). Beijing: China Fangzheng Press.

[99] Bajun, & Shiren. (1985). *Questions and answers on the Seventh Five-Year Plan proposals* (p. 52). Hangzhou: Hongqi Press.

transportation. ... In line with the requirements to accelerate transportation construction, it was necessary to develop the automotive manufacturing industry as an important pillar industry, aiming for significant development, while also advancing the manufacturing of locomotives, ships, and aircraft. During this period, the concepts of "combined transport" or "integrated transportation" began to take shape. In 1987, in *Current Situation and Future of Transportation Development at Home and Abroad* edited by China Communications and Transportation of Association and the Institute of Comprehensive Transportation under the State Economic Commission, six major issues within the existing transportation system were identified. The book proposed solutions that included, "replacing rail with road for short-distance passenger transport, using waterways instead of rail for long-distance freight transport, and increasing the proportion of investment in transportation, particularly in road and waterway projects".[100] This approach was in line with the guidelines of the *Seventh Five-Year Plan.*

Thus, from a macro perspective at that time, the economic focus was on "pursuing scaling production primarily through intensive development and improvement". In the field of transportation, highways and waterways gained prominence, while the proportion of investment in railways decreased. After 1985, the strategy of "tapping the potential and expanding capacity" frequently mentioned in speeches by leading officers at all levels of the CPC Central Committee and MOR, almost becoming a buzzword for the railway industry during that period.

In February 1986, Li Peng, then Vice Premier of the State Council, pointed out at the National Railway Work Conference that "to establish a comprehensive, complete, and structurally sound transportation system in our country, railway transportation should hold a dominant position. The railway sector needs to do a good job in renovating existing lines, further exploiting its potential, and playing its leading role".[101] Li instructed that the railways should "focus on enhancing transportation capacity" to address the shortage in railway transportation and "to build new railway lines with priority while also primarily focusing on reforming the existing lines during the Seventh Five-Year Plan period".[102] Therefore, the focus on exploiting potential, supplemented by the construction of new lines, became the basic railway policy of the state during the Seventh Five-Year Plan period. Following this, the leadership of the State Council continued to underscore this policy.

In October 1986, at the Railway Construction Conference in East China, Wan Li, then Vice Premier of the State Council, stressed the strategic principles of railway construction in East China. According to him, we need to depend on reform to give

---

[100] Yao Yashan et al. (1987). *Current situation and future of transportation development at home and abroad* (p. 44). Beijing: Scientific and Technical Documentation Press.

[101] Huang Fengchu, & Wang Lijiang. (1986, February 27). Li Peng's remarks on the Seventh Five-Year Plan at the National Railway Work Conference. *People's Daily.*

[102] Fu Shanglun, & Luo Chaoqing. (1986, October 26). Addressing the incompatibility between railway transportation and economic development: The acceleration of railway construction in East China, with over 1700 kilometers of new lines, double tracks, and electrified railways to be built. *People's Daily.*

full play to people's enthusiasm and creativity; second, we need to rely on renovation to fully exploit the potential of existing equipment; third, we need to focus on construction, concentrate our efforts on fighting decisive battles, establish a sense of time and efficiency, and prioritize both economic and social benefits. Reform and renovation were the primary measures for tapping the potential, with construction placed as a secondary focus. In March 1987, when inspecting the Dayao Mountain and Nanling tunnels on the Hengyang-Guangzhou Railway, Li Peng reiterated, "infrastructure development of the railway should yield economic benefits through the prioritization of investment in key lines that are cost-effective and can quickly yield results, following the path of tapping the potential".[103] In the first year of the Eighth Five-Year Plan, 1991, railway policy remained unchanged. Hence, from 1986 to 1991, the strategy of "tapping the potential and expanding capacity" served as the fundamental policy for Chinese railways. In other words, to address the situation where railway transportation was increasingly becoming a "bottleneck" in national economic development, the main strategy determined by the state was to "tap the potential and expand capacity" by maximizing the internal potential of railway equipment and management to increase transport capacity.

According to the national railway policy, "Railway construction should primarily focus on tapping the potential and renovation, following a path of pursuing scaling production through internal development and improvement, and building urgently needed new lines to prepare the necessary conditions for the economic revitalization of the next decade".[104] At that time, Ding Guangen, then Minister of Railways, said, "Efforts should primarily focus on tapping the potential, and some new lines and factories are necessary".[105] The strategic deployment for MOR during the Seventh Five-Year Plan period was, "Focusing on capacity expansion, adhering to reform, developing technology, strengthening management, advancing both material and cultural progress, and effectively handling three key tasks, i.e. improvement of safety, railway work style, and efficiency, as well as achieving success in two critical battles (referring to the reinforcement and renovation of the 16,000 kilometers of eastern railways and the railway locomotives and rolling stock industries)".[106] In 1988, Li Senmao, who succeeded Ding as Minister of Railways, stressed that "we should strive to tap the potential of transportation. Given the pressure from the increase in transport volume, it is impossible to form new capacity in

[103] (1987, March 10) The completion of the Dayao Mountain Tunnel Project is just around the corner, as constructors are working tirelessly on the final 70 meters. *People's Daily*.

[104] (1985). A Blueprint for the Development of China's Railways (the Third Draft). Archived in Southwest Jiaotong University, T-272.

[105] Ding Guangen. (1985, November 15). Prioritizing reform and striving to improve railway transport capacity: Speech of Minister Ding Guangen at the Closing Ceremony of the First Railway Transportation Modernization Management Seminar on August 28, 1985 (Excerpt), *People's Railway Daily*.

[106] (1986, Feb 19). The national railway work conference held in Beijing. *People's Railway Daily*.

the short term, so the focus must remain on tapping the potential of existing transportation equipment".[107]

In retrospect, the strategy of "tapping the potential and expanding capacity" was a direct cause of the railway's transformation from a "weak link" to a "bottleneck" in the national economy. Any entity's development relies on both internal and external factors, with the latter acting through the former. However, Marxist Dialectical Materialism teaches us that both internal and external factors are indispensable. Since 1979, such a strategy of "tapping the potential and expanding capacity" was a reluctant action of the railway sector. In other words, as the central government was unable to provide additional investment in railway infrastructure, the railways nearly lost the ability to "expand transportation capacity and improve equipment levels". During this period, there was a greater emphasis on the railways' internal issues, while the critical problem was that the hardware lagged far behind.

Certainly, we should view the issue historically. This also involves the state and the general public's understanding of the nature of the railways, which is particularly evident in the initial explorations of railway reform.

2.3 Decentralization and Revenue Sharing to Advance the Railway Reform

While "tapping the potential and expanding capacity", the railway industry was also exploring the path of reform. As mentioned earlier, the reforms during this stage lacked industry-specific characteristics, and in retrospect, these reform measures only marginally influenced the inherent operational practices of Chinese railways. A more significant issue encountered across different levels of administration, from the central government to MOR, was the apparent ambiguity in the understanding of the railway sector's nature—whether it was considered a business entity or a government department. In other words, whether the railway sector was an industrial sector or a functional administrative department, and were railway products (transportation) regarded as commodities or services?

These perceptions inevitably exerted a profound influence on railway reform. Overall, during this stage and for an extended period, the railway sector was treated as an industrial sector, and the reform plans implemented were not different from those performed in any other industrial sector. The basic logic was that Chinese railways, as represented by MOR, were viewed as a large, profit-oriented state-owned enterprise, thus all reforms applicable to state-owned enterprises were equally applied to the railway sector. Thus, the general strategies for state-owned enterprise reform were applied to the railway reform in the initial stage. In terms of

[107] (1988, Oct. 10). Enhancing confidence, strengthening reforms, and tapping the potential to overcome difficulties. *People's Railway Daily*.

outcomes, the reforms of state-owned enterprises during this period were successful, while the railways seemed to experience the opposite.

In the initial stage, the reform of the Chinese railways was characterized by "Decentralization and Revenue Sharing".

Revenue Sharing refers to the reform measures primarily involving changes in financial management systems, which included three different economic accounting systems for the distribution of benefits and taxes between the state and the railways: the enterprise fund system (1978–1980), the profit retention system (1981–1982), and the system of the lump-sum contracting of the progressive increase of the post-tax profit (1983–1985). Compared to the subsequent "Input-Output Contract Responsibility System" i.e. the Contract Responsibility System introduced after 1986, the system of the lump-sum contracting of the progressive increase of the post-tax profit can be seen as a form of "Partial Contract System".

Decentralization entailed pilot projects aimed at expanding the autonomy of enterprises. The following sections are dedicated to separate elaboration of the two aspects.

2.3.1 *Reforms in Revenue Sharing*

Firstly, the enterprise fund system was reinstated. This system allowed state-owned enterprises to allocate a portion of their profits to a special fund based on conditions stipulated by the state. Essentially, it was a mode of profit distribution between the state and state-owned enterprises. The railway sector was the first to implement this system in China, with an initial form being the Director's Fund introduced in 1950 on the China-Changchun Railway. In 1951, the railway transportation enterprises initiated an incentive fund system, where, apart from infrastructure investment, 30% of profits exceeding the planned target were allocated to funds for welfare and bonuses for outstanding workers. The following year, the experience from railways was extended to other state-owned enterprises, manifested in both an incentive fund system and an excess profit-sharing system. For railway enterprises, the allocation of the incentive fund was adjusted to 3.5% of planned profits and 15% of excess profits, with the same intended use. However, due to subsequent political and economic changes, this distribution method was abolished after 1966.

In 1978, to align with economic reforms, the State Council approved the *Regulations of the Ministry of Finance on Trial of the Enterprise Fund System in State-Owned Enterprises* issued by the Ministry of Finance, marking the comprehensive resumption of the trial implementation of the enterprise fund system. According to the *Regulations*, upon fulfilling all annual plan targets and supply contracts, state-owned enterprises could allocate a portion of the profit at the rate of

5% of the total annual wages of employees to the enterprise fund, primarily for collective welfare facilities and socialist competition bonuses to employees.[108]

In accordance with relevant regulations, MOR formulated the *Provisions on the Trial Implementation of Enterprise Funds for Railway Transportation Enterprises*, which outlined eight economic and technical indicators for evaluation. These indicators were consistent with national standards but were explained in a "railway-specific" manner. The eight indicators were: "1. Output (converted ton-kilometer); 2. Variety (volume of coal shipments); 3. Quality (turnaround time of freight car); 4. Profit (revenue from transportation); 5. Consumption (coal and fuel consumption per 10,000 ton-kilometers of locomotive operation); 6. Labor Productivity (overall labor productivity in transportation); 7. Cost (cost per 10,000 converted ton-kilometers); 8. Working Capital (working capital consumed per 10,000 converted ton-kilometers)".[109] This "reinterpretation" was essentially a reclassification of the 28 indicators used to assess railway administrations at that time.[110] To address the issue of an excessive number of evaluation indicators, on October 17, 1979, the Ministry of Finance issued the *Notice on Improving the Method for Allocating Enterprise Funds from State-Owned Enterprises*, which reduced the eight indicators to four. For the railway industry, the newly proposed indicators were "Output (converted ton-kilometers), Variety (volume of coal shipments), Efficiency (turnaround time of freight car), and Profit.

In reality, there are issues with these relevant evaluation indicators in the railway sector. For example, "The products of railway transportation are measured in passenger-kilometers and ton-kilometers. Their quantity depends on the level of national economic demand for railway transportation. We require that transportation plans be as realistic, objective and balanced as possible. Exceeding planned targets or completing plans ahead of schedule may not necessarily generate positive outcomes".[111] "In addition to assessing the absolute target numbers achieved, it is also necessary to examine whether there is a backlog of related materials in the national economy and whether there are any adverse reactions from the shipping departments. This relates to the principles, policies and effectiveness of railway transportation, which should be closely inspected".[112] In summary, first, the volume of output generated by the railway industry is passive, determined by national and

[108] (1980). Regulations of the Ministry of Finance on trial of the enterprise fund system in state-owned enterprises. In Institute of Law, Chinese Academy of Social Sciences (Ed.), *Selected economic laws of the People's Republic of China* (Volume 2) (p. 274), Beijing: China Financial & Economic Publishing House.

[109] Zhang Qiusheng. (1993). *On economic accounting of railway transportation* (p. 48). Beijing: China Railway Publishing House.

[110] Transportation Economics Teaching and Research Office of Northern Jiaotong University. (1978). *Eight economic and technical indicators for the railway transportation sector* (p. 12). Beijing: People's Railway Publishing House.

[111] Statistics Department of Nanchang Railway Administration. (1979). Views on the eight current economic and technical indicators and their evaluation in china's railways, *Railway Transport and Economy*, (2), 38–39.

[112] Ibid.

social needs rather than the railway itself. Second, an increase in output, variety or even profit in the railway sector may not necessarily be beneficial. Ultimately, it depends on how well it aligns with national and social needs.

Furthermore, the reinstatement of the enterprise fund system was merely a transitional measure in response to the looming economic collapse in our country, where comprehensive reforms had yet to be fully implemented.[113] The primary limitations were as follows. Firstly, the amount allocated to the enterprise fund was not directly related to profits which were only used as performance evaluation indicators. Secondly, the enterprise fund was primarily designated for the collective welfare of employees. In the *Regulations of the Ministry of Finance on Trial of the Enterprise Fund System in State-Owned Enterprises*, the competent departments, rather than the enterprises themselves, could withdraw a proportion of the fund from profits exceeding national annual targets—10% for the transportation sector.[114] For instance, MOR oversaw the railway system, imposing restrictions on the financial autonomy of railway enterprises.

Nonetheless, the reinstatement of the enterprise fund system was a manifestation of the central government's Four-Word Guiding Principle of "Adjustment, Reform, Rectification, and Improvement". It had positively influenced the motivation of both enterprises and employees. Furthermore, in the railway transportation sector, its reinstatement also brought economic accounting systems back on track, "indicating a shift in the relationship between the railway transportation industry and the state from a purely production-based model to a production-operation-oriented one".[115] In other words, the railway sector needed to start re-recognizing its corporate identity and could not continue to simply position itself as a government functional administrative department.

The enterprise fund system for railways began in 1978 and was replaced by the profit retention system after approximately three years in 1981.

The reinstatement of the enterprise fund system was aimed at enhancing the profit retention system, which had been unified under a specific directive. The *Regulations of the Ministry of Finance on Trial of the Enterprise Fund System in State-Owned Enterprises*, issued in 1978, stated that "departments that have already implemented profit retention system shall not withdraw enterprise funds from profits exceeding planned targets". On July 13, 1979, the *Regulations on Implementing Profit Retention System for State-Owned Enterprises*, promulgated by the State Council, required that "all enterprises operating under independent economic accounting system and generating profits can retain a portion of their profits based

[113] Han Yingjie, & Xia Qingcheng. (1995). *New research on profit distribution system of state owned enterprises* (p. 42). Beijing: China Economic Publishing House.

[114] (1980). Regulations of the Ministry of Finance on trial of the enterprise fund system in state-owned enterprises. In Institute of Law, Chinese Academy of Social Sciences (Ed.), *Selected economic laws of the People's Republic of China* (Volume 2) (p. 274), Beijing: China Financial & Economic Publishing House.

[115] Li Xuesong. (1999). *Game theory and economic transformation: On China's railway reform* (p. 81). Beijing: Social Sciences Academic Press.

on the proportions set by the state",[116] marking the start of pilot reforms for profit retention system. On January 22, 1980, the State Council approved the *Trial Measures for Profit Retention System in State-Owned Industrial Enterprises* jointly issued by the National Economic Commission and the Ministry of Finance. This document stipulated that "the original measures of total profit retention shall be revised into measures that combine base profit retention with incremental profit retention. Enterprises can use the funds retained from base profits and incremental profits for development funds, employee welfare funds, and employee incentive funds".[117] In December of the following year, the *Several Provisions for the Implementation of the Profit Retention System and the System of Enterprises Responsible for Their Own Profits and Losses in State-owned Industrial and Transport Enterprises* further clarified that "the state may implement various forms of systems of profit retention and contracted responsibility for profit and loss of enterprises and their competent departments based on different circumstances, including measures of 'base profit retention combined with incremental profit retention', 'total profit retention', 'fixed profit submission with excess sharing or retention', 'fixed subsidies for losses with deficit sharing or retention', and 'retention on excessive planned profit'".[118] It further required, "For departments and enterprises with stable production and profits, the measure of base profit retention combined with incremental profit retention or the measure of total profit retention should be implemented. Additionally, the measure of total profit retention is straightforward and simple to implement, which can better reflect the interests of the country, enterprises, and employees, and should be actively promoted".[119]

MOR, regarded as a sector with "stable profit", implemented the measure of total profit retention. After the founding of PRC, the railway sector was among the first to implement the economic accounting system. This approach initially stemmed from the application of the management experience of the China-Changchun Railway.[120] In 1951, MOR issued the *Order on the Implementation by Railway Administrations of the Economic Accounting System*, mandating that all railway

---

[116] (1980). Regulations on implementing profit retention system for state-owned enterprises. In Institute of Law, Chinese Academy of Social Sciences (Ed.), *Selected economic laws of the People's Republic of China* (Volume 1) (p. 382), Beijing: China Financial & Economic Publishing House.

[117] (1983). Trial measures for profit retention system in state-owned industrial enterprises jointly Issued by the National Economic Commission and the Ministry of Finance. In Institute of Law, Chinese Academy of Social Sciences (Ed.), *Selected economic laws of the People's Republic of China* (Volume 2) (p. 220), Beijing: China Financial & Economic Publishing House.

[118] (1983). Several provisions for the implementation of the profit retention system and the system of enterprises responsible for their own profits and losses in state-owned industrial and transport enterprises. In Institute of Law, Chinese Academy of Social Sciences (Ed.), *Selected economic laws of the People's Republic of China* (Volume 2) (p. 224), Beijing: China Financial & Economic Publishing House.

[119] Ibid.: 225.

[120] Harbin Railway Administration. (1988). *Compilation of materials on the China Changchun Railway*. Beijing: China Railway Publishing House.

administrations nationwide would adopt the independent economic accounting system starting from July 1, 1951. The core objective was to "manage enterprises in a planned manner".[121] From 1955 to 1956, the Ministry gradually promoted the economic accounting management approach, which was implemented across all railway administrations, sub-administrations, and various departments, including locomotive depots, vehicle depots, track divisions, signal and communication divisions, construction divisions, major marshaling stations, freight stations, and railway sections. "The implementation of the economic accounting system marked a historic and fundamental shift from a supply-based production system to an economic accounting system at all levels. However, due to the limited autonomy of enterprises in production and operations, the economic accounting system during this period was merely a production responsibility system".[122]

With the advent of the reform and opening up, MOR issued the *Trial Measures for Reforming the Financial Management System of Railway Transportation* in November 1979, which stipulated that railway administrations should operate as relatively complete economic accounting entities with specific financial management requirements, including the rational allocation of transportation revenues and a more comprehensive accounting of transportation costs, linking revenues and expenditures while calculating profits. Ye Keming, then Director of the Ministry's Finance Bureau, noted that "if the issue of profit distribution is not resolved, it would be challenging to expand the autonomy of enterprises. Through internal settlements, we recognized the differences in profit levels among administrations while ensuring their cost expenditures were covered. Only after accurately calculating profits could we discuss profit distribution, and only then could we motivate enterprises and individuals, under the premise of strengthening ideological and political work, through certain material benefits".[123] This approach aimed to both share profits and decentralization. "It strengthens the economic responsibilities of each administration while also expanding their autonomy. In addition to completing transportation tasks, railway administrations are required to meet profit and cost plans. Within the scope of their cost plans, they could decide how to allocate funds based on production needs and economic outcomes, aiming to achieve maximum economic effect with minimal expenditure".[124] This can be seen as preparations for profit retention reform.

In June 1981, the Ministry of Finance officially approved the trial implementation of the total profit retention system across the national railway industry for

---

[121] Locomotive General Administration of MOR. (1952). *How the Vehicle Department of the Changchun Railway Administration implements the Economic Accounting System* (p. 1). Beijing: People's Railway Publishing House.

[122] Zhang Qiusheng. (1993). *On economic accounting of railway transportation* (p. 41). Beijing: China Railway Publishing House.

[123] Ye Keming. (1979, November 22). Issues on independent economic accounting of Railway Administrations. *People's Railway Daily.*

[124] (1979, November. 15). Independent economic accounting to be trialed next year by National Railway Administrations. *People's Railway Daily.*

enterprises under MOR. Under the system, these enterprises including transportation, industrial, and supply and marketing ones were permitted to retain 21% of their total profits. Compared to the enterprise fund system, the profit retention system has several advantages.

Firstly, profit retention is based on a share of profits—higher profit means a larger share. In contrast, the enterprise fund system withdraws a fixed 5% of total wages from profits. The fundamental difference lies in the fact that the enterprise fund system is a production responsibility system, while the profit retention system is an economic responsibility system, thus incentivizing enterprises and employees to strive for higher profits.

Secondly, the profit retention system enhances the financial autonomy of the railway sector. On the one hand, "enterprises have control over the retained profit funds. Funds allocated for production development should align with national economic adjustments and primarily be used for energy conservation, production increase of short-term products, management of waste gas, waste water and industrial residue, and occupational safety and protection".[125] On the other hand, MOR has gained a certain level of financial authority to approve profit retention, profit contracting, loss sharing, and other internal distribution forms, as well as to set retention ratios. Railway administrations are also granted partial financial authority, which enables them to allocate incentive and welfare funds within certain limits to motivate internal departments and employees.[126] Due to these advantages, "the total profit achieved by the entire railway system including transportation, industrial, and supply and marketing enterprises rose from 2.47 billion yuan in 1980 to 3.09 billion yuan in 1982, an increase of 25.1%, with both remitted profits and retained earnings growing accordingly".[127]

Initially, the profit retention system was set to be implemented for three years but was only in effect for two. In 1983, enterprises under MOR began trialing a reform approach of substitution of tax payment for profit remittance.

The so-called "substitution of tax payment for profit" describes the process of "converting the profits remitted by state-owned enterprises to the taxes according to the tax types and tax rates established by the state. This approach allows the after-tax profits to remain entirely under the control of the enterprises, eventually reinforcing the distribution relationship between the state and state-owned enterprises through the mechanism of taxation".[128] Specifically, "all profitable large and medium-sized state-owned enterprises (including financial and insurance

[125] Ye Keming. (1981, June 15). Responses to journalists on several issues regarding the trial implementation of the Total Profit Retention System across the national railway industry. *People's Railway Daily.*

[126] Zhang Qiusheng. (1993). *On economic accounting of railway transportation* (p. 51). Beijing: China Railway Publishing House.

[127] Xu Zenglin. 1999. *Fifty Years of the Railroads of People's Republic of China (1949–1999)* (p. 192). Beijing: China Railway Publishing House.

[128] (1983). *Responses from the Ministry of Finance regarding the implementation of Profit-to-Tax Reform.* In the Profit-to-Tax Reform Office of the Ministry of Finance, *Compilation of documents*

organizations) shall pay income tax at a rate of 55% based on their realized profits. Part of enterprises' profits after paying income tax shall be remitted to the state and the remaining part retained by the enterprises according to the profit retention ratio approved by the state".[129] Profits are remitted to the state in four ways, of which MOR is subject to "lump-sum contracting of the progressive increase of the post-tax profits (the Partial Contract System)".[130]

Specifically for the railway system, the plan for substitution of tax payment for profit generally includes two parts. First, the profits and income taxes paid to the state by the entire railway industry under MOR are centrally collected and remitted by MOR. Second, MOR adopts a tiered verification method for its subordinate enterprises, implementing different approaches based on each specific circumstance. "Each railway administration adopts a method of setting an annually incremental base for after-tax profits, and shares the profit when actual profits exceed the approved level; China National Railway Locomotive and Rolling Stock Industry Corporation and the China Railway Signal & Communication Corp (CRSC) adopt the Partial Contract System; the Railway Materials Corporation (industrial sector) and Zhengzhou Loading and Unloading Machinery Plant pay regulatory taxes; the Railway Materials Corporation remits a fixed proportion of profits; the Railway Engineering Corporation (industrial sector) and the Beijing Engineering Machinery Plant managed by China Railway Construction Corporation, after paying income tax at a rate of 55%, receive a fixed subsidy when their retained profits fail to reach the approved level, and share profit when their retained profits exceed the approved level; for planned loss-making enterprises directly under the Ministry, loss subsidies are granted according to the rate approved by the Ministry, with no compensation for excess losses and profit sharing for reduced losses. Operating losses should be rectified within a specified time frame to achieve profitability".[131] This plan essentially changes the dual management of railways by the central and local governments that was restored after 1975, in other words, local governments would no longer interfere with the management and operation of railways.

Then, how was the 55% income tax rate determined? In fact, this rate was calculated based on the previous profit retention ratio. In other words, "the profit retention for enterprises across the country would essentially be maintained at the

on the conversion of state enterprises from profit to tax (p. 102). Beijing: China Financial and Economic Publishing House.

129 (1983). Measures for the Trial Implementation of Profit-to-Tax Reform in State-Owned Enterprises. In the Profit-to-Tax Reform Office of the Ministry of Finance, *Compilation of documents on the conversion of state enterprises from profit to tax* (p. 2). Beijing: China Financial and Economic Publishing House.

130 Ibid.: 3. The other three methods are "fixed proportion submission", "regulation tax submission", and "fixed quota submission" (for mining enterprises).

131 (1983, May 29). Issues concerning the implementation of Profit-to-Tax Reform in railway enterprises: An Interview with Tan Hequan, Deputy Director of the Finance Bureau of MOR. *People's Railway Daily.*

existing level".[132] Before the substitution of tax payment for profit, the profit retention ratio in the railway industry was 21%. Following the reform, enterprises had to pay an income tax at a rate of 55%, a business tax at 15%, and an incremental profit tax at 2%, resulting in a profit retention ratio of 23%. The final profit retention ratio was slightly higher than that before the substitution of tax payment for profit. However, taking a 2% annual incremental rate from 1982 as the benchmark, the proportion of profits and taxes remitted by the railway industry reached over 80% by 1985.

In that case, what are the advantages of the substitution of tax payment for profit? According to the initial conception, the substitution of tax payment for profit "can stabilize the distribution relationship between the state and enterprises through tax laws. This provides guarantee for the state's fiscal revenue and furnishes enterprises with a statutory income source. If production grows steadily and economic efficiency continues to improve, both state fiscal revenue and the benefits that enterprises derive from it can increase stably. The state levies taxes and enterprises pay taxes in accordance with the law, which can prevent enterprises from arguing over tax base and tax rate every year. Enterprises that perform well will receive more, while those that perform poorly will receive less or nothing. To some extent, this can eliminate the drawbacks of "eating from the same big pot", which means indiscriminate egalitarianism, and is conducive to strengthening the responsibility system for enterprise management. After the substitution of tax payment for profit, enterprises shall pay taxes to both the central and local governments regardless of their affiliation, which can help them break away from their dependence on local governments and departments. Enterprises can operate in accordance with economic laws and organize production rationally".[133] It can be seen that the original intention of the substitution of tax payment for profit was part of the overall plan for economic system reform, aimed at enhancing the autonomy of enterprises, developing the economic responsibility system, and thereby promoting economic progress.

The overall reform of the financial management system reflected the conscious, planned, and stepwise reform of the central government. From restoring the enterprise fund system to profit retention, the ultimate goal was the substitution of tax payment for profit. However, when applied to the railway industry, such substitution failed to encompass adequate analysis of specific issues and, to some extent, led to the mistake of overgeneralization. In fact, the challenges faced by the railway industry became increasingly severe. The formula for calculating income tax in the railway industry is "taxable income multiplied by the applicable tax rate". As stipulated in *the Regulations on Income Tax for State-Owned Enterprises of the People's Republic of China* in 1984, the specific calculation formula is "taxable income = total profit + profit received from other enterprises - the amount of deductible

[132] (1983). Broadcast and television speech on Profit-to-Tax Reform by Wang Bingqian. In the Profit-to-Tax Reform Office of the Ministry of Finance, *Compilation of documents on the conversion of state enterprises from profit to tax* (p. 2). Beijing: China Financial and Economic Publishing House.

[133] Ibid.: 88.

items". The profit calculation formula for the railway industry is "realized profit = transport profit + industrial and ancillary profit + other profits".

There are two obvious issues here. First, can railways be considered as a single or purely commercial entity? If so, how can they fulfill their significant political functions? Second, if railways are treated as enterprises, are they profit-making? In 1982, a report by a World Bank delegation stated, "Chinese railways are said to be highly profitable. Their 1979 revenue doubled operating costs, including depreciation — though Western railways don't regard the latter as actual profit.... Freight rates appear high in comparison to costs, yet remain uniformly priced nationwide.... In a non-market economy, the interplay between railway pricing and costs defies straightforward interpretation. For instance, the relative costs of steam locomotive coal versus diesel fuel bear no proportion to their true economic costs, with unclear causation.... Without comprehensive data, limited analysis can be made: short-distance freight rates fail to cover costs; the unified national pricing implies cross-region subsidies for underdeveloped parts of the railway network, which, due to satisfactory distribution of income to some local users, may lead to uneconomic decisions".[134]

This shows that Chinese railways' dual nature—both economic entity and political instrument—leads to distinct management, operation, and overall financial situation unparalleled in conventional enterprises. Any simplistic either-or judgment risks gross oversimplification. Nevertheless, railway industries were treated as merited "large, profitable state-owned enterprises" in various reforms from 1978 to 1985. This calculation method excluded the largest cost in the railway industry, which is the huge expenditure on infrastructure development. However, as mentioned earlier, the most pressing issues for Chinese railways at that time were the unfinished railway network and outdated technical capabilities, and the renovation of existing lines and traction power could not truly resolve these issues. Due to constraints on overall economic strength, the country repeatedly reduced investment in the railway industry, and the limited profit retention of railways was far from sufficient to cover infrastructure expenditure. The substitution of tax payment for profit seemed to give railways more autonomy, but in reality, the pressure on the railway industry actually increased.

2.3.2 *Reforms in Decentralization*

During the reform of the financial management system, the reform to enhance the autonomy of enterprises within the railway industry was quite limited.

The policy of Decentralization and Revenue Sharing was a general practice among state-owned enterprises in China during the early 1980s, applicable to

[134] World Bank Mission. (1982). *China: The development of a socialist economy—Annex VI: Transportation Sector* (p. 59). Beijing: *China Financial & Economic Publishing House.*

railway reform without exception. This policy originated from Deng Xiaoping's significant theoretical innovations. At the closing ceremony of the Working Conference of the Central Committee of CPC held in December 1978, Deng delivered an important speech entitled *Emancipating the Mind, Seeking Truth from Facts and Moving Forward in Unity*. This speech is one of the landmark documents marking China's entry into a new era of reform. Deng pointed out, "When it comes to emancipating our minds, using our brains, seeking truth from facts, and uniting as one in looking into the future, the primary task is to emancipate our minds", and "One important condition for getting people to emancipate their minds and using their heads is genuine practice of the proletarian system of democratic centralism. While unified and centralized leadership remains essential, it must be rooted in full measure of democracy to achieve correct centralization". In the economic field, this is specifically manifested as the "issue of economic democracy", which means that "the current over-centralization in China's economic management system necessitates planned yet bold decentralization. Failure to delegate power would not be conducive to fully leveraging the initiative of the country, localities, enterprises, and individual workers, nor to modern economic management and labor productivity. Localities, enterprises, and production teams, should be granted more autonomy in operation and management.... The most urgent task at hand is to expand the autonomy of factories, mines, and production teams so that each factory and production team can exert every effort to unleash their proactive and creative spirit".[135]

After the founding of PRC, the blurring of the lines between government and business functions was determined by specific historical conditions. Notably, the newly established government at that time faced political isolation and economic blockades from Western countries, represented by the United States. This situation made the leaders of the Party and the country realize that whether they could quickly establish an industrial system, especially a heavy industry system (including military industry), was the top priority related to the survival of the country, and also the most important task to change the backward economic structure and improve the level of industrialization as soon as possible. Therefore, "the emergence of Chinese state-owned enterprises was aimed at implementing the strategy of prioritizing heavy industry development. After the selection of the traditional development strategy, the government nationalized urban industrial and commercial enterprises to control economic surpluses and accumulate funds in compliance with this strategic goal. In this logic, state-owned enterprises were an endogenous institutional arrangement. In the context of incentive incompatibility, information asymmetry, and responsibility mismatch between the owners and operators of state-owned enterprises, depriving enterprises of their autonomy in production and operation was also an endogenous governance approach to prevent operators from infringing

[135] Literature Research Office of CPC Central Committee. (1994). *Selected works of Deng Xiaoping (Volume II)* (p. 255). Beijing: People's Publishing House.

upon the rights and interests of owners",[136] especially for the Chinese railway industry.

As early as October 1978, Sichuan Province began pilot projects for Decentralization and Revenue Sharing. On July 13, 1979, the State Council issued the *Notice on Launching Pilot Projects According to Five Documents on Reform and Management System*. "These documents combined economic interests, economic effectiveness, and economic responsibilities, as well as the interests of the state, enterprises, and individuals, and enhanced the autonomy of enterprises, especially financial autonomy, allowing enterprises to have more flexibility to accomplish more complex tasks".[137] In the *Several Provisions on Giving State-owned Enterprises a Big Say in Management*, the six major powers on "staffing, finance, materials, production, supply, and sales" that enterprises cared about were delegated and regulated. By 1980, over 6600 enterprises had been designated to implement pilot projects for enhanced autonomy.

The first pilot for decentralization in the railway industry was the Shanghai Railway Administration. "In May 1979, the Shanghai Railway Administration implemented a pilot project to enhance the autonomy of enterprises in operation and management at the Shanghai Vehicle Depot. In 1980, it was expanded to organizations such as the Shanghai East Vehicle Depot and Shanghai Locomotive Depot. The pilot project for enhancing autonomy at the grassroots level first focused on the fulfillment of enterprise responsibilities to the state. At the same time, key points for autonomy enhancement were stipulated: under the premise of ensuring basic service, enterprises had the authority to develop supplementary plans, undertake external commissioned tasks, and plan for increases in production and revenue. Under the premise of completing the task of remitting profits to the state, enterprises had the right to utilize and dispose of the remaining profits, and seal up or lease surplus fixed assets. Not to exceed the authorized staffing level, enterprises had the right to adjust internal organizational structures and staffing.

Depending on different situations, a profit retention system was implemented: locomotive and vehicle depots tested the retention of increased profits. With the actual profit of the previous year before the pilot project as the base, 20% of the increased portion was retained to establish three funds: production development fund, welfare fund, and bonus fund".[138] During this period, all the enterprises involved in the pilot projects accomplished record-high tasks. In 1980, the profits remitted by the five pilot depots increased by 150.2%, resulting in an additional revenue of 2.06 million yuan for the state, and a total retention of 646,000 yuan by

---

[136] Lin Yifu, Cai Fang, & Li Zhou. (2014). *Full information and the reform of state-owned enterprises* (p. 12). Shanghai: Truth & Wisdom Press, Shanghai Sanlian Bookstore, Shanghai People's Publishing House.

[137] (2015). *Notice on launching pilot projects according to five documents on reform and management system*, In Wang Zhenchuan (ed.), *Yearbook of the new period of China's Reform and Opening up (1979)* (p. 599). Beijing: China Democracy and Law Publishing House.

[138] Compilation Committee of the *Chronicle of Shanghai railways*. (1999). *Chronicle of Shanghai railways*. Shanghai: Shanghai Academy of Social Sciences Press.

the five pilot depots, achieving the goal of "providing more revenue for the state, more retention for the enterprises, and more benefits for the employees".

Overall, since 1978, the railway industry had implemented reforms to enhance enterprise autonomy in the following key areas: (1) Organizational restructuring, primarily involving the adjustment and establishment of railway administrations and subdivisions. After the reform, Chinese railways were restructured from 20 railway administrations and 62 subdivisions into 12 railway administrations and 56 subdivisions. The basic principles of the adjustment were "to facilitate the smooth flow of major transportation routes, to match cargo and train flow patterns, and to support the development of new technical equipment for railway transportation", which ultimately resulted in a coexistence of large, medium-sized, and small railway administrations. (2) Reform of the Factory Director Responsibility System. This mainly involved rectifying the problems in various types of railway enterprises, improving Party leadership, and strengthening the command of production and operation. (3) Establishment of industry-specific organizations. Key organizations established included CREC, China National Railway Locomotive and Rolling Stock Industry Corporation, China Railway Signal & Communication Corporation Limited, China Railway Materials Group Corporation, China Civil Engineering Construction Corporation, China Railway Foreign Service Corporation, etc. In 1983, the Railway Engineering Corps was also disbanded, and the Railway Engineering Command was established, which was later renamed as China Railway Construction Corporation. (4) Enhancement of management autonomy for enterprises under MOR. Generally speaking, MOR delegated 53 management authorities to its subordinate enterprises, including planning, finance, labor, materials, personnel management, and other authorities, which somewhat boosted the vitality of enterprises.

Even today, it remains challenging to accurately evaluate the railway reforms in their initial stages. This difficulty mainly stems from the lack of precise assessment and validation of various reforms. Strictly speaking, within the railway industry, there was no discussion or attempt to answer questions such as why greater autonomy was needed, how it could be strengthened, in what areas it should apply, how its impacts could be measured, and how previous measures should be adjusted and justified, along with the reasons for such adjustments. Perhaps it was not just the railway industry, the entire country lacked sufficient theoretical preparation for reform. Although the macro goal of reform was to achieve "Four Modernizations", the specific direction of reform was still being explored. In-depth discussions on reform on the academic level were only in their infancy.

Therefore, on the one hand, we recognize that consistent progress was reported in the railway industry. After arduous efforts by railway practitioners, various key indicators grew at a considerable rate, with relevant transportation and profit-making targets completed as scheduled or even exceeded annually and quarterly. On the other, however, it is undeniable that the railway system was increasingly becoming a weak link in the progress of the national economy. In terms of its strategies, the reform of the railway industry was reactive, lacking its own industry-specific characteristics and creativity based on industry peculiarities.

In this context, the railway industry was continuously compelled by national and social demands to seek its reform path. In 1984, China began to comprehensively implement the policy of "Conversion of Capital Allocations to Loans". This policy marked a significant reform in China's infrastructure investment sector and a major reform in the financial sector. It can be said unequivocally that this reform played an undeniable historical role in establishing a socialist market economic system in China. However, for Chinese railways at that time, this reform brought significant pressure. The practical situation compelled the railway industry to undertake autonomous reforms, leading to the Contract Responsibility System.

Chapter 3
Proactive Change: Overall Exploration of Reform (1986–1991)

After the Third Plenary Session of the Eleventh Central Committee of the CPC, China embarked on the path of reform and opening up. During this period, the railway industry, while undergoing rectification, began to experiment with reforms and attained some progress. In the years after 1980, freight increased by 40–50 million tons annually, growing at a steady rate of around 4%. However, compared with the faster pace of economic growth, the development of the railway industry in China was relatively slow. It is calculated that for every 1% increase in the national gross industrial output value, the corresponding railway freight was required to increase by 0.5%. From 1980 to 1985, the ratio of the growth rate of the national gross industrial output value to the growth rate of railway freight was 1:0.31,[1] indicating that during the Sixth Five-Year Plan period, railway development lagged significantly behind industrial development. Therefore, by around 1985, the main issue facing the railway industry was its inadequate transportation capacity to meet national and social demands. The railway industry had to intensify efforts to reform and implement the Contract Responsibility System across the entire industry.

3.1 Background of Exploration in Package Reform

3.1.1 Rail Transportation under Severe Strain

After the reform and opening up, the primary issue facing the railways was the railway's severely inadequate capacity to meet these needs, specifically the conflict between freight volume and transport capacity. This conflict was particularly evident

[1] Policy Research Office of MOR. (1987). Basic situation and countermeasures of current railway transportation. Archived in Southwest Jiaotong University, No. 1-11-1.

X. Zhang, *China's Railway Expansion in 40 Years of Economic Reform and Opening Up*, Understanding China,
https://doi.org/10.1007/978-981-95-0333-9_3

in the development of the national economy, with overall tension in passenger and freight transport. In 1985, railway freight "could only meet 74% of the planned cargo transport".[2] Various industries reported that "currently there is a shortage of national transport capacity, and railway transport is especially under strain".[3] "Due to strained railway and highway transport in Sichuan, there is an accumulation of supplies bound for regions outside Sichuan, and there is an encouraging sign in water transport".[4] In Shanxi, coal produced in the Datong coal mines could not be transported in time, leading to serious spontaneous combustion issues. For example, "with railway transport unable to keep up, a large amount of coal piled on the ground for a long time will inevitably self-combust", and "according to some colleagues in the coal industry, railway transport serves as the key to prevent coal self-combustion".[5] Freight transport was under strain, and passenger transport faced the same challenge. To be more specific, "the passenger traffic on busy main lines often exceeds capacity by more than 30%. Even during holidays, despite various emergency measures such as adding temporary trains, expanding train formations, and canceling or reducing sleeper carriages, the demand could not be met, leading to acute overcrowding, not to mention improving the quality of passenger services".[6]

The tension in railway transportation had drawn the attention of the State Council, MOR, and all social sectors. On March 28, 1985, at the Third Session of the Sixth National People's Congress, Song Ping, State Councilor and Director of the State Planning Commission, highlighted the issues in national economic development in the *Report on the Draft National Economic and Social Development Plan for 1985*. The first issue he mentioned was that "energy, transportation, particularly electricity supply and railway transportation, are still under grave strain, and the contradiction of insufficient raw material supply is becoming increasingly prominent". He particularly emphasized the problem in railway transportation, noting that "the tension in the railway main lines has reached a rather serious level, with a large amount of goods piling up waiting for transport and passengers being subjected to acute overcrowding".[7] At the National Symposium on Transportation Work held in May of the same year, then Vice Premier Li Peng candidly stated, "Currently, due to the strain on transportation, the railway sector has become a seller's market, and

---

[2] Ibid.

[3] Xing Fengbing. (1986, February 28). Authorities should give the green light to horizontal integration. *People's Daily*.

[4] Luo Maocheng. (1986, June 15). No spring water flowing to the east in vain: Sichuan's shipping industry is recovering. *People's Daily*.

[5] (1986, October 26). Mining less coal if it can't be shipped out: refuse energy waste caused by spontaneous coal combustion. *People's Daily*.

[6] Science and Technology Bureau of MOR. Suggestions for strategic countermeasures for the development of Chinese Railways. Archived in Southwest Jiaotong University, T-36.

[7] Songping. (1986, April 13). Report on the draft national economic and social development plan for 1985. *People's Daily*.

everyone seeks favour from the transportation departments".[8] In the same month, at the Seminar on Railway Development Strategy, experts believed that "currently all modes of transportation in our country are inadequate. Transportation has become a prominent weak link constraining national economic progress",[9] with railways being the most critical. "The contradiction between railway capacity and volume is becoming increasingly serious. Due to insufficient capacity, the backlog of goods awaiting transport has exceeded 100 million tons, with coal alone accounting for over 40 million tons. Many factories and mines' operation depended on transport capacity, which has drastically impeded economic development. As for passenger transport, the difficulties in buying tickets and getting seats, and the severe overcrowding of trains have reached unprecedented levels, becoming a serious social issue. Therefore, it is urgent to systematically study and strengthen our transportation infrastructure, especially to quickly enhance railway transportation capacity, as it is a vital issue for achieving a fourfold increase in the national economy in this century".[10]

The railway has always been dubbed as the "pioneer" of social economy, which has been a source of immense pride for those working in the industry. However, after several years of restructuring and reform, it has now become a "dead weight" of the national economy. This challenge lies before the Party, the country, and even the Railway Man, pressing for reform and development of the railways. "For a long time, the railway has been a weak link in the national economy, and to change this situation, comprehensive and in-depth reforms are necessary".[11]

3.1.2 The Coercion of the Policy of "Conversion of Capital Allocations to Loans"

"Conversion of Capital Allocations to Loans" is a significant reform in China's infrastructure investment, as well as a pivotal one in the financial sector. As early as 1952, the Central Financial and Economic Committee issued two documents, namely, *The Interim Measures for Basic Construction Allocations* and *The Infrastructure Development Financial Plan Submitted for Central Review and Approval and Reaffirmation of the Advance Allocation Principle*, which marked the establishment of the investment system for infrastructure allocations in China. Until 1979, the only mechanism for investment and financing in construction projects in China was financial allocations. This investment method played a significant role during certain historical periods.

---

[8] Notice on forwarding Vice Premier Li Peng's speech at the National Symposium on Transportation Work. Archived in Southwest Jiaotong University, T-342.

[9] Proceedings of the seminar on railway development strategy (internal information). Archived in Southwest Jiaotong University, T-235-1.

[10] Ibid.

[11] Opinions of MOR on railway reforms (TZY No. 41). Archived in Southwest Jiaotong University, T-7.

After the new period of reform and opening up, the approach of investment through government funding gradually became insufficient to meet the new demands. In particular, enterprises were using investment funds at no cost, lacking a self-regulation mechanism where rights, responsibilities, and benefits were integrated, leading to problems such as wasteful use of national investment funds and low investment efficiency. In 1979, in view of the various drawbacks of government funding, the State Council approved the *Report on the Trial Implementation of Loans for Investment on Infrastructure Development* and *the Pilot Regulations on Granting Loans for Infrastructure Development* proposed by the National Planning Commission, National Construction Committee, and Ministry of Finance, which shifted basic investment mode from government grants to bank loans. The *Report on the Trial Implementation of Loans for Investment on Infrastructure Development* stated that "Pilot bank loans for infrastructure investment will play a positive role in fully mobilizing the initiative and enthusiasm of various regions, departments, and enterprises, strengthening infrastructure management, changing the current 'long, scattered, and chaotic' situation of infrastructure, and improving investment effectiveness".[12]

The Pilot Regulations on Granting Loans for Infrastructure Development emphasized that borrowing entities should be independent economic entities that bear full financial responsibility for loan applications and repayment of principal and interest, as well as the reconstruction and expansion of themselves.[13] After approximately 4 years of pilot testing, the National Planning Commission, Ministry of Finance, and the China Construction Bank jointly issued the *Interim Provisions on Changing Investment on All Infrastructure Development Covered by the State Budget from Appropriation to Loan* in December 1984, deciding that starting from 1985, all infrastructure investments within the national budget would be converted changed from appropriations to loans. "After the Conversion of Capital Allocations to Loans, the original channel of 'investment directly allocated from the national budget' will be canceled accordingly".[14] The method and purpose are as follows, "The state will allocate the total amount specified in the investment package agreement to the China Construction Bank, which will make payments according to the progress of the project and actual needs. Without exceeding the total investment amount, disbursements were not subject to annual limits. Policies should be formulated to encourage the shortened construction periods, lower investment, early production, and achievement of design capacity".[15]

The railways mainly adopted a fiscal management model of separating revenue and expenditure before 1985. In this model, the infrastructure investment of the railway was funded by the state on the one hand. This investment method had the

[12] Department of System Reform and Regulations, State Development & Planning Commission of PRC. (1989). *Overview of the ten-year plan system reform (1979–1988)* (p. 229). Beijing: China Planning Press.

[13] Ibid.: (233).

[14] Ibid.: (275).

[15] Ibid.: (21).

following basic characteristics: First, the scale of infrastructure was determined by the state, with the majority of investment allocated by the state. Designs, planning, and other operations required prior approval from state planning and construction departments. Second, infrastructure investment was mainly funded through national appropriations and directly allocated within the national budget. Third, the majority of the key materials for construction were distributed by the central government. Fourth, design and construction for infrastructure were also controlled primarily at the central level, though MOR has its own design and construction teams.

According to statistical data, before 1979, investments from the national budget accounted for over 98% of the total investment in railway infrastructure. From 1979 to 1984, this proportion was also around 80%.[16] It can be said that the rapid development of railway construction in the first 30 years of PRC was indeed due to the state's funding for railways. On the other hand, the profits of the railways were basically turned over to the state, with 85% remitted to the state.[17] The remaining 15% was mainly used for employee welfare and bonuses. This fiscal management system, with revenue and expenditure managed separately, has clear drawbacks. "Even if the railway earns more, it cannot reform or develop itself; with fewer earnings, it has to ask for investment and projects from the central government. This not only affects the operational enthusiasm of the enterprise but also encourages extravagant spending. For a large interconnected system like the railway that spans nearly the entire country, it is certainly difficult to thrive if its own earnings are not linked to its own development and everything relies solely on state support".[18] Therefore, this system was not conducive to mobilizing the enthusiasm of railway personnel thereby necessitating structural reform. As Li Peng, then Vice Premier of the State Council, mentioned at the National Symposium on Transportation Work, "Given the limited infrastructure investment from the state, if we were to inappropriately expand the scale of infrastructure, the harm it would bring to the national economy would be significant. We shall act within our means. Our task is to make good use of the existing funds, plan carefully, improve investment efficiency, lower costs, and ensure that investment is frugal, construction periods are short, and efficiency is high. Therefore, we should carry out reforms in infrastructure".[19] One of the reforms implemented at the time was the Conversion of Capital Allocations to Loans.

However, the Policy of Conversion of Capital Allocations to Loans has, to some extent, pushed the railways into a corner. Firstly, it has reinforced their responsibilities

---

[16] Statistics Center of MOR. (2008). *Collection of national railway statistical data* (internal resources, p. 226). Beijing: China Railway Publishing House.

[17] The tax reform starting in 1983 simply changed the method of transferring profits to MOR in the form of taxes, but did not substantially change the practice of railway profit remittance.

[18] Lv Yinhua. (1986, April 20). The road of operation and management with unique characteristics of Chinese railways. *People's Railway.*

[19] (1985). Notice on forwarding Vice Premier Li Peng's speech at the national symposium on transportation work (TBZ No. 1181). Archived in Southwest Jiaotong University, T-342.

which are beneficial for the improvement of investment efficiency by transforming national investment into bank loans, but this policy inevitably intensified the financial pressure on the railways. The state issues construction plans and loan quotas, and the railways borrow from the China Construction Bank. Different from the direct allocation of government funds, loan requires repayment of both principal and interest, and imposes a heavy burden on the railways. For example, in 1985, "the country initially arranged for our ministry to allocate a fund of 3.781 billion yuan for the conversion of capital to loans".[20] According to the *Interim Provisions on Converting All National Budgetary Infrastructure Development Investment into Loans*, the annual loan interest rate for MOR was 3.6%, meaning that the Ministry needed to pay 136 million yuan in interest annually, plus repayment of the principal. Railway construction, especially new line construction, involves large-scale projects with significant initial investment, long construction periods, and slow investment recovery, thus placing a heavy burden on the railways in repaying the loans.

The Policy of Conversion of Capital Allocations to Loans converted subsidies into loans, but the complementary reforms lagged behind, leaving the railways in a passive position. First, the revenue management model for railways was not reformed. The policy required the railways to repay the principal and interest on the loans, but where did this revenue come from? The railway provides transportation services, so the revenue for the railway industry mainly comes from transportation service fees, namely, the income from passenger and freight transport. Before the implementation of the policy, most of the railway's profits were remitted to the state. In 1985, the investment system was changed to the model of Conversion of Capital Allocations to Loans, but the "regulations on railway taxes and profits were not correspondingly adjusted, leading to substantial debts for the railways. With slow capacity formation and mostly social benefits, the railways themselves generated little revenue and were unable to repay the debt due to large investments and a long cycle of construction. Furthermore, a significant portion of the loans to the railways was arranged as bank loans".[21] Therefore, after the implementation of the policy, the profit distribution model where most of the railway's profits were remitted to the state would also undoubtedly change. How to propose a more beneficial and balanced "comprehensive reform" plan that would give the railways more initiative was an urgent issue for the railway authorities to address.

Second, the transportation price control model related to railway revenue was not reformed. The pricing authority for railway transportation services did not rest with the railway department but was controlled by the state. In January 1988, the State Council issued the *Interim Regulations on the Management of Important Production Materials and Transportation Prices*, which kept on stressing that railway transportation prices and fee standards were managed, set, and adjusted by the National Price Bureau. Of course, as the railways has to provide for public welfare,

---

[20] Speech materials at the conference of railway sector in august 1985 (leading officers' speech). Archived in Southwest Jiaotong University, ([1985] No. 1-0-2).

[21] Sun Xing. (1986, July 18). The origin of the Contract Responsibility System of railways and its practical and theoretical importance. *People's Railway.*

it is impossible for the market or railway departments to make decisions entirely on their own under the socialist system. However, it also needs to be set at a reasonable level. The management reform of China's railway transportation prices lagged behind significantly, leading to long-term operating losses for the railways, which was especially evident during the price surge around 1988 (to be elaborated later). Third, the tax rate for the railways was changed. After the founding of PRC, railways were classified as large state-owned enterprises, and under the separated fiscal management model, the railways did not have overall input-output accounting. For a long time, the railways functioned as a purely profit-remitting sector providing substantial taxes and profits to the state. The state imposed heavy taxes on the railways. Before the end of 1985, the railway business tax was 15%, much higher than the 3% for road transportation, aviation, and postal services.[22] This excessively high business tax did not significantly impact the railways when most of their profits were remitted to the state. However, once independent input-output accounting was required, it would have a significant impact on railway profits.

In all, the Conversion of Capital Allocations to Loans placed the heavy burden of railway infrastructure development on MOR, while the necessary reforms did not match the steps taken, thus cornering the railways. As Li Senmao, then Minister of Railways, summarized the origin of the "Contract Responsibility System" in 1990, "Starting from 1985, the state stipulated that funding should be changed to loans. However, the amount and duration of the loans were not guaranteed, and the interest rates were high and it was difficult to repay. To find a way out, the railways requested approval to implement a contract system across the industry, envisioning using the surplus accumulated from their own efforts to accelerate railway renovation and development, thereby enhancing the vitality of the railway sector".[23]

3.1.3 Experience from the Pilot Reform of the Guangzhou-Shenzhen Railway

Before the implementation of the Contract Responsibility System, the railway sector had been experimenting with small-scale reforms. For example, MOR had previously introduced the economic contracting responsibility system in the transportation, industrial, and infrastructure systems. These reforms accumulated experience and yielded positive results. "Enterprises with more liberated thinking and significant reform showed notable outcomes, with substantial increases in transportation production and construction, and significant improvements in economic benefits. This stimulated the enthusiasm of enterprises and employees, bringing vitality and

[22] Ibid.

[23] Summarizing experience, deepening reform and improving the railway economic Contract Responsibility System: Comrade Li Senmao's speech at the Railway Leading Officers' Meeting. Archived in Southwest Jiaotong University, ([1990] No. 1-0-2).

vigor".[24] Notably, the successful pilot reform of the Guangzhou-Shenzhen Railway Company served as a strong catalyst on the introduction of the Contract Responsibility System.

The Guangzhou-Shenzhen Railway Company is the first enterprise in the Chinese railway industry to implement the Contract Responsibility System based on the "Scheme of Optimization of Input for Better Output" and the "Self-sustaining Approach to Building New Railways". The Company officially began operations in early 1984. The main line under its jurisdiction, the Guangzhou-Shenzhen Railway, starts from Guangzhou in the north and passes through stations such as Shitan to Luohu Bridge in Shenzhen, with a total of 23 stations and a length of 147 kilometers. The state provided no financial support to the reform of the Guangzhou-Shenzhen Railway but adopted a "package" policy, that is "self-management, bearing profits and losses independently, raising funds by itself, and a self-sustaining approach to building new railways".

The specific policies given by the State Council are: "(1) The reform of the Guangzhou-Shenzhen Railway shall have complete independent accounting, with passenger and freight operating revenue as the company's income, offsetting costs and expenditures, and calculating profits and losses independently; (2) it shall implement the lump-sum contracting of the progressive increase of the profits (the profit-sharing arrangements) with a base of 20 million yuan to be increased by 2.32% annually, with the remaining profit retained as accumulation; (3) it shall pay 12.97 million US dollars in foreign exchange each year, and excess can be fully retained; (4) it shall have independent investment and management autonomy, including rights for project design approval, contract awarding, construction loans, fund utilization, external business negotiations, and equipment and technology introduction; (5) it shall increase freight rates by 50%".[25]

After the establishment of the Guangshen Railway Company, a series of new measures were taken to expand the market: First, the management system was reformed. The company adopted a managerial accountability system, introduced a recruitment system for technical personnel, and implemented contract systems for some workers. Responsibility was assigned to individuals specifically, and performance evaluations were strengthened. For infrastructure, the company implemented a uniform management system for planning, construction, and operation. The Guangshen Railway Company, as both the investment and operation entity for the Guangzhou-Shenzhen Railway electrification project, facilitated the integration of transportation services and infrastructure development, thus saving investment and improving construction quality. Second, the Company closely followed market demand and improved service quality. "Leveraging its advantages of location in a special economic zone and having independent rights to negotiate with foreign firms, the Company engaged in extensive contacts with foreign and Hong Kong

[24] Opinions of MOR on railway reforms (TZY No. 41). Archived in Southwest Jiaotong University, T-7.

[25] Yang Peixin. (1990). The Contract Responsibility System: the Essential Path to Successful Enterprises (p. 68). Beijing: China Economic Publishing House.

business institutions stationed in Hong Kong to introduce technology and funds and exchange ideas on transportation business".[26] It investigated market needs and provided services accordingly. "According to the fluctuation of demand and supply, the Company built cargo yards, increased direct passenger coaches from Guangzhou to Kowloon, added more passenger coaches between Guangzhou and Shenzhen, and used returning empty trains bound for Hong Kong for cargo transportation",[27] which gave passengers great convenience and increased revenue. Third, the Company diversified its business by establishing affiliated transportation firms, infrastructure construction firms, and external service firms.

Just one year later after the series of reforms, the Company experienced a significant increase in economic benefits. Zhang Chaoqun, then Deputy Secretary of the Party Committee of the company, said, "The company's transportation revenue reached 153 million yuan, and after taxes and profits, the net profit was 87.9 million yuan, which was 46% higher than the targets set in the annual plan".[28] After three years of pilot testing, the Guangzhou-Shenzhen Railway's "transportation revenue, profit, and net profit growth rates reached 49.7%, 57.2%, and 80.7%, respectively. Notably, with 360 million yuan accumulated entirely from its own operations and a small amount of loans, the company completed a new 147-kilometer second line of the Guangzhou-Shenzhen Railway. MOR praised the company as the 'trailblazer'"[29] for revitalizing the national railway system. Thus, it became known as the "Guangzhou-Shenzhen Experience" or "Guangzhou-Shenzhen Model", which was promoted throughout the railway industry.

Around 1985, the Guangzhou-Shenzhen Railway became a prominent topic of discussion. The "Guangzhou-Shenzhen Experience" can be summarized in three points: (1) Comprehensive contracting shall be implemented to enhance the enterprise's initiative and proactivity, making it a self-reforming and self-developing investment entity; (2) The "state taking the biggest share" approach should not be the fixed one to adopt, allowing the enterprise to have financial autonomy commensurate with its role as an investment entity; (3) The enterprise operators should be assigned the authority to exercise their management rights without obstruction.[30] Although these points might seem like the enhancement of the policy of

---

[26] Wang Lijiang. (1984, November 14). The new economic entity shows vitality, and the economic benefits of Guangshen Railway Company are very encouraging. *People's Daily*.

[27] Yan Shigui. (1986, July 25). The independent operation and self-development of Guangshen Railway Company gains net profit of 280 million yuan in two years. *People's Daily*.

[28] Xu Miaozhong, & Huang Chifeng. (1985). *Selected compilations on experience of business management in Shenzhen Special Administrative Region* (p. 167). Yunnan: Yunnan people's Publishing House.

[29] Zhang Chaoqun, et al. (1988). How to revitalize large and medium-sized enterprises from the practice of Guangshen Railway Company. In Compilation Group of the Guangdong Reform and Opening Up Theoretical Seminar (Ed.), *Guangdong Reform and Opening-up Studies* (p. 280). Guangdong: Guangdong people's Publishing House.

[30] Ibid.

Decentralization and Revenue Sharing, they were actually not. Objectively speaking, the trial reform was indeed already at the forefront of national reform.

For the railway sector, the pilot establishment of the Guangzhou-Shenzhen Railway Company revealed an opportunity. It allowed the railways to seek a more market-oriented and enterprise-based development model, aiming for greater autonomy and, thus, greater development potential. "The Guangzhou-Shenzhen Railway experience provides insight into the national railways, which are seeking reform solutions to their difficulties. It vividly demonstrates that even a highly centralized basic industry like the railways can operate autonomously, bear profits and losses, and undergo self-reform and self-development. The key is to be open-minded, reform the management system, and address the issues of separation between income and expenditure and the imbalance between profits and development, thereby enabling the railways to gradually transition toward enterprise-based operations".[31] The reforms of the Guangzhou-Shenzhen Railway "received applause like a great performance",[32] clearly showing that reform not only improves capacity and alleviates the conflict between transportation demand and capacity shortages to some extent but also significantly increases revenue. This provided the entire railway system with the courage and experience to explore reforms and implement comprehensive contracting across the rail network.

MOR's summary of the "Guangzhou-Shenzhen Experience" is one of the most important materials for justifying the "Contract Responsibility System". However, the unique circumstances of the Guangzhou-Shenzhen Railway, especially the advantages from the favorable economic and geographic conditions of Shenzhen Special Economic Zone and Hong Kong, have not been fully assessed. These conditions cannot be replicated nationwide. Perhaps due to the urgent need for a solution, MOR had to quickly propose a plan to address new situations and difficulties, which has led to an overly optimistic view of the Contract Responsibility System.

In summary, after the reform and opening up, Chinese railways have seen significant improvements through reorganization and partial reforms. However, the contradiction between the railway capacity and the national and public demand for railway transportation has become more pronounced. The Conversion of Capital Allocations to Loans reform has placed a heavy burden on the railways, while reforms such as those implemented on the Guangzhou-Shenzhen Railway, have given the Railway Man hope of overcoming the difficulties.

---

[31] Lv Yinhua. (1994). *China's railways advance in reform* (p. 31). Beijing: China Railway Publishing House.

[32] Lin Xi, & Liu Guosheng. (1986, July 3). A strong resolution to reform: A chronicle of reform of Guangzhou Railway Administration. *People's Daily*.

3.2 Execution and Adjustment of "Contract Responsibility System"

3.2.1 Full Implementation of Contract Responsibility System

On March 31, 1986, the State Council issued a document approving MOR's reform plan for railways, commonly known as the "Contract Responsibility System", encompassing Optimization of Input for Better Output and the Self-sustaining Approach to Building New Railways. This System involved a model of investing in and constructing railways on the basis of contractual agreements.

The Contract Responsibility System was a comprehensive reform proposal initiated by MOR. As a specific embodiment of the economic system reform focused on enhancing enterprise vitality, it was decided at the Third Plenary Session of the 12th CPC Central Committee and represented an active exploration of economic reform by the Ministry. At that time, with widespread social consensus and a vigorous reform atmosphere nationwide, the Ministry actively explored comprehensive reform proposals. The Railway Man believed that "comprehensive and in-depth reform was essential" and that "without reform, it would be impossible to adapt to new pressures".[33] Ding Guangen, then Minister of Railways, emphasized that reform was of paramount importance and that railway development shall "place reform as the top priority. To improve transportation capacity, we shall adhere to the approach of reform". "To achieve the goals of the *Seventh Five-Year Plan.* reform shall be adhered to, new paths shall be explored, and the importance of the railway shall be elevated to a new level".[34]

On May 20, 1985, MOR drafted the first version of an "urgent report" in order to advance railway reform, requesting the central government to increase investment in railways to address the sharp conflict between the shortage in railway capacity and the increase in transportation volume. On July 1, Ding Guangen, then Minister of Railways, once again reported the grave situation of the railways to national leaders and suggested that if intensive investment was not feasible, the Ministry should be allowed to implement a contracting system. National leaders instructed MOR to propose a contracting plan and conduct extensive evaluations. On August 6, the Ministry submitted the plan, which, after evaluation, was reported to the central government on September 20 and received high attention from the Central Committee of the Communist Party of China. After hearing Ding's report on the reform, Deng Xiaoping remained silent for a while and then said, "Let the railways break the ice".[35] On October 14, General Secretary Hu Yaobang, endorsed the Contract Responsibility System and supported its implementation the following

[33] Opinions of MOR on railway reforms. Archived in Southwest Jiaotong University, T-7.

[34] Wang Lijiang. (1986, Feb, 20). Ding Guanguan mentioned that we must adhere to reform and explore new ways in railroad work at the national railroad work conference. *People's Daily.*

[35] Li Wei. (2012, March 23). Railway reform review: Three difficulties causing failure three times. *Oriental Morning Post.*

year.[36] Under the direct supervision of the Central Committee and with the support of various ministries under the State Council, the Contract Responsibility System was officially approved in March 1986, symbolizing a proactive move initiated by MOR to seize control. The State Council recognized the Ministry's reform rationale and deemed the contracting plan as "a positive and promising reform to increase the railway's self-improvement and development vitality", promising not to change it for five years.[37] Then, starting in 1986, the railways began to fully implement the Contract Responsibility System.

What does Contract Responsibility System cover? In short, it encompasses Optimization of Input for Better Output, as well as the Self-sustaining Approach to Building New Railways. MOR proposed six contracting tasks involving transportation, rolling stock production, expansion of railway construction scale, increasing transportation capacity, infrastructure development investment, and rolling stock acquisition costs. It specifically outlined passenger and freight volumes to be reached during the Seventh Five-Year Plan, the volume and growth rate for each organization, the number of locomotives, passenger coaches, and freight cars to be manufactured, the mileage of double-track, electrified railways, and new railways to be completed, the transportation capacity to be achieved by 1990, the amount of investment in basic construction and acquisition costs for rolling stock during the Seventh Five-Year Plan, the taxes to be paid to the state, and the allocation principle for converted ton-kilometer wages in the transportation department starting from 1986. The plan of the Contract Responsibility System also proposed policies and regulations on material supply, railway freight rates, basic depreciation rates, and major repair depreciation rates. Upon approval by the State Council, these indicators became mandatory.

(1) Transportation task: During the period of Seventh Five-Year Plan, passenger volume reached 6.4 billion person/trip, with an average annual increase of 58 million person/trip, compared to an average annual increase of 40 million person/trip during the period of Sixth Five-Year Plan. In 1990, passenger volume reached 1.4 billion person/trip. The total freight volume during the period of the Seventh Five-Year Plan was 7.1 billion tons, with an average annual increase of 65 million tons, compared to an average annual increase of 38 million tons during the period of the Sixth Five-Year Plan. In 1990, freight volume reached 1.6 billion tons. The total coal transport volume was 2.95 billion tons, including 850 million tons of coal transported from Shanxi. (2) Rolling stock production task: During the Seventh Five-Year Plan period, railway factories manufactured a total of 5,000 locomotives (830 electric locomotives, 2,260 internal combustion locomotives, and 1,910 steam locomotives), 116,000 freight cars, and 10,860 passenger coaches. (3) Expanding railway construction scale and increasing transport capacity: During the Seventh

[36] Sun Xing. (1986, July 18). The Origin of the contract responsibility systeming of railways and its practical and theoretical importance. *People's Railway Daily*.

[37] Notice of the state council approving the five departments of the state planning commission's decision on MOR's adoptation of the economic contract responsibility system (GF. No.40 [1986]). Archived in Southwest Jiaotong University, T-87.

Five Year Plan period, 2,687 kilometers of double track railway and 3,660 kilometers of electrified railway were completed, and 3,809 kilometers of new railways were constructed (including joint venture railways but excluding local railways). By the end of the Seventh Five-Year Plan, the targeted transport capacities were: 1.6 billion person/trip for passenger transport and 1.7 billion tons for freight transport. (4) Infrastructure development investment and rolling stock acquisition costs: The basic construction investment for the period of the Seventh Five-Year Plan was 33.2 billion yuan (with 25.2 billion yuan provided by railway profits and an additional 1.5 billion yuan added during execution), and the rolling stock acquisition cost was 10 billion yuan, totaling 43.2 billion yuan. (5) Tax payment: During the period of the Seventh Five-Year Plan, railway enterprises paid urban maintenance and construction tax, as well as value-added tax (or business tax, product tax) according to current regulations. Transportation enterprises paid business tax at a rate of 5%. Newly added taxes were exempted or deducted from the payable business taxes. (6) From 1986 onwards, the railway transportation sector implemented a system where the wage was calculated based on converted ton-kilometers.[38] These indicators were quite higher compared to those of the Sixth Five-Year Plan period. By 1990, freight volume had increased by 25% compared to 1985, and passenger volume had increased by 26% compared to 1985. The calculated turnover volume reached 1.4 trillion converted ton-kilometers by 1990, a 33% increase from 1985. Basic construction investment over five years amounted to 33.2 billion yuan, with 46% allocated for the renovation of old railways. Manufacture of rolling stocks increased by 73%, freight cars by 60%, and passenger coaches by 75% compared to the the period of the Sixth Five-Year Plan.[39]

The Contract Responsibility System brought two main changes to railway.

First of all, the railways gained more initiative, which was manifested in two aspects. On the one hand, the railways gained greater autonomy in terms of national operations. With responsibility for the optimization of input for better output and unified revenue-expenditure management, the railways, aside from paying a certain percentage of business tax and contributing a fixed amount to the energy and transportation construction funds, were allowed to retain all income tax and post-tax profits for use by MOR.[40] The railways began to adopt the self-sustaining approach to build new railways, no longer relying on state investment. To promote railway construction, the state granted the railways some preferential policies including reducing business tax for transport enterprises from 15% to 5%, exempting new taxes, setting income tax and post-tax profits as railway construction fund, fixing the baseline contribution to the Energy and Transportation Key Construction Fund based on 1985 income with future income growth exclusion, and a certain amount of funds subsidized by the Ministry of Finance (later all of the funds in the Key

---

[38] Ibid.

[39] (1986, Feb 19). The national railway work conference held in Beijing. *People's Railway Daily.*

[40] Zhao Weichen, Vice Director of the National Economic Commission: The contract responsibility systeming will definitely accelerate railroad construction. (1986, May 14). *People's Railway Daily.*

Energy and Transportation Construction Fund was returned for the use by MOR). MOR was also exempted from repaying principal and interest on certain loans as stipulated in the policy of Conversion of Capital Allocations to Loans during the period of the Sixth Five-Year Plan, with possible deferred principal repayment and lower interest rates for bank loans.

Domestic bank loans (including 2.25 billion short and middle-term loans and 10 billion loans on locomotives) had a 10-year repayment period, with no principal repayment in the first 5 years and preferential rates. Joint venture railways followed local freight rates, and new railways with losses had special rates. It was recommended to include the adjustment of railway freight rates in the Seventh Five-Year Plan for price adjustments, and to appropriately lean on MOR at the adjustment of these rates. The railway depreciation rate and major repair depreciation rate were increased from 3.5% to 4%.[41] These favorable policies seemed very attractive to the railway industry at the time. They not only provided tax incentives, partial loan principal and interest exemptions, easing the railway's burden, but also made some exceptions in pricing the freight rates. Both the government and the Railway Man believed that these measures could ensure funding for railway construction and facilitate the positive development of the railway industry.

In addition, the transportation administrations and industrial enterprises under MOR also gained more autonomy. During the development of Chinese railways, characteristics of "highly centralized dispatching command, coordinative actions among transportation departments, semi-military approach to order execution" emerged, with "centralization" referring to highly centralized dispatching command, ensuring the implementation of national will and meeting the transportation needs of the nation. However, the highly centralized management system failed to recognize the autonomy and initiative of subordinate railway enterprises. The Contract Responsibility System delegated responsibility, authority, and benefits to railway administrations and enterprises, featuring "finishing the contract task, handing in the preset profit, retaining the excess part, and covering the shortfalls independently",[42] and granting them greater development initiative.

Second, the income of each railway employee was closely connected with the profit of railway business. Besides other reforms, the Contract Responsibility System also changed the distribution mechanism. The railway transportation department implemented a system where wages were calculated based on converted ton-kilometers, encouraging more work for more payment and further motivating

---

[41] *Notice of the state council approving the five departments of the state planning commission's decision on MOR*'s *adoptation of the economic contract responsibility system* (GF. No.40 [1986]). Archived in Southwest Jiaotong University, T-87. *Summarizing experience, deepening reform and improving the railway economic contract responsibility system: Comrade Li Senmao's speech at the Railway Leading Officers' Meeting*. Archived in Southwest Jiaotong University, ([1990] No. 1-0-2).

[42] Taking capacity expansion as the center, deepening railway reform and moving forward along the path guided by the 13th Party Congress: Ding Guanguan's speech at the National Railway Work Conference. (1988). *Railway Transport and Economy*, (3), 1–10.

railway enterprises and employees to complete transportation tasks. The contracting plan included policies such as "retention of bonuses for revenue overachievement, profit-sharing on excess earnings, dividends from cost savings, performance-based wage adjustments and mandatory profit remittance quotas".[43] Consequently, the Contract Responsibility System addressed a fundamental issue in railway development at the time—its lack of vitality, where the outcomes of railway development were not related to the efforts of railway workers, nor were personal incomes linked to labor contributions. The System resolved this fundamental issue through the optimization of input for better output in advance, thereby addressing the crucial problem of their motivation. With the implementation of the System, the railway industry gained greater autonomy. Employees were motivated, as their income was directly linked to their efforts. Additionally, every railway employee could clearly feel that their work was closely related to railway development, which stimulated their initiative, enthusiasm, and creativity. Only by effectively mobilizing the enthusiasm of more employees to participate in production and management can greater economic benefits be achieved.

The specific method of this System is as follows: First, MOR contracts with the state. Then, MOR sets the contracting targets for various departments such as transportation, industry, and construction, and signs contracts with them. Each department further divides targets and signs contracts with its subordinate departments. The process continues all the way down to the smallest working groups and teams. Since railway work was centered around transportation, the Contract Responsibility System began with the transportation department and then extended to industrial, construction, and other departments. On May 6, 1986, at a national railway administration directors' meeting, MOR allocated the annual targets to each railway administration. After taking on the tasks, they were required to allocate them to their sub-administrations by May 29. In December 1986, MOR signed contracting agreements with 12 subordinate railway administrations.

As the economic artery of the country, railways generate both economic and social benefits. The implementation of the Contract Responsibility System, however, may focus on the interests of contractors at the expense of national interests. To balance economic and social benefits, as well as partial and overall interests, in April 1986, Ding Guangen, then Minister of Railways, issued "The Three Chapters of the Covenant" at a ministerial meeting: First, we should adhere to the socialist direction and firmly practice the principle that the people's railway serves the people, always putting national interests and social benefits first, and never engaging in practices that favor local interests to the detriment of the general public. Second, it is important to follow the principle of "meticulous planning, prudent investments, operational improvement, and transportation volume expansion". This means that our priorities should be on increasing revenues while minimizing expenses, building railways with diligence and frugality, and maximizing investment benefits.

[43] Making good use of the contract responsibility system to win achievements and be a good leader. (1986, July 20). *People's Railway Daily.*

Third, we shall strictly enforce discipline, act with integrity, abide by rules and regulations, and strictly prohibit arbitrary price hikes, unreasonable charges, exploition of railway resources for personal gain, and indiscriminate issuance of bonuses.[44] After implementing the Contract Responsibility System, the three chapters serve as the fundamental principles for the railway sector to balance the overall and local development, as well as economic and social benefits. They are also critical measures to ensure the success of railway reform. Ding emphasized, "The three chapters are set at a high standard, grounded in the realities, and are indeed achievable".[45]

The years of 1986 and 1987 marked not only the first two years for the railway sector to implement the Contract Responsibility System but also a process of advancing through exploration. The guiding principles were grounded in reality, emphasizing innovation and continuous refinement of strategies while moving forward. The awareness of development was embedded in the fine adjustments, with incremental steps adding up to significant strides, thus gradually enhancing reforms and improving the economic contract responsibility system.[46] Based on the original economic management system, MOR made fine adjustments to the policy of profit sharing and power delegation to the railway administrations, encouraging them to increase revenue. To achieve a balance between revenue and expenditure, in these two years, the plans for infrastructure investment and purchases of locomotive and rolling stocks were both lower than the average for the period of the Seventh Five-Year Plan, while the number of loans used was higher than the average.[47] At the same time, some indicators improved compared to the original plan, with freight volume increasing by 4.3%, converted turnover by 8.3%, and transportation revenue by 9.6%.[48] Due to time constraints, the income distribution could not be completely refined, so we could only make "fine adjustments" based on the previous measures for unified settlement unit prices set by railway administrations. Work was contracted out first, followed by organizing personnel to formulate distribution methods. Therefore, the Contract Responsibility System was actually a reform initiated from practices that gradually improved the distribution rules. To this end, the Party Leadership Group of MOR established several foundational principles and put forward the "three links" and "nine principles".

[44] (1986, May 11th). Three Chapters of the Covenant is the action code of the Contract Responsibility System, which we should firmly support, vigorously publicize and take the lead in abiding by. *People's Railway Daily.*

[45] Ibid.

[46] Political Department of MOR. Key work of the railway in 1987 (TB [1987] No. 1). Archived in Southwest Jiaotong University, T-52.

[47] Summarizing experience, enhancing reform and improving the railway economic contract responsibility system: Comrade Li Senmao's speech at the Railway Leading Officers' Meeting. Archived in Southwest Jiaotong University, ([1990] No. 1-0-2).

[48] (1986, May 14th). This year's road-wide goals were implemented in all road bureaus. *People's Railway Daily.*

The "three links" refer to "the links between benefits of each railway administration and its retained profits, the benefits and vitality of self-transformation and self-development, and between the input and output".

The "nine principles" are:

> "First, check the needs of the railway infrastructure construction funds allocated by the state this year (referring to 1986, hereinafter the same);
>
> Second, ensure compliance with the tax payment obligations to the state and the repayment of principal and interest to the bank;
>
> Third, enhance the vitality of the railway administrations, with retained profits higher than last year's figures. Net profits will be remitted to the Ministry at a fixed amount based on last year's, without further increases;
>
> Fourth, when formulating the settlement unit price, fully consider the cost differences among railway administrations caused by objective economic factors such as geographical conditions and line structure;
>
> Fifth, strive to ensure that the growth of retained profits of each railway administration this year is not lower than the growth rate from 1984 to 1985;
>
> Sixth, formulate policies to encourage overproduction and implement a multi-tiered pricing policy for overproduction. Specifically, in the first tier, the settlement unit price remains at the original level for the volume of the previous year's target. In the second tier, the settlement unit price increased for the volume exceeding the previous year's target and reaching this year's target. In the third tier, all taxes paid on the excess volume of this year's target will be retained by the railway administration;
>
> Seventh, in terms of retained profit distribution, shift from the previous retroactive payments to some administrations to advance payments, turning 'hidden subsidies' into 'explicit subsidies;
>
> Eighth, allocate newly added traffic volume based on the amount of infrastructure and renovation funds invested by the state in each administration, linking input to output;
>
> Ninth, the adjustment of the settlement unit price this year should be aligned with the pace of reform next year".[49]

These nine principles aim to motivate railway workers while ensuring the completion of tasks under the Contract Responsibility System.

After the contracting tasks were allocated, MOR established an economic responsibility contract reform team in June 1986. The reform team conducted widespread investigations, research, calculations, and comparisons among relevant administrations. They also held discussions with various railway administrations to understand the situation at the local level, ultimately formulating a specific implementation plan.[50]

---

[49] (1986, May 9). Implementing the contract tasks of 1986 to ensure a victorious start in the Seventh Five-Year Plan, *People's Railway Daily.*

[50] (1986, September 24). The expert consultation meeting held by MOR on the *Implementation measures for the economic contract responsibility system of Railway Administrations (Draft). People's Railway Daily.*

On October 12, 1986, the State Development Planning Commission, the State Economic Commission, the Ministry of Finance, the Ministry of Labor and Personnel, and MOR jointly issued the *Implementation Measures for the Contract Responsibility System for Wage-Content Based on Converted Ton-Kilometers in Railway Transportation Departments*. The *Measures* stipulated that, based on the converted ton-kilometers completed and total wages expended in 1985, after deducting wages that should not be expended and adding wages as stipulated by relevant national regulations, the wage per 10,000 converted ton-kilometers would be calculated. Then, MOR would determine the lump-sum responsibility coefficient for wage compositions based on the actual situation of each railway transportation enterprise. For enterprises that failed to meet the contract targets set by MOR, their lump sum wages would be deducted by a certain proportion. This regulation came into effect on January 1, 1986, which led to the introduction of the Contract Responsibility System for income distribution reform, where employee income became directly related to their performance.

Starting from the second year for the implementation of the Contract Responsibility System in 1987, MOR, began to sign four-year contracts with some railway administrations for the period from 1987 to 1990, after research, calculations, and consultations with them.[51] Due to different conditions, the specific targets for each administration varied. After assigning the mandatory plans for each year to the administrations, MOR recognized that the completion of these plans and their growth rates would differ from year to year. Additionally, MOR also made contracting arrangements for engineering group corporations. Within each engineering group corporation, the Contract Responsibility system for its construction projects was also put into practice.

3.2.2 Milestone of the Contract Responsibility System

As the first package of reform plans initiated by MOR, the Contract Responsibility System aimed to overcome the constraints hindering the railway sector from contributing effectively to national economic development and to advance the modernization of the railways.

As a result, the Railway Man looked forward to this reform. MOR viewed the approval of the Contract Responsibility System by the State Council as "a great demonstration of the Party and the State's support and concern for the railway work, an important decision to accelerate the development and construction of the railways, and a significant event that the three million railway workers have long been

[51] MOR. Approval reply regarding the economic contract plan of Guangzhou Railway Administration for the years 1987 to 1990 (TB [1987] No. 264). Archived in Southwest Jiaotong University, T-66.

hoping for to change the passive and backward state of the railways".[52] The implementation of the Contract Responsibility System left railway workers both excited and nervous. The prospect of embarking on an all-round reform was exhilarating, as it inspired hope among railway workers. "Many colleagues who had previously perceived the railways as trapped in a seemingly endless cycle of challenges and difficulties began to see a glimmer of hope, alleviating the long-held sense of oppression caused by the railways' passive and backward state. They felt that there was finally a way out, and wanted to move forward again". Railway workers' confidence in achieving the goals set for the period of the Seventh Five-Year Plan soared. "Throughout the railway system, there emerged an unprecedented intellectual vitality and morale, leading to a thriving momentum in the transportation, production, and construction, as well as various other tasks".[53]

However, a sense of nervousness prevailed due to the challenging targets set under the Contract Responsibility System. "The targets set by MOR under national contract for the volume of passenger and freight traffic, the output of locomotive and rolling stock, the investment in and the scale of infrastructure, as well as the required transportation capacity, which was all ambitious and necessitated significant efforts to be invested". Moreover, "our goal in undertaking this contract is to use our own accumulated earnings to accelerate the pace of railway construction and technological transformation… The vitality of the railways' self-transformation and self-development depends on how much we can accumulate through our efforts".[54] The excitement and nervousness gave a sense of responsibility and mission among railway workers, who felt that "to bear the pressure that weighed a thousand pounds, every person has a mission to shoulder one pound".[55] This combination of a sense of mission, urgency, and responsibility has mobilized railway workers to take proactive actions. "When accepting tasks, the directors of administrations were all very proactive, believing that the heavier the task, the more fulfilling it would be. They were all determined to leverage their intelligence and talents, and work hard to accomplish great things".[56]

However, the "Contract Responsibility System" was not without its challenges. Amidst the scarcity of railway carriages and tickets at that time, the system not only involved the reform of management concepts and systems but also implicitly

---

[52] MOR. Notice on conscientiously implementing the *Notice of the State Council forwarding the proposal of the state planning commission and four other departments on implementing the economic contract responsibility system in MOR* (TB [1986] No. 340). Archived in Southwest Jiaotong University, T-87.

[53] (1986, April 20). Laying a solid foundation for achieving the Seventh Five-Year Plan goals. *People's Railway Daily.*

[54] (1986, April 25). Recognizing the new situation, clarifying the new tasks, and adapting to the new requirements. *People's Railway Daily.*

[55] (1986, May 23). Efforts to implement the fine adjustments plan, racing against time to increase transportation and revenue. *People's Railway Daily.*

[56] (1986, May 30). The Contract Responsibility System indicators assigned to 62 Railway Administrations. *People's Railway Daily.*

addressed the issue of the "gray income" among certain railway employees. The former required a shift in management concepts and a thorough renovation of outdated management systems, while the latter necessitated punishing illegal and irregular practices. Additionally, while some people supported the Contract Responsibility System, their habits formed under the old system over decades proved difficult to change, or their understanding of the System was inadequate, thereby hindering the implementation of the system. The semi-militarized characteristic of Chinese railways, to a certain extent, ensured the implementation of the System. At the same time, MOR strengthened several aspects of its work: Firstly, it intensified ideological and political work to urge railway employees to accelerate the transformation of their mindsets and clear ideological obstacles for the smooth progress of the system. Secondly, it accelerated institutional reforms to safeguard the implementation of the Contract Responsibility System. Thirdly, it continuously rectified misconduct and intensified efforts to combat illegal and irregular practices. These measures collectively provided a solid foundation for the successful advancement of the Contract Responsibility System.

With the implementation of the Contract Responsibility System, the railway sector quickly witnessed its significant progress. The first notable transformation was the mobilization of enthusiasm and initiative among railway employees. They began to genuinely understand "the drawbacks of egalitarianism and the benefits of performance-based distribution". For instance, the Xuzhou Track division "ended the equal distribution mode of bonuses, linking the floating wages with contract bonuses, job allowances, and an additional ten yuan from each employee's salary. This system rewarded diligence and excellence while punishing laziness and underperformance. Workers who completed tasks well and made significant contributions received higher bonuses, up to over fifty yuan per month, while those who were reluctant to engage in difficult, dirty, or tiring work only received a few yuan or even nothing. Bonus now truly reflects its essence, and there is no more guaranteed fixed incomes regardless of performance".[57] The economic lever effectively motivated the majority of workers. After the practice of the Contract Responsibility System, the creativity and ingenuity of employees at the Guangzhou Railway Administration were fully motivated, sparking a wave of increase in performance indicators.... Employees at the Guangzhou Railway Administration demonstrated a strong sense of ownership and actively contributed ideas to the Contract Responsibility System. In just two months, transportation and production divisions received nearly ten thousand suggestions from employees, with over five hundred being adopted by relevant departments. These strategies were projected to increase revenue by over ten million yuan when applied to production".[58] The Guangzhou Railway Administration also saw remarkable benefits. "Shortly after the system's implemen-

[57] (1986, August 6). Suffering from backtracking to equalitarianism and enjoying the benefits of adhering to distribution according to work after the Contract Responsibility System. *People's Railway Daily.*

[58] (1986, June 25). Employees at Guangzhou Railway Administration contribute nearly 10,000 suggestions for the Contract Responsibility System. *People's Railway Daily.*

tation, the Guangzhou Railway Division began to see results within just a few months. Despite fierce competition in passenger and freight transportation, the volume of cargo shipped increased by 17.9%, the converted turnover volume grew by 9.6%, and transportation revenue rose by 10.3% in the first half of this year compared to the same period last year".[59]

The second notable transformation was the positive change in the railway's work style. "In early August, some corporate executives in Shanghai cited facts to reporters, praising the ongoing transformation of the railway's 'boss mentality' since the implementation of the Contract Responsibility System. ... In the past... due to the railway's monopoly, the 'boss mentality' among officers and employees was entrenched, and shippers also had a mindset of relying on the railway and being afraid of offending it, thus becoming accustomed to the difficulties in accessing the railway services. However, after implementing the Contract Responsibility System, with increased tasks and intensified competition, the 'boss' began to change its stance. Workers from the railway administrations, sub-administrations, and relevant stations visited customers door-to-door to seek feedback, connect with cargo sources, and even conducted on-site investigations at company warehouses, helping and guiding the clients to increase efficiency in loading and transportation. In just a few months, the company (Note: referring to the Shanghai Commercial Warehousing & Transportation Co.) loaded over 450 more carriages than planned, adding 110 million yuan of value to Shanghai's income".[60]

The third transformation involved fostering business awareness and cultivating a mindset focused on profitability. Under the System, railway administrations "first prioritized renewing concepts and strengthening business awareness".[61] "To survive and develop in a competitive environment, the China Railway Major Bridge Engineering Group Co., Ltd. established a Business Development Department, and corresponding institutions were also set up in various departments and factories, extending their business to East, Central, and South China to stay informed of market trends and participate in bidding".[62]

After implementing the Contract Responsibility System, Harbin Railway Administration's Binjiang Station emphasized the need for a shift in mindset, stating, "In terms of business, we should learn from privately owned restaurants. When customers arrive, we should warmly greet them and find seats for them. Would they leave under such hospitality? Similarly, when shippers arrive, we should also greet them warmly, and help them solve difficulties, so worry no more about returning

[59] (1986, July 22). Vigorously promoting conceptual renewal and strengthening business awareness. *People's Railway Daily.*

[60] (1986, August 17). Contract Responsibility System eliminating the "railway boss" mentality. *People's Railway Daily.*

[61] (1986, July 22). Vigorously promoting conceptual renewal and strengthening business awareness. *People's Railway Daily.*

[62] (1988, January 17). Succeeding in market entry and competition. *People's Railway Daily.*

customers".[63] Driven by the Contract Responsibility System, the organizations worked hard to "increase revenue and reduce expenditures". Some employees said, "Saving expenditures is equivalent to increasing profits, a way to increase revenue without investment or taxation. It's been said for many years that every penny saved counts, but under the 'indiscriminate egalitarianism' system, the phenomenon of 'talking about saving while wasting persisted. Now things are different. Some workers from the Shenyang Railway Administration put it bluntly, 'Under the new System, spending money means dipping into your own pocket'. Railway personnel are starting to make the most of resources and minimize waste now!"[64]

The Contract Responsibility System initially progressed smoothly. Ding Guangen, then Minister of Railways, summarized, "1986 and 1987 marked the two years of significant economic reforms centered on the Contract Responsibility System in the railway sector. These were also years of determination, hard work, and a major turnaround for the railways". "The Contract Responsibility System is both a necessary step and the inevitable path for railway development. This fundamental breakthrough in the railway's economic system has revitalized its development and spurred all-round reforms within the sector".[65] From 1986 to 1987, the annual average passenger and freight turnover increased by over 88 billion ton-kilometers, 30% higher than the average annual growth during the period of the Sixth Five-Year Plan. In 1987, the daily average number of loaded wagons exceeded 70,000. The freight density per kilometer of railway reached 23.6 million converted ton-kilometers, an increase of 13% compared to 1985. Vehicle manufacture capacity also grew rapidly, with the average output of newly manufactured diesel locomotives, electric locomotives, passenger coaches, and freight cars in 1987 increasing by 29%, 32%, 14%, and 9% respectively compared to 1985, alleviating the acute shortage of locomotives to some extent. Employee incomes also improved, with the average salary of railway employees reaching 1,833 yuan in 1987, an increase of 24.8% over 1985. At the same time, significant achievements were made in railway construction. Over these two years, 1,181 kilometers of new railways, 892 kilometers of double-track lines, and 490 kilometers of electric railways were built, exceeding the average construction speed of the period of the Sixth Five-Year Plan.[66] Without the System, the railways would have paid 16.2 billion yuan to the country in 1986 and 1987. However, under the System, the total amount of taxes paid by the railways and funds invested in infrastructure development, locomotive and rolling

[63] Zhou Ziping, & Zhang Shugui. (1986, August 15). New developments in the Contract Responsibility System: A chronicle of the transformation of business concepts at the Binjiang Station of the Harbin Railway Administration. *People's Railway Daily.*

[64] Li Runqing. (1986, June 11). Saving expenses is also a form of increasing revenue. *People's Railway Daily.*

[65] (1988). Taking capacity expansion as the center, enhancing railway reform and moving forward along the path guided by the 13th Party Congress: Ding Guanguan's speech at the National Railway Work Conference. *Railway Transport and Economy*, (3).

[66] Ding Guanguan's speech at the National Railway Work Conference. Archivesd in Southwest Jiaotong University, T-32.

stock purchases reached 21.6 billion yuan over these two years, effectively contributing one-third more to the country than before the implementation of the System. This fully demonstrates the remarkable effectiveness of the Contract Responsibility System.

3.2.3 Major Challenges and Adjustments Faced by Contract Responsibility System

At the 13th National Congress of the Communist Party of China held in November 1987, it was concluded that China was in the primary stage of socialism. The Congress also established the basic guiding principle for building socialism with Chinese characteristics during this stage, "to make economic development our central task, adhere to the Four Cardinal Principles, and persevere in the reform and opening up". Against this backdrop, the reform of the railway sector continued to progress. In 1988, the Contract Responsibility System of the railway sector advanced, with a comprehensive contracting system being implemented. MOR signed three-year contract agreements with all regional railway administrations, encouraging diversified operations. The Contract Responsibility System also began to extend to railway enterprises. By 1988, major entities such as the China National Railway Locomotive & Rolling Stock Industry Corporation, the General Adminstration of Capital Construction, China Railway Signal & Communication Corporation Limited, and the Research & Design Institute completed their own Contract Responsibility System reform. Other enterprises, including companies related to logistics support, followed suit shortly thereafter.

In 1988, the Contract Responsibility System encountered unforeseen issues that posed significant challenges. These challenges stemmed both from the broader national economic environment and the railway sector itself. The first challenge was inflation, which seriously weakened the railway's ability to accrue capital for construction. During the period of the Seventh Five-Year Plan, the liberalization of prices, wage reforms, and the evaluation of professional titles and large-scale investments in infrastructure led to a substantial increase in the demand for currency in Chinese society, triggering inflation. According to statistics, the national retail price index rose by 18.5% in 1988 compared to 1987. Nearly 80% of goods saw price increases, with an average rise of 30%, particularly in products like household appliances, motorcycles, and fuel. In 1989, the price index further surged by 17.8%, marking a major challenge for China's reform and opening-up process. Inflation posed significant challenges for the railway sector. In 1988, escalating prices led to a notable increase in railway expenditures compared to the previous year, "as the sector had to absorb the rising costs of steel and diesel fuel internally".[67] Consequently, the prices of production materials soared sharply, leading to higher

[67] Ibid.

costs, increased bank loan interest rates, adjustments in import exchange rates, and a rise in various taxes and rates. These factors collectively resulted in a substantial decline in the railway's capacity to accumulate internal capital. Throughout the period of the Seventh Five-Year Plan, railway expenditures increased by over 23.37 billion yuan due to the rising prices,[68] equivalent to the entire investment in railway infrastructure during the Sixth Five-Year Plan.

At the same time, the transportation income—the main source of revenue for the railway sector—remained stagnant due to low freight rates. From 1985 to 1986, the government began to deregulate prices for unplanned industrial goods gradually, allowing enterprises to set prices based on market conditions. Although the railway sector was classified as a large state-owned enterprise, the prices for the services they provided—transportation service—were still regulated, and the railways had no right to set prices. The passenger fare, set in June 1955, remained unchanged except for a slight increase in short-distance tickets under 100 kilometers in 1985. In 1987, the suburban fare per kilometer was 0.015 yuan, while the cost of rail operations that year was 0.04063 yuan, indicating "a sharp discrepancy between railway fare and actual costs, which resulted in significant losses in passenger transportation".[69]

Freight rates were even lower. By the end of 1987, the revenue per ten thousand ton-kilometers, when calculated based on a combination of market prices and state-planned prices, represented only 10% to 20% of its value in 1955. Compared to road freight rates, it was even more evident that railway freight rates were low. For instance, in short-distance transportation, "before the price adjustment, the average cost per ton-kilometer by rail was 0.017 yuan, while by road it was 0.17 yuan—a tenfold difference".[70] Railway freight rates have always been regulated by the country. Although the government may adjust these rates based on market conditions, such adjustments inevitably involve numerous procedures and often lag considerably behind market changes. Despite significant price increases by 1988, the State Council did not approved adjustments to passenger fares until September 1989, with freight rates following suit in March 1990.

Thus, the disparity of "fixed freight rates against dynamic market prices, and low freight rates against high commodity prices" led to a sharp decline in railway revenue, making it difficult to self-fund its maintenance and construction. In 1988 and 1989, compared to 1985, profits of the railway sector decreased by 180 million yuan and 1.86 billion yuan respectively. However, after the fare adjustments in 1990, revenues increased by approximately 10 billion yuan compared to 1989, and profits

[68] Summarizing experience, enhancing reform and improving the railway economic contract responsibility system: Comrade Li Senmao's speech at the Railway Leading Officers' Meeting. Archived in Southwest Jiaotong University, ([1990] No. 1-0-2).

[69] Zhu Kuicheng. (1989, February 22). Strengthening reform should take this step. *People's Railway Daily.*

[70] Notice on forwarding Vice Premier Li Peng's speech at the National Symposium on Transportation Work (TBZ [1985] No. 1181). Archived in Southwest Jiaotong University, T-342.

rose by 6 billion yuan.[71] These additional incomes were immediately allocated to key construction projects, and combined with increased loans, such funds somewhat alleviated the financial strain on railway infrastructure development. All these underscored the critical role of transportation prices in the railway's own capital accumulation.

The significant challenges posed by inflation to the railway industry highlighted the fundamental contradictions of the Contract Responsibility System. Under this system, however, the railway industry lacked the authority to set transportation fares. The goal of the System was to move the railways towards marketization. Although a dual-track price system was in place at the time, in which the prices of major production materials were controlled by the state, the quantity and variety of materials under state control decreased significantly. As an enterprise, the railways had to procure a large amount of production materials from the market at prevailing prices, in addition to those controlled by the state. However, the fares of the transportation services provided by the railways were not determined by the railways themselves and could not be adjusted according to market prices. This contradiction fundamentally reflects the complexities inherent in railway reform in China.

The Chinese railway system, as a multifaceted entity integrating construction, transportation, and industrial manufacturing, necessitated market-oriented operations. Yet, given the relatively underdeveloped infrastructure in China at the time, the railway, as a type of infrastructure requiring capital-intensive investment but yielding slow returns, cannot be categorized as a general business entity, and its complete marketization was quite challenging. This ongoing contradiction has persisted throughout the railway reform process, causing the reform efforts to fluctuate. As renowned economist Li Yining noted a few years later, "The railway management departments are government authorities, and for them, the issue is not about whether to push the sector to the market. Their role is to manage according to laws, regulations, and policies". In contrast, "for the railway administrations and sub-administrations, which are railway transportation enterprises, the questions of whether and how to push them into the market do arise. Marketization is necessary; thus, how can they facilitate this process? The two essential conditions are: firstly, the railway enterprises should gradually become independent operators responsible for their own profits and losses, and move onto the track of self-development and self-regulation; secondly, it is crucial to establish a rational pricing relationship, particularly between transportation fares and other prices, to create a more reasonable price structure. With these two requirements met, integrating railway transportation enterprises into the market will be feasible".[72] In the 1980s, however, it was evident that these requirements, especially the latter, were not adequately fulfilled.

The second challenge was ensuring operational safety. In transportation, safety is the top priority. Although safety was stressed during the implementation of the

[71] Statistics Center of MOR. (2008). *Collection of national railway statistical data* (internal resources, p. 271). Beijing: China Railway Publishing House.

[72] Wu Guoliang. (1992, April 2). Railways and market mechanisms: An interview with Li Yining, the Deputy to the National People's Congress and renowned economist. *People's Railway Daily*.

Contract Responsibility System, the main focus remained on economic efficiency. As a result, the principle of "safety first" was not firmly established, nor was it effectively implemented. More importantly, the aging of railway equipment had long remained unresolved, and the continuous increase in freight volume under the Contract Responsibility System further exposed existing equipment issues, resulting in frequent safety incidents. In the first quarter of 1988 alone, there were four serious safety accidents, "causing significant losses to the country and its people, tarnishing China's reputation both domestically and internationally, and drastically setting back railway operations".[73] As a result, Ding Guangen, then Minister of Railways, resigned in accountability, marking the first such resignation in the history of railway development in China. Subsequently, some suggested that the frequent safety incidents were closely linked to the implementation and abolition of the Contract Responsibility System, "These frequent safety incidents not only damaged the image of Chinese railways but also hindered the progress of railway reform initiated in 1986". How should the balance be struck between maximizing investment returns and ensuring railway safety?

The third challenge was the high-performance targets of the Contract Responsibility System, coupled with failure to correspondingly elevate the remuneration for railway workers, making it difficult to fully mobilize their enthusiasm. Under the Contract Responsibility System, the sector has maintained a relatively high annual growth rate without state investment. Initially, MOR made minor adjustments to various performance targets, and the large-scale exploration of reforms sparked enthusiasm among railway workers. However, as the targets became more demanding, the lack of improvement in workers' welfare in the pursuit of these targets dampened their morale. "Over the past three years of implementing Contract Responsibility System, the average wage of railway workers has fallen below the average of the energy and transportation sectors. After accounting for inflation, the average wage growth for railway workers in 1987 turned negative (-0.4%)". "In the past, people took pride in being a railway worker. Now, the new saying among railway workers is 'the more we take on, the harder it gets'".[74] This does not imply that the Contract Responsibility System itself was wrong, but rather that the external and internal conditions are too strained, and workers themselves saw little personal benefits. In other words, the workload was high and the remuneration was low. While short-term motivation could be maintained through ideological and political work, over the long run, the morale of railway workers is inevitably dampened.

The fourth challenge was the issue of work ethics in the railway system. During periods of high demand for railway transportation, the limited availability of freight cars created opportunities for rent-seeking, and it was not uncommon for railway employees to use their access to freight cars for personal gain.

[73] (1989). Summarizing experiences, facing challenges, striving to move forward: Summary of Vice Minister Tu Yourui's speech at the New Year's Teleconference. *Railway Situation Briefing*, (1).

[74] Zhang Guorong. (1989, March 12). The railway's struggles. *People's Railway Daily.*

Li Senmao, the then Minister of Railways, reflected on the lessons learned from the implementation of the Contract Responsibility System, stating, "The Contract Responsibility System in railways was a new initiative. Due to a lack of experience, we were not fully prepared for the external changes and their impact on the system, nor did we anticipate the new problems that arose after the decentralization and revitalization. Many lessons are worth reflecting upon, including a lack of effective oversight that led to serious issues with railway work ethics, overly ambitious targets that diminished the motivation of contractors, changes in the external conditions that exacerbated funding shortfalls and hindered the completion of construction tasks, the insufficiently refined contracting mechanism, and increased reliance on loans that resulted in an excessive debt burden".[75] The emergence of these issues led to "growing doubts both within and outside the railway sector regarding the Contract Responsibility System".[76]

In the face of these challenges, the State Council affirmed its commitment to the practice of the Contract Responsibility System. At the transportation safety conference held by the State Council on June 4, 1988, then Premier Li Peng emphasized in his speech that "the 'Contract Responsibility System' in the railway sector should be endorsed and carried forward, but it requires further improvement to address its current limitations". Li Peng argued that the Contract Responsibility System should be tailored to the unique characteristics of the railway sector, described as "highly centralized dispatching command, coordinative actions among transportation departments, semi-military approach to order execution". He proposed that "railway administrations and sub-administrations could adopt a Contract Responsibility System, according to which they are responsible not only for fulfilling transportation tasks but also for managing transportation revenues and remitting accumulated funds to higher authorities. At the grassroots level, however, management and assessment approaches should be based on comprehensive performance indicators, and direct linking of station income to individual employee earnings should not be encouraged, as this would facilitate unified scheduling and command across the entire railway system. Most importantly, the railway shall ensure the timely transportation of key materials outlined in the national plan, as it is crucial for overall economic stability. There should be a shared understanding throughout the railway sector that partial or departmental interests should never override national and overall interests".[77] His remarks highlighted the specific nature of the railway sector, noting the difficulty of applying purely economic metrics to measure both economic and social benefits, as well as overall and partial interests. Therefore, he called for further refinement to the Contract Responsibility System. Additionally, Li urged the railway authorities to "reform in labor and wage systems to address issues of 'equal

[75] Summarizing experience, enhancing reform and improving the railway economic contract responsibility system: Comrade Li Senmao's speech at the Railway Leading Officers' Meeting. Archived in Southwest Jiaotong University, ([1990] No. 1-0-2).

[76] Ibid.

[77] Xu Yaozhong, & Yang Like. (1988, June 8). Advancing reforms in the transportation sector and strengthening comprehensive management of safety. *People's Railway Daily*.

distribution regardless of performance' ('da guo fan' in Chinese), advocating for widening the income gap between key personnel and general staff".[78] He emphasized that "like any new initiative, this current reform still has its shortcomings. Thus, in the next stage of contracting, the Contract Responsibility System should maximize benefits while correcting its flaws, further enhancing and refining its practices".[79]

In response to the directives of the State Council, MOR decided to adjust and improve the Contract Responsibility System. Meanwhile, starting from the Third Plenary Session of the 13th Central Committee of the CPC in 1988, China's reform and opening up entered a stage of "governance and rectification, along with advancing reforms", marking a significant adjustment in the national economy. As a result, the railway sector had also stepped into such a stage, during which it focused on summarizing the experiences and lessons learned from the Contract Responsibility System in order to further improve it. An investigation team for advancing reform, established by MOR, conducted in-depth research at the grassroots level, and issued seven measures to perfect contracting in the course of the governance and rectification, that is, adjusting contracting indicators for certain departments, addressing new issues arising in the financial settlement, making appropriate adjustments to the planned economic system, improving the distribution of "the contracted wage quotas measured in ton-kilometers", adhering to the reward-and-penalty system for the paid use of freight cars, vigorously developing diversified operations, and improving the contracting methods for railway transportation departments at the grassroots level.[80] Throughout the railway sector, transportation safety was the top priority. At the same time, efforts were being made to improve work ethics, tighten expenditures, and launch a large-scale campaign for "increasing production while reducing consumption, and boosting revenues while cutting costs". The principle of "three safeguards and three constraints" (safeguarding planned projects while restricting unplanned ones, safeguarding productive construction while restricting non-productive construction, and safeguarding key construction while restricting general industrial construction), was being upheld, all aimed at firmly controlling the scale of investment.[81]

In 1990, MOR issued the *Opinions on Further Consolidation and Deepening Reform of Railways*, outlining a plan to continue the railway system's rectification and reform over a three-year period (1990–1992). The focus of this initiative was on: "(1) adjusting structures and clarifying relationships; (2) strengthening the foundations, ensuring quality, maintaining safety, and improving railway work ethics;

[78] Ibid.

[79] Xu Yaozhong. (1990, June 22). Premier Li Peng urges further strengthening and improvement of Contract Responsibility System reforms during discussion with representatives of the Railway Leading Officers' Meeting". *People's Railway Daily*.

[80] The Editorial Center of Chinese Railway History (1996). *Memorabilia of Chinese railway* (1876-1995) (p. 436). Beijing: China Railway Publishing House.

[81] (1988, January 20). Continuing to bear the burden and strive towards the railway strategic goals of the Seventh Five-Year Plan. *People's Railway Daily.*

(3) renovating equipment and accelerating the rectification of maintenance backlogs; (4) tapping the potential and expanding capacity to enhance the overall transportation capability of limiting sections along main lines".[82] This reform aimed primarily at internal structure adjustments, safety assurance, rectification of railway work ethics, equipment maintenance, and the improvement of capacity in the "bottleneck" sections. From 1990 onwards, the settlement method directly linking economic interests to the performance of railway stations at the grassroots level was replaced with a comprehensive management and assessment system of performance indicators centered on the fulfillment of mandatory plans, continuously improving the Contract Responsibility System.[83]

At the end of 1990, the Contract Responsibility System for the *Seventh Five-Year Plan* came to an end. On January 12, 1991, the State Council issued a notice stating that during the period of the Eighth Five-Year Plan, the railway sector would continue operating under the existing contracting system until the new one was approved.[84] In 1991, MOR adjusted and improved the contracting system based on the *Seventh Five-Year Plan*, adhering to the principle of "creating a stable operating environment for each railway administration to enhance the vitality of their transportation operations".

The adjustments included: Firstly, the separation between operations and constructions. The railway authorities would contract infrastructure construction and operations separately.[85] Railway administrations would only manage operations, while MOR would assign construction tasks through project contracts. Secondly, changes in the contracting hierarchy. In line with the characteristics of railway production and operations, MOR served as the center for investment and decision-making, railway administrations and their branches were centers for operations, and stations and depots were centers for production. Economic contracts in the railway transportation system would extend only to the branches, with stations and depots managed and assessed based on comprehensive indicators primarily for task completion, safety, quality, and performance. This marked a shift from the previous practice of contracting down to the station and depots. Thirdly, boost in motivation of sub-administrations. The increase in "contracted wage quotas measured in ton-kilometers" approved by the State Council was allocated to each railway administration, with the Ministry retaining minimal amounts. Additionally, measures were taken to raise settlement prices, adjust the base for profit retention,

[82] (1990, March 21). Recognizing the situation, lifting spirits, and uniting in striving to fully accomplish this year's tasks: Summary of Li Senmao's speech at the National Railway Planning Conference. *People's Railway Daily*.

[83] (1989, October 22). Recognizing the situation, lifting spirits, and advancing railway work amid governance and rectification efforts. *People's Railway Daily*.

[84] The State Council: Notice on the temporary execution of all contractual policies between MOR and the State during the Seventh Five-Year Plan. Archived in Southwest Jiaotong University, T-15.

[85] (1992, January 2). Focusing on quality and economic efficiency to steadily advance railway work: Li Senmao's deployment for railway work in 1992 at the National Railway Teleconference. *People's Railway Daily*.

and modify the content and indicators of the contracts. Fourthly, incentive and restriction mechanisms in income and distribution. The adjustments introduced appropriate incentives while also establishing comprehensive mechanisms for accountability. These changes increased the motivation of sub-administrations and employees while alleviating pressure on them. However, such adjustments did not address the most significant issue of the Contract Responsibility System, that is, the railway's inability to generate sufficient internal funds to cover the substantial funding gap for construction. Under the circumstances at the time, without large-scale construction of new railway lines, it would have been impossible to significantly increase capacity and effectively resolve the stark imbalance between transportation capacity and actual volume.

In summary, starting in 1986, the railway sector implemented an economic contract responsibility system, embarking on a path of large-scale reforms through exploration and experimentation. After more than six years of bold efforts, the railway sector successfully fulfilled its contracting tasks. During the period of the Seventh Five-Year Plan, transportation revenues increased by 10.2 billion yuan, coupled with an additional 10 billion yuan from adjustments in passenger and freight fares, leading to a total transportation revenue of 150.2 billion yuan. This represents a 96% increase compared to the period of the Sixth Five-Year Plan and an increase of 20.2 billion yuan over the 130 billion yuan predicted at the introduction of the contract system in 1986. At the same time, economic management in the railway sector transitioned from a production-based model to a production-operation-oriented one, forming a new operational structure centered on transportation with diversified operations. This shift transformed the traditional transportation organization and management methods, marking the beginning of its adaptation to the socialist market economy.

The Contract Responsibility System was the railway industry's exploration under the impetus of China's reform policies. While it was generally in line with China's reform policies and pace at the time, the railway industry showed significant initiative in the reform, marking its first comprehensive attempt at transformation. The reform achieved notable results, completing the contracting goals and propelling the railway industry to take its first step towards the market, which merits full recognition. However, despite fulfilling the contracting goals, the Contract Responsibility System failed to achieve its original intention—to fundamentally resolve the contradiction between transportation capacity and traffic volume.

In early 1992, following Deng Xiaoping's Southern Tour Speeches and the subsequent convening of the 14th National Congress of the Communist Party of China, China's reform and opening up entered a new phase. The railway industry proceeded to promote reforms and refine economic contracting, but the focus has shifted to market-oriented reforms to accelerate the transformation of internal business mechanisms. The Contract Responsibility System has been virtually abolished, marking a new stage in the development of railway reforms.

3.3 New Model of "Self-sustaining Approach to Building New Railways"

3.3.1 "Self-Sustaining Approach to Building New Railways"—A New Model for Railway Construction Funding Reform

Construction fund is one of the critical issues that should be addressed in railway construction. As mentioned earlier, prior to 1985, Chinese railway infrastructure was mostly funded by the country. In 1985, the country comprehensively implemented the policy of Conversion of Capital Allocations to Loans, transforming the investment in railway infrastructure into bank loans. In 1986, the railway industry began to implement the Contract Responsibility System, which led to significant changes in the financing system for railway infrastructure development, marking the beginning of the new model of the Self-sustaining Approach to Building New Railways. As the name suggests, the term refers to relying on railway revenue to fund railway construction. This model differs significantly from previous ones in two ways:

Firstly, railway construction funds no longer relied on state allocations or loans but were primarily sourced from the capital accumulations of the railway industry. Before implementing the policy of Conversion of Capital Allocations to Loans, as mentioned before, railway construction funds were allocated through state budgets. After its implementation, these funds were converted into loans from the China Construction Bank, with the state setting loan quotas for the railway industry. With the advent of the Contract Responsibility System, the railway construction funds were no longer sourced from bank loans, but mainly from the funds accumulated by the railway industry itself. Profits generated after taxes were retained and utilized for construction. The country also ceases to allocate funds specifically for railway construction and issues construction funding plans, with the railway industry relying on its own accumulated funds as the primary source of funding for railway construction. During the period of the Seventh Five-Year Plan (1986–1990), the railway industry's self-accumulated funds for infrastructure development totaled 25.44 billion yuan, accounting for 73.9% of the total investment of 34.433 billion yuan for national railway construction.[86] This underscores the pivotal role of self-accumulation in railway construction funding during this period.

Secondly, railway construction became directly tied to its operational performance. During the period of the Contract Responsibility System, the country canceled direct funding for the railway, and the amount of funds available for railway

[86] Summarizing experience, enhancing reform and improving the railway economic contract responsibility system: Comrade Li Senmao's speech at the Railway Leading Officers' Meeting. Archived in Southwest Jiaotong University, ([1990] No. 1-0-2). Statistics Center of MOR. (2008). *Collection of national railway statistical data* (internal resources, p. 228). Beijing: China Railway Publishing House.

construction was determined by its profit accumulation. The railway's operational performance directly affects the funding available for construction. Although the country had issued funding plans for railway construction, these were contingent upon the operational profits of the railway. If the railways' own capital reserves were substantial, the targets set in the national plan could be exceeded; otherwise, reductions in investment were inevitable, as shown in several instances of infrastructure funds being scaled back in the latter years of the Contract Responsibility System. To some extent, the change in the main source of funding for railway construction is the embodiment of the Contract Responsibility System in railway construction investment, that is, the "Input Quota Contract Mechanism" in the "Input-Output Contract Responsibility System".

Of course, in addition to its own reserves, the railway construction also receives certain loan quotas from the state, allowing the railway to borrow money from banks and pay interest in order to make up for the shortage of construction funds. During the period of the Sixth Five-Year Plan (1981–1985), the railway industry borrowed 720 million yuan in loans, while from 1986 to 1991, it utilized a total of 2.732 billion yuan of domestic loans,[87] representing a relatively small proportion of overall infrastructure investment.

To accelerate Chinese railway construction with the Self-sustaining Approach to Building New Railways, MOR explored diversified funding channels, including attracting local and foreign investments. Key initiatives included joint ventures between MOR, local governments, and enterprises, aiming to achieve "overall planning, combined regulation and control, tiered responsibilities, and joint construction".[88] For instance, the Beijing-Kowloon Railway broke the monopoly of state investment, and introduced funding from both the central government and provinces along the route, "marking a new path and model for jointly constructing trans-provincial main lines".[89] From 1981 to 1985, local governments and enterprises invested merely 915 million yuan in railway construction, whereas from 1986 to 1991, this figure soared to 4.058 billion yuan,[90] a 4.4-fold increase over the entire period of the Sixth Five-Year Plan. During this latter period, a total of 1,410 kilometers of local railways were constructed.

In addition to leveraging local government and enterprise investment, Chinese railways also accelerated the utilization of foreign capital. The Chinese railway sector began to use foreign capital in small amounts starting in 1981. From 1981 to 1985, a total of 1.068 billion yuan of foreign capital was used, with an average annual utilization of 213.6 million yuan. After 1986, the amount of foreign capital utilized gradually increased. On March 1, 1987, the Office of MOR for Using

[87] Statistics Center of MOR. (2008). *Collection of national railway statistical data* (internal resources, pp. 228–232). Beijing: China Railway Publishing House.

[88] *Key* points of railway work *in 1992*. Archived in Southwest Jiaotong University, T-18.

[89] (1991, September 10). Persist in reform and explore a new model of joint venture railway construction. *People's Railway Daily.*

[90] Statistics Center of MOR. (2008). *Collection of national railway statistical data* (internal resources, pp. 228–232). Beijing: China Railway Publishing House.

Foreign Capital and Introducing Technology was established to better introduce foreign capital and technology. From 1986 to 1991, a total of 1.832 billion yuan of foreign capital was utilized, averaging 366 million yuan per year.[91] These foreign funds mainly included World Bank loans and yen loans from Japan. Foreign loans were mainly used for purchasing advanced railway technology and equipment.

Over the six years following the full implementation of the Contract Responsibility System, the total investment in railway reached 43.544 billion yuan, of which MOR contributing 39.486 billion yuan. The average annual investment and the Ministry's share increased by 3.341 billion yuan and 2.848 billion yuan respectively compared to the period of Sixth Five-Year Plan, marking growth rates of 85.3% and 76.3%.[92]

The Self-sustaining Approach to Building New Railways is an important part of the Contract Responsibility System reform, which eased the pressure on the country under the extremely tight financial situation, and also as a follow-up reform of the policy of Conversion of Capital Allocations to Loans, reducing the strain on MOR. However, the Approach was in the wake of the basic railway policy that focused on "giving priority to tapping the potential, supplemented with new line construction", which cannot provide funds for large-scale new line construction.

Firstly, the Self-sustaining Approach to Building New Railways is a tangible expression of the "new line construction as a supplement" policy during the 1970s and 1980s. As mentioned earlier, since the late 1970s, the basic policy of Chinese railways has been "giving priority to tapping the potential, supplemented with new line construction", which places the construction of new railway lines in a secondary position. At that time, the Chinese government believed that the railway's weakness in the national economy stemmed not from a lack of lines, but from the underutilization of existing lines' transportation capacity. Therefore, the focus of railway work was on technological transformation and efficiency gains from management, that is, tapping into the potential. Under this policy, the construction of new lines was placed in an auxiliary position, so the idea that railways can rely on their own capital accumulation to solve the necessary funding needs for new line construction was believed within industry and beyond. As such, the investment in new line construction is inevitably limited, which is also the basic premise for the railway to adopt the Self-sustaining Approach to Building New Railways and carry out the Contract Responsibility System. With the railway freight rate not significantly increased, the capital accumulation of the railway remained limited. If the country does not have additional capital investment, it can be expected that there will be no large-scale new line construction during the Contract Responsibility System period, which means that the railway mileage will not have a significant increase. Later developments also confirmed this point. In 1991, the national railway mileage increased by only 2,606 kilometers compared with that in 1985, a mere 4.7% increase.[93]

[91] Ibid.

[92] Ibid.: 228–230.

[93] Ibid.: 29.

Furthermore, the "Supplementation with New Line Construction" policy has led to the initial compression of funding for new line construction whenever the railway industry faces urgent situations. Through years of efforts on "potential tapping", the transportation volume and frequency of Chinese railways are among the top in the world. Operating at overcapacity, facing serious equipment aging, and a significant need for technical upgrades, along with frequent safety issues around 1988, the railways urgently required safety investments. Where did these investments come from? The only solution was to compress funding for new line construction, resulting in an acute shortage of funds for such projects. In July 1989, MOR convened a conference focused on the reduction of infrastructure investment. "As freight rates have not been adjusted, bank loan interest rates have risen, and our debt repayment has increased, further widening the funding gap for railway construction",[94] the Ministry noted. In 1989, railway construction funding decreased by over 40%, even affecting the key projects. Although subsequent adjustments to freight rates increased revenue and partially replenished infrastructure investment, the overall period of the Seventh Five-Year Plan saw a reduction of RMB 1.6 billion in railway infrastructure funding compared with that in the original plan. In the latter stages of the Contract Responsibility System, MOR felt the inadequacy of railway construction funds. In 1990, Li Senmao, then Minister of Railways, stated at a Railway Leading Officers' Meeting that despite increased loans from the state, "there is still a funding gap of nearly RMB 7 billion in terms of capital balance".[95] When announcing the 1991 plan, the Ministry also pointed out that "the railway industry still faces a stark shortage of funds"[96] in 1991. Although the Ministry barely fulfilled its contracting plan of infrastructure investment under the Contract Responsibility System,[97] it could not raise more funds to support the rapid development of railways.

Moreover, despite completing the investment plan, due to inflation, the actual mileage of new or double tracks constructed with the allocated funds fell far short of the original projections. From 1986 to 1990, it was originally planned to build 3,809 km of new lines and 2,687 km of double tracks, totaling 6,496 km. However, only 1,797 km of new lines and 2,420 km of double tracks were completed, totaling 4,217 km, achieving respectively only 47%, 90%, and 65% of the original targets.[98]Strictly speaking, although the 'Contract Responsibility System' met the

[94] (1989). Plans on adjustments and compressions in the construction investment on the Railway Construction Conference. *Railway Situation Brief*, (47).

[95] Summarizing experience, enhancing reform and improving the railway economic contract responsibility system: Comrade Li Senmao's speech at the Railway Leading Officers' Meeting. Archived in Southwest Jiaotong University, ([1990] No. 1-0-2).

[96] Notice on the issuance of the 1991 Railway Plan. Archived in Southwestern Jiaotong University, T-501.

[97] The target was 25.2 billion yuan from self-accumulated fund, and actual investment exceeded the target by over 200 million yuan.

[98] Summarizing experience, enhancing reform and improving the railway economic contract responsibility system: Comrade Li Senmao's speech at the Railway Leading Officers' Meeting. Archived in Southwest Jiaotong University, (No. 1-0-2 [1990]).

financial plan for infrastructure development, it did not achieve the targeted basic construction mileage.

Thus, an apparent paradox arose: despite consistently surpassing annual targets, the railway industry still lagged far behind the needs of the national economy. Relying solely on the self-accumulated funds for construction was merely a make-shift, let alone providing funds for faster development.

3.3.2 "Three Major Campaigns" of Railway Construction

After implementing the Contract Responsibility System, the railway industry adopted the Self-sustaining Approach to Building New Railways, where funds for new line construction primarily came from self-accumulation, thus providing the railway with greater autonomy. However, it was still required to adhere to national plans for new line construction. "The national plan serves as the foundation for macro-level control. The scale, structure, and benefits of railway infrastructure should conform to the overall national plan, prioritizing social and economic benefits. Railway development shall cater to the needs and coordinate with the entire transportation system. The scale of railway infrastructure should be commensurate with national strength. Autonomy and financial resources shall not lead to the arbitrary launch of projects that exceed the national construction scale or prioritize corporate interests over national interests in investment, project decisions, and investment returns. Following the national plan is not a restriction but a means for better railway development, serving national economic development and maximizing social and transportation benefits".[99]

Given the scarce railway construction funds, during the Contract Responsibility System period, railway construction focused on regions with the most pressing national economic and transportation needs. "Currently, there are three major areas where the national railway system is most strained and inadequate. The first is the transportation of coal from Shanxi province, the second is access to Guangzhou, and the third is the connection to eastern China. Opening up these three corridors will play a crucial role in alleviating railway transportation tensions, promoting national economic development, and meeting people's living needs".[100] These three regions thus became the foci of construction efforts in the mid-to-late 1980s, collectively known as the "Three Major Tasks" among the Railway Man—the Northern Campaign for the Datong-Qinhuangdao Railway, the Southern Campaign for the Hengyang-Guangzhou Railway, and the Central Campaign for railways in the east China.

[99] Shen Zhijie. (1986, April 18). Railway infrastructure construction should proceed according to national plans. *People's Railway Daily.*

[100] Policy Research Office of MOR. (1987). Basic situation and countermeasures of current railway transportation. Archived in Southwest Jiaotong University, No. 1-11-1.

3.3.2.1 The Northern Campaign for the Datong-Qinhuangdao Railway

The Datong-Qinhuangdao Railway, spanning 653 km from Datong in the west to Qinhuangdao in the east, is China's first double-track electrified railway for heavy-haul transportation and a vital corridor for coal transportation from Shanxi. After reform and opening up, China's economy gained traction, and energy demand soared. Eastern and northern China experienced severe power shortages, compelling numerous factories to adopt a three-days-on, four-days-off operational schedule due to electricity scarcity. A direct cause of these shortages was the lack of coal at power plants, leading to power rationing. This energy deficit drastically hindered national economic development. Meanwhile, Shanxi, abundant in coal, faced a dilemma: mined coal piled up like mountains, some even spontaneously combusting, forcing Datong, the coal capital, to limit production based on transportation capacity. Transporting coal from Shanxi to the regions with critical needs became a pressing national priority after the reform and opening-up, highlighting the urgent requirement for the construction of a dedicated railway.

Heavy-haul transportation represents the progress of railway freight. Thus, after thorough technical assessments, the construction of a dedicated heavy-haul railway for coal transportation has been prioritized. Designed with an annual transport capacity of 100 million tons, the Datong-Qinhuangdao Railway could fundamentally resolve the issue of transporting coal from Shanxi Province, considering that the total annual outward transport of coal from Shanxi was 140 million tons in the early 1980s.[101] In September 1983, the State Council held an executive meeting and made the strategic decision to build the Datong-Qinhuangdao Railway and expand coal export channels in Shanxi, designating the project as a key national construction project during the period of the Seventh Five-Year Plan.

The construction of the Datong-Qinhuangdao Railway was carried out in two phases. The first phase, costing nearly 4 billion yuan, connected Datong to Dashizhuang and further linked Duanjialing on the Beijing-Qinhuangdao Railway via connecting lines. It encompassed four major projects: the line introducing the Datong-Qinhuangdao Railway into the Datong transport hub, the Hanjialing-Chawu section, the Chawu-Dashizhuang section and its connections with the Beijing-Qinhuangdao Railway, and the line leading into the third-phase coal terminal in Qinhuangdao. This phase covered the construction of a double-track electrified railway, with a total main line length of 410.788 kilometers. In the second phase, the line, a single-track electrical railway with allowance for future double track, extended from Dashizhuang to Qinhuangdao with a main line length of 242.232 kilometers.

In January 1985, the construction of Datong-Qinhuangdao Railway officially commenced. In November, to provide unified leadership for the construction efforts, MOR established the Datong-Qinhuangdao Railway Construction Office. This

[101] Policy Research Office of MOR. (1987). Basic situation and countermeasures of current railway transportation. Archived in Southwest Jiaotong University, No. 1-11-1.

Office was primarily responsible for the management of the design, scientific research, technology and equipment importation, construction, and overall development of the Datong-Qinhuangdao Railway. The Railway was constructed with joint efforts from various administrations and engineering organizations under MOR, including the First, Third, Sixteenth, Seventeenth, Eighteenth Engineering Groups, Tunnel Engineering Co., and Electrification Engineering Co., as well as the Beijing Railway Administration and Harbin Railway Administration. In total, over 70,000 workers were involved in the project. Composed of 28 key control projects, the first phase was immensely challenging, fraught with complexities arising from geological conditions.

On December 28, 1988, the first phase of the Datong-Qinhuangdao Railway was officially opened to traffic. Premier Li Peng and State Councilor Zou Jiahua, who were in office at the time, attended the opening ceremony. The congratulatory message from the State Council stated, "The completion of the first phase of the Railway has created new experiences in the construction of heavy haul transportation railways and the electrification and double-tracking of new lines in China, marking another significant achievement in Chinese railway construction". The opening of the first phase of the Datong-Qinhuangdao Railway "marks a gratifying step forward in modernizing Chinese railway heavy haul technology, equipment, and transportation organization and command. This is of great significance for developing the coal base in Shanxi and promoting national economic development".[102] After a year of trial operation, the first phase of the Datong-Qinhuangdao Railway was handed over to the Beijing Railway Administration for operation in December 1989. The second phase of the project, with an investment of over 2 billion yuan, officially commenced in June 1988 and was fully completed on December 21, 1992.[103] The opening of the Datong-Qinhuangdao Railway significantly alleviated the tension in transporting coal from Shanxi. Following the completion of the first phase, the Railway transported over 20 million tons of coal during the trial period in 1989, 33.8 million tons in 1990, 36.66 million tons in 1991, and 42.599 million tons in 1992. From August 1, 1992, the Railway operated six heavy-haul trains daily, each with a single locomotive pulling 72 wagons carrying 6,000 tons of coal, setting a new record for single locomotive traction on the entire railway network.[104] After the completion of the second phase, the Railway transported 46.659 million tons and 50.855 million tons of coal in 1993 and 1994 respectively, approaching the initial design target of 55 million tons, playing a crucial role in addressing coal transportation needs in China.

---

[102] (1988, December 30). Congratulatory message from the State Council on the opening of the first phase of the Datong-Qinhuangdao Railway. *People's Railway Daily.*

[103] Zhu Haiyan. (1992, May 21). Greeting the *wulong* into the Bohai Sea. *People's Railway Daily.*

[104] Lei Fengxing. (1992, September 10). The Datong-Qinhuangdao Railway boosts coal transportation from shanxi, demonstrating its power. *People's Railway Daily.*

3.3.2.2 Improvement in Transport Capacity of Railways in the East China

East China has experienced the fastest economic development after the reform and opening up. Accounting for 22% of the national total, the population in this region's five provinces and one municipality including Shanghai, Jiangsu, Zhejiang, Jiangxi, Anhui and Fujian benefit from the railway network managed by the Shanghai Railway Administration. In 1985, the total industrial and agricultural output value of these five provinces and one municipality accounted for 31% of the national total. However, the railway mileage in this region only accounted for 10% of the national total, and the double-track rate was only 13%, far below the national average double-track rate of 19%. "The railway transportation capacity is struggling to keep pace with the economic development of East China, leading to overcrowding and frequent delays in cargo loading. Among the four railway ports with the most prominent transportation capacity constraints, three of them are in East China, i.e., Fuliji (in Anhui Province), Yingtan (in Jiangxi Province), and Liling (in Hunan Province)".[105] Therefore, improving railway transportation capacity in East China has become an urgent task.

MOR organized experts for a thorough evaluation. In August 1986, Ding Guangen, then Minister of Railways, presided over a meeting in Shanhaiguan, where he asked the experts to specifically evaluate the railway construction in East China. The experts believed that it was necessary to strengthen the technological renovation of the Beijing-Shanghai Railway, expand its capacity, and actively use modern technology to increase the transportation capacity of East China. However, the transportation pressure in East China could not be truly resolved by merely renovating the Beijing-Shanghai Railway. Given many technical difficulties in modernizing the Beijing-Shanghai Railway that would require some time to address, the experts emphasized the "urgent need to quickly open a second corridor in East China since relying solely on the renovation of the Beijing-Shanghai Railway could not immediately transform the network in East China into one that could meet the imminent transportation demands".[106]

MOR prioritized "Improving Transport Capacity of Railways in the East China" as one of the Three Major Tasks. During the period of the Seventh Five-Year Plan, an investment of 7 billion yuan was planned for "Improving Transport Capacity of Railways in the East China", accounting for one-fifth of the total railway investment for the period. In October 1986, then Vice Premier Wan Li chaired a meeting on railway construction in East China. Wan Li pointed out, "East China is one of the more economically developed regions in our country… The economic zone in East

[105] Cao Zhaoqin, & Pan Yanfei. (1986, September 16). MOR, along with experts from various fields, has developed a new blueprint for the East China' Railway. The plan strengthens the construction of the Beijing-Shanghai line, opens a second channel in East China, and builds the Zhejiang-Jiangxi double-track line. The situation where railway transportation and economic development are highly mismatched will greatly improve after five years. *People's Daily*.

[106] Ibid.

China symbolizes the progress of the entire nation in the world". He emphasized that "good transportation is essential for the economic development in East China".[107] The State Council approved MOR's construction plan for Improving Transport Capacity of Railways in the East China, which included 17 major and medium-sized projects, constructing 473 kilometers of new lines, 740 kilometers of double-track lines, and 513 kilometers of electrified railways.[108] The project involved strengthening the first Beijing-Shanghai corridor, opening the second north-south corridor, and accelerating the construction of the east-west corridor. Strengthening the first Beijing-Shanghai corridor included completing the Xuzhou hub, building the Fuliji interchange, expanding the marshalling yard of Nanjing East Railway Station with additional departure tracks, and constructing the third railway from Nanjing to Nanjing East Railway Station. Through these upgrades, the freight on the Beijing-Shanghai Railway was expected to increase from 45 million tons in 1986 to 70 million tons. The second north-south corridor was the largest and most investment-intensive construction project in the East China campaign, including constructing new railways such as the Shangqiu-Fuyang Railway, the Fuyang-Huainan Railway, and the Xuanzhou-Hangzhou Railway, as well as building the Huainan-Hefei-Wuhu double-track railway, connecting the Wuhu railway hub, and upgrading existing lines such as the Huolonggang-Xuanzhou Railway and the Changxing-Hangzhou Railway. Once the entire line was operational, materials from the north could travel directly from Shangqiu to Hangzhou, with the annual transportation capacity of the Shangqiu-Fuyang Railway and Yuxikou Railway Crossing reaching 16 million tons.

Upgrading the first east-west corridor included the construction of the double-track railway on the Shanghai-Hangzhou route, the new Qiantang River Bridge, and the double-track railway on the Zhejiang-Jiangxi route. After the Shanghai-Hangzhou double-track railway was completed, its transportation capacity would increase from 17 million tons in 1986 to 32 million tons, and the pressure on the Zhejiang-Jiangxi Railway would be partially alleviated. The opening of the second east-west corridor included the construction of the Daye-Shahejie-Nanchang-Xiangtang Railway and the upgrading of the Nanchang-Jiujiang section of the Nanxun Railway. Once this corridor was opened, the distance from Wuhan to the southeastern coast would be shortened by 350 kilometers compared with the detour via Zhuzhou, and the annual freight of 4 million tons transported by the Zhejiang-Jiangxi Railway through Lilingkou could be diverted. Additionally, the electrification of the Yingxia Railway was accelerated to increase the annual transportation capacity entering Fujian from 9.6 million tons in 1986 to 18 million tons.

In terms of organizational model, a coordination group, composed of officers and experts from MOR and the five provinces and one municipality in East China, was

---

[107] MOR. Notice on studying and implementing the *Key Points of Comrade Wanli's Speech on Accelerating Railway Construction in East China* (TB [1986] No. 1239). Archived in Southwest Jiaotong University, T-315.

[108] (1988). Improving Transport Capacity of Railways in the East China fully launched. *Railway Information and Recommendations*, (1).

set up for "Improving Transport Capacity of Railways in the East China". This group collaboratively arranged every stage and addressed various issues during construction. With concerted efforts, continuing difficulties concerned with land acquisition cost standards, navigational standards, and key engineering construction plans were resolved one by one. In December 1986, MOR established the Headquarters of Eastern China Railway Construction, and China Railway Siyuan Survey and Design Group Co., Ltd. was charged with the task of construction organization plans. The plans cover five major targets: the reinforcement of the Beijing-Shanghai Railway from Xuzhou to Shanghai; the second north-south corridor from Shangqiu to Hangzhou; the east-west corridor from Shanghai via Hangzhou to Zhuzhou; the railway from Wuhan via the Dashan Railway to Xiangtang; and the railway from Yingtan to Xiamen, plus the Waiyang-Fuzhou Railway.

The construction design for these five major tasks was undertaken by the Shanghai Railway Administration, the Engineering Contracting Company of MOR, and China Railway Siyuan Survey and Design Group Co., Ltd.[109] In June 1987, the East China Railway Construction Command held a meeting in Wuxi, and in July, the East China Railway Construction Headquarters held its first meeting in Suzhou with the participation of the governors and mayors of the five provinces and one municipality. These two meetings reviewed the construction scale, investment, projects, and duration of the tasks of Improving Transport Capacity of Railways in the East China, and discussed and approved the organizational plans.

Contract packages were awarded to nine institutions, including the Fourth Engineering Group Co., the Second Engineering Group Co., the Third Engineering Group Co., the Fifth Engineering Group Co., the Fourteenth Engineering Group Co., the Sixteenth Engineering Group Co., the Twentieth Engineering Group Co., the Bridge Engineering Group Co., and the Electrification Engineering Group Co. of MOR, and a total of 100,000 construction personnel were mobilized. During the construction, a "joint venture between railways and local governments" was established. MOR concentrated its financial resources on maintaining main lines, ensuring smooth operations, prioritizing key projects, and safeguarding safety facilities. Local railways and some passenger and cargo transport facilities were mainly funded by local governments or co-funded by both the railways and local governments.[110] Coordination among the five major tasks was essential, with stations and hubs, as well as lines, being interrelated. Collaboration was required between the owner, designer, and contractor, between transportation and construction departments, and between the railways industry and local governments. The tasks of Improving Transport Capacity of Railways in the East China progressed smoothly. By the end of 1991, 8.3 billion yuan had been invested. Upon completion of 473 km of new railways, 513 km of electrified railways and 620 km of double-track lines, the traffic congestion in East China was effectively alleviated. Compared with 1985,

[109] (1986, September 30). To finish the construction plan for East China's Railway in half a year. *People's Railway.*

[110] (1988). Progress was made in Improving Transport Capacity of Railways in the East China. *Railway Information and Recommendations*, (15).

the number of passenger coaches from East China to other regions increased from 38 pairs to 47 pairs daily by 1990, and freight cars increased from 85 pairs to 111 pairs. The freight transported into East China through various railway ports increased by 24.55 million tons, a 40.3% increase from 1985, while freight turnover increased by 22.9 billion ton-kilometers, a 19.4% increase from 1985.[111]

3.3.2.3 The Southern Campaign for the Hengyang-Guangzhou Railway

The Beijing-Guangzhou Railway stretches 2,300 kilometers, serving as a major transportation artery running through China from north to south. The section from Beijing to Hengyang had long been double-tracked, but the one of 520 kilometers between Hengyang and Guangzhou was a single-track line, built before 1936. This section had lower standards, many curves, steep gradients, and rudimentary equipment, resulting in limited transport capacity. In 1982, the downward freight through the Pingshi Railway Crossing was only 11 million tons, meeting only about 40% of the demand, thus this section emerged as a bottleneck for north-south transportation.

Beginning in 1978, the country commenced the construction of the Hangzhou-Guangzhou Railway. Due to insufficient investment, construction was intermittent, and by 1985, 1.278 billion yuan had been invested, accounting for 39.8% of the total, with which 54% of the earthwork, 72.7% of the tunnels, 40.1% of the major and medium bridges, 16.6% of tracks on main lines, and 19.3% of tracks on temporarily occupied lines were completed. Additionally, 50.07 kilometers of double-track railway was open to traffic.[112]

In December 1985, Wan Li and Hu Qili went to Guangzhou for an on-site meeting. They decided to concentrate the prime resources on strengthening the construction of the Hengguang Railway, requiring the entire line to be operational by the end of 1988 and all supporting construction to be completed before 1990. The Hengguang Railway Construction Leading Group was established at this meeting, with Ding Guangen, the then Minister of Railways, as the head, responsible for resolving major issues in the construction process.[113]

MOR designated the "the Hengyang-Guangzhou Railway" as one of the Three Major Tasks of railway construction during the period of the Seventh Five-Year Plan. The Hengyang-Guangzhou Railway is 526.6 kilometers long, shortening the original route by 26.4 kilometers. It is a national first-class railway, with 155 kilometers from Chenzhou to Shaoguan being an electrified railway with a designed capacity of 50 million tons per year. The entire project is of immense scale,

[111] Xue Guibao. (1991, December 26). Decisive victories were made in Improving Transport Capacity of Railways in the East China. *People's Railway.*

[112] (1988). Experiences on the construction of Hengyang-Guangzhou Railway. *Railway Information and Suggestions*, (14).

[113] (1988, September 7). Chronology of major events in the construction of the Hengyang-Guangzhou Railway, *People's Railway.*

including 61 tunnels and 139 large and medium-sized bridges. On December 24, 1985, MOR held a meeting in Beijing to discuss the construction of the Hengyang-Guangzhou Railway, decided to establish a construction command headquarter for it, adjusted the construction deployment, and increased the construction workforce from 23,500 to over 50,000.[114]

The construction of the Hengyang-Guangzhou Railway is complex. First, the double track crosses the Nanling Mountains, with complex terrain and well-developed karst features, making construction difficult. Second, there are many key projects with high technical requirements. The Dayao Mountain Tunnel, with a total length of 14.295 kilometers, was the longest double-tracked tunnel in China at that time and is also one of the top ten longest ones in the world. Third, there were significant conflicts between construction and transportation. The Hengyang-Guangzhou Railway is one of the busiest routes, and the construction and renovation should be carried out with minimal disruption to traffic flow. Fourth, the project covers a wide range of areas measuring about 29,000 acres, and land acquisition and demolition of existing structures are arduous. The workload of the station reconstruction and demolition is particularly substantial.[115]

On May 6, 1987, the tunnelling of the Dayaoshan Tunnel, a key project of the double-track railway, was completed. Wanli and Hu Qili, accompanied by Ding Guangen and others, attended the completion ceremony at the Dayaoshan Tunnel exit. On August 4, 1988, the double track was fully laid in the Dayaoshan Tunnel. On December 6, 1988, the 155-kilometer electrified railway from Chenzhou to Shaoguan opened to traffic, and the Hengyang-Guangzhou Railway was declared fully operational 36 days ahead of schedule. On the day the new line was put in operation, a trail run was conducted with a 3,600-ton long freight train at the speed of 45 km/h.[116] On December 16, 1988, Li Peng attended the opening ceremony at the new Shaoguan station. In his speech, he pointed out, "The construction of this double track was challenging and had to conform to high technical standards. Its completion signifies that our railway construction has reached a high level, reflecting the spirit of our builders who adhered to reform, relied on technologies, and worked hard, setting a successful example for the technological transformation of our main lines".[117]

On December 27, 1989, the Hengyang-Guangzhou Railway passed the national inspection after over a year of safe operation, and was handed over to the Guangzhou Railway Administration for use starting January 1, 1990. In addition to the Three Major Tasks, construction of the Beijing-Kowloon Railway and the

---

114 (1988). Experiences on the construction of Hengyang-Guangzhou Railway. *Railway Information and Suggestions*, (14).

115 (1988, September 7). Chronology of major events in the construction of the Hengyang-Guangzhou Railway, *People's Railway*.

116 (1988, September 7). The Hengyang-Guangzhou Railway came into service 36 days ahead of schedule. *People's Railway*.

117 (1988, December 18). The adress of Comrade Li Peng on the Inaugural Ceremony of the Hengyang-Guangzhou Railway. *People's Railway*.

Nanning-Kunming Railway also commenced. From 1986 to 1991, the railway added 2,606 kilometers of operational mileage, extending the total railway mileage to 11,190 kilometers.[118] Consistent with the Contract Responsibility System, reforms were also carried out for "large and medium-sized railway construction projects such as newly constructed railways, the addition of second lines, the renovation of existing lines, electrification, the construction or expansion of independent hubs and large bridges. The Infrastructure General Bureau (supervisory department) shall sign agreements with the relevant construction institutions".[119] This contracting approach allowed for effective management and control. The Datong-Qinhuangdao Railway "adopted the Contract Responsibility System for all engineering projects and technical equipment development and technology transfer projects, including bidding, negotiation, and budget-based contracting, through which funds, quality, and duration were effectively controlled, and three shifts in mindsets were promoted, i.e., breaking the mindset of spending lavishly on key projects and embracing the concept of hierarchical contracting and cost-effectiveness through careful calculation; abandoning the mindset of relying on large numbers of people and blind rushing, and establishing a management concept of scientific organization and balanced production; discarding the mindset of merely rushing to meet deadlines and delivering work at any cost, and establishing a concept of quality first and reputation above all in competition".[120]

In response to the immense pressure of inadequate capacity to meet national economic development needs, Chinese railways actively embraced comprehensive reforms after 1986, leading to rapid development. By 1991, freight and freight turnover had increased by 17.0% and 35.0%, respectively, compared with those in 1985.[121]

China's railway remains to be the "bottleneck" of the national economy, or a more pronounced one, despite its tellingly great development of itself. From 1985 to 1991, the gross national product grew by 141.6%, while the railway operating mileage increased by only 4.7% during the same period.[122] The total industrial output value grew by 134.5% from 1985 to 1991, but railway freight only increased by 17.0%, with an elasticity coefficient of 1:0.13.[123] The gap as compared with the

---

[118] Statistics Center of MOR. (2008). *Collection of national railway statistical data* (internal resources, p. 29). Beijing: China Railway Publishing House.

[119] MOR. Supplementary provisions of the Investment Contract Responsibility Agreement on the project of railway infrastructure development (TJZ [1985] No. 952). Archived in Southwestern Jiaotong University, T-288.

[120] Datong-Qinhuangdao Railway Construction Office of MOR. (1988, December 25). Experience on the construction of Datong-Qinhuangdao Railway. *People's Railway*.

[121] (1990, November 30). Notable achievements were made on railway passenger and freight transport during the Seventh Five-Year Plan period. *People's Railway*.

[122] Statistics Center of MOR. (2008). *Collection of national railway statistical data* (internal resources, p. 3, p. 28). Beijing: China Railway Publishing House.

[123] Comprehensive Division of National Economic Statistics. (2009). *Compilations of statistics of the past 60 years since the founding of PRC* (p. 9). Beijing: China Statistics Press. Statistics Center

normal coefficient level of 1:0.5 is quite large, and the difference between the 1991 one and that of 1:0.29 in 1985 has widened, indicating that the mismatch between railway and industrial development in 1991 was more grave.

Han Zhubin, then Minister of Railways, pointed out that by the end of 1991, "with the continued rapid development of the national economy, the demand for railway transportation has increased, making the already inadequate railway transport even more strained, and the contradiction of insufficient capacity has further intensified. Such inadequacy in capacity has become a 'bottleneck' that constrains the national economy from reaching new heights". He noted that "many regions have seen industrial and agricultural production growth rates of 17% or even 20% this year, placing higher demands on the railways. However, railway freight during the period of the Seventh Five-Year Plan could only increase by 4.3% annually. As a result, the contradiction between freight and capacity has become more prominent, and the railways continued to face a disadvantaged position".[124]

The increasingly prominent "bottleneck" situation in railways is not due to a lack of railway development, but rather because of the rapid growth of the national economy and the relatively slow progress of the railways, which has widened the gap between the two. The Contract Responsibility System was introduced under the basic railway policy of focusing on tapping the potential and supplementing it with new line construction, but the potential was always limited. By the early 1990s, the existing potential of the railways had been exhausted to its limit after more than a decade of tapping the potential, especially after several years of implementing the Contract Responsibility System. It was necessary to increase the capacity, i.e., to expand the railway network to meet the demands of the rapidly growing national economy. The Contract Responsibility System shifted the task of road construction onto the railway construction, which required huge amounts of funds. However, the accumulation of the railway itself was limited, and it was unrealistic to rely on the railway's own accumulation for the practice of "Self-sustaining Approach to Building New Railways and Maintain Old Ones" in a short period of time. Therefore, while the Contract Responsibility Systems could help complete the assigned tasks, it could not put the railways on the fast track of progress. To break free from the "bottleneck", the railway industry needs to find alternative solutions.

---

of MOR. (2008). *Collection of national railway statistical data* (internal resources, p. 17). Beijing: China Railway Publishing House.

[124] Han Zhubin. (1996). The reform and development of China's railways. Archived in Southwest Jiaotong University, X-211–15.

Chapter 4
Marketization: Development Accelerated in the Deepening of Reform (1992–2003)

In early 1992, Deng Xiaoping delivered south tour speeches. In October of the same year, the 14th National Congress of the Communist Party was held, and the goal of building a socialist market economy system was established. This marked the entry of China's reform and opening-up as well as modernization drive into a new stage of development. In November 1993, the Third Plenary Session of the 14th Central Committee published the *Decision of the Central Committee of the Communist Party of China on Some Issues Concerning the Establishment of the Socialist Market Economy*, which concretized goals of the economic system reform and basic principles proposed at the 14th National Congress and became the action plan for China's economic system reform in the 1990s.

Entering the 1990s, the problems with the railway system became more pronounced. On the one hand, railway transportation was far from meeting the needs of social and economic development, and the "bottleneck" effect of the railways became increasingly evident. On the other hand, the drawbacks of the closed-loop railway system began to show, with the former advantages such as "highly centralized dispatching command, coordinative actions among transportation departments, semi-military approach to order execution" turning into constraints on railway progress. Although there was a clear direction for reform, the question that persisted was how the railway system could be more market-oriented. This required the railway department to explore reform practices based on a shift in mindset and starting from real-time scenarios.

Although under historical inertia, the promotion of railway reform is not so easy, and the achievements of reform cannot be revealed overnight. However, in the more than 10 years after 1992, the railway departments had gradually made attempts and efforts to reform the transportation management system, change the operation mechanism, reform the railway investment system and develop diversified operations, with the goal of establishing a modern enterprise system. This has effectively promoted the construction and development of railways in this period. In particular,

X. Zhang, *China's Railway Expansion in 40 Years of Economic Reform and Opening Up*, Understanding China,
https://doi.org/10.1007/978-981-95-0333-9_4

the reform of "Separation of Railway Network and Transportation" is an important attempt to solve the institutional drawbacks and structural contradictions that have long plagued the development of railways, while the implementation of the reform of "Separating Railway Sideline Businesses from Core Businesses" is to spin the non-core sectors off the railways, ensuring the professionalism of the main business, and effectively bring the railway to the market.

At the same time, the new pace of railway's opening up during this period has also accelerated. First of all, the exploration of a diversified investment and financing system has effectively alleviated the shortage of railway construction funds. As a result, China's railway construction ushered in the process of greater development. From 1992 to 2003, the investment in railway construction continued to increase, and so did the mileage of railway operations, which also stimulated the growth of the national economy. Secondly, the opening of railways to market beyond the railway industry and to international market has also been fully rolled out. With the opening up from the inside to the outside, local governments began to participate in the construction and management of railways, with joint venture railways and local railways developed. In addition, during this period, the opening of railways from the domestic market to the international market, including the introduction and utilization of international capital and advanced technology and equipment, the development of Sino-foreign joint ventures and cooperative production, the development of international trade and economic cooperation, and the expansion of international transportation, were gradually promoted, leading to significant progress in these areas.

4.1 Railway Reform for Modern Enterprise System

4.1.1 Determining the Direction of Market-Oriented Reform in the Railway Sector

Since 1992, China's economy has entered a stage of rapid progress, which demanded for a substantial uplift of China's railway transportation capacity. However, despite the establishment of a socialist market economy that has propelled China's overall economic growth to new heights, the railway sector has not seen comparable advancements. As a result, railway transportation is facing considerable strain and remains a critical area for improvement. In 1992, China's GDP increased by 12% compared with 1991, and the national gross industrial output value increased by more than 20%, while the railway freight volume only increased by 3%. At the beginning of 1993, the assessed demand in China was 120,000 vehicles per day, but the railway could provide about 75,000 vehicles, which could only meet about 60% of the actual demand. The capacity of the Beijing-Guangzhou, Beijing-Shanghai, Beijing-Shenyang, Lanzhou-Lianyungang, Zhejiang-Jiangxi and other main lines and major hubs were saturated, and long-distance passenger coaches on the main lines were seriously overcrowded, with more than 20 strain-constrained sections. The contradiction between traffic volume and transportation capacity was acute, intensifying the difficulty in the transportation of goods and passengers. According to the forecast of the railway department in 1993, the railway freight volume would reach 1.7 billion tons in 1995, which required the railway freight capacity to be increased by 60 million tons annually in the last three years of the Eighth Five-Year Plan period (i.e. 1993–1995).[1]

The railway transportation capacity was so seriously insufficient that it could not meet the actual demands of economic and social development and further highlighted the "bottleneck" effect of railway on national economic progress. According to the report of *People's Daily*, from 1992 to 1993, the elasticity coefficient of national gross industrial output value and railway freight volume was as low as 1: 0.2.[2] Due to the shortage of railway transportation capacity, transportation of many important materials related to the national economy and people's livelihood had to be "determined by capability". In particular, the coal in Shanxi, Shaanxi and western Inner Mongolia could only be stored in the open air awaiting transportation. Over time, the coal even weathered and spontaneously combusted, while factories in the eastern coastal areas often had to limit or even stop production due to a lack of coal and electricity. According to the calculation of relevant departments at that time, the energy shortage resulting from insufficient railway transportation capacity led to a loss of nearly 400 billion yuan in output value and 50 billion yuan in profits and taxes every year.[3]

In terms of passenger transport, overloading was quite serious, with typical excesses ranging from 50% to 70%, and in some cases, surpassing 100%. According to statistics, approximately 800,000 people had to stand during their trips on trains every day nationwide. Some people even joked that it used to be "standing all the way to Shanghai", but now it is "standing wherever you go". During the Spring Festival travel rush in 1993, MOR took a series of emergency measures and increased the number of passenger coaches by more than 4,600, reduced the number of freight cars by more than 5,000, and used all available boxcars and spare carriages. The number of carriages for one train increased from more than a dozen to 30 at most, but even so, it was a drop in the bucket. Almost every train was overloaded by more than 100%. Three people crowded around the one-foot-square tea table, and more than 10 people were crowded in the one-square-meter toilet. Under the seats and on the luggage racks, the train was filled with people, literally. Many passengers couldn't

---

[1] Han Zhubin. (1994). Thoroughly implement the principle of the 14th National Congress of the Communist Party of China, accelerate the pace of reform and opening up, and strive for the historic development of railways: Speech at the National Conference of Leading Officers in the Railway Sector. In Wu Changyuan (ed.), *The important manuscripts on the reform and development of Chinese railways in 1993* (pp. 56–57). Beijing: China Railway Publishing House.

[2] Zhang Guorong. (1993, February 22). When Chinese Railway out of "bottleneck" (First Part). *People's Daily.*

[3] Ibid.

buy train tickets in time, leaving stranded people everywhere.[4] Severe overcrowding led to a series of problems. During the Spring Festival travel rush in 1993, on Train No.52 from Guangzhou to Chengdu, six passengers jumped out of the train due to dizziness. The conductor and passengers managed to pull five people back from the window, but one passenger who jumped from the toilet window unfortunately fell to death. The sordid air in the carriage also caused many passengers to become ill during their journey. The 5th charter crew of the 51st train rescued more than 40 passengers in the dining car in just one trip. The carriage was so crowded with people, that the passengers couldn't use the toilet. When Train No. 51 arrived in Guangzhou, the carriage was filled with abandoned shoes and luggage—passengers could no longer wear their shoes as their feet were swollen for long time standing, and some of them had to "let go" of their belongings as their luggage was contaminated with waste.[5]

Due to a severe shortage of railway capacity, the railway system has to operate under long-term overload to meet the needs of national economic development. In 1993, China's railway load intensity reached 25.794 million ton-kilometers per kilometer, 3.4 times that of the United States, 3.28 times that of India, and 1.98 times that of Japan, ranking first globally. There were 20,000 freight cars and more than 2,000 passenger coaches in operation every day nationwide. The average interval time between main railway lines was less than 10 minutes with the shortest interval being 6 minutes. The exceptionally high traffic density was rare in the world.[6] Many people are often amazed that Chinese railways, despite having lower-quality technology, can achieve economic benefits comparable to modern railways in developed countries. However, this is actually at the expense of the interests of passengers and cargo owners, resulting in idle production capacity, a long-term decrease in railway safety standards, and inefficient national economic operations. While it can be sustained for a while, it is not a viable long-term solution.

In short, after 1992, with the rapid progress of China's economy, the serious lag in railway development became increasingly apparent, and the railway became a major "bottleneck" in the development of the national economy. If the situation could not be significantly improved, it would be difficult to sustain the rapid growth of national economy, and there was a risk of missing excellent opportunities in the 1990s. Therefore, it became a pressing need of the time as well as a hope for the people to transform the railway system from a constrained model to an adaptive one to address the severe lag in railway development. So, how could the situation be reversed? This became a thorny problem for the CPC Central Committee, the State Council and MOR at that time.

There were many reasons for the increasingly prominent "bottleneck" effect of Chinese railways in the early 1990s, but the fundamental issue lay in the problems of operation mechanisms and management systems. For a long time, due to the

[4] Ibid.

[5] Gan Lin, & Deng Congyi. (1993, February 25). Railways in Southwest China are moaning: Records of transporting Sichuan servicemen this spring. *People's Daily*.

[6] Zhang Guorong. (1993, February 22). When will Chinese railways out of "bottleneck" (First Part). *People's Daily*.

influence of planned economic system and the characteristics of railway departments, Chinese railway was semi-militarized and highly centralized in operation and management. However, with the gradual transition of China's economy from the planned economy system to the socialist market economy system in the early 1990s, the original management and operation mechanism of centralizing management power at the railway authorities and leaving little autonomy for subordinate railway administrations became much more unsuitable for the market economy.[7] Therefore, in order to adapt to the progress of socialist market economy, the railway system shall be reformed.

Consequently, how to carry out the reform? First of all, an overall reform principle was needed as a guide. In response to the decrees of the State Council, MOR put forward the general principle of "making great strides in railway reform, raising transportation to a higher level, and making great progress in construction", to explore effective solutions for the lagging development of Chinese railways and deepen its reform. From July 14 to 18, 1992, MOR held the National Conference of Leading Officers in the Railway Sector in Beijing to find ways to alleviate the tension of railway transportation and solve the problem. At this conference, MOR initially formed the general guideline of reform: the two major systems of railway—industry and infrastructure construction would be gradually liberalized and invigorated. In addition to the highly centralized transportation organization and command dispatch, the transportation system would be liberalized and invigorated to meet the demands of the market economy.[8]

From January 8 to 11, 1993, MOR held another National Conference of Leading Officers in the Railway Sector in Beijing, focusing on studying and implementing the guidelines determined at the 14th National Congress of the Communist Party of China, discussing the basic ideas of how to carry out railway reform under the new situation of establishing a socialist market economic system, and pushing railway reform and progress to a new stage.[9] Han Zhubin, then Minister of Railways, delivered a speech entitled *Enforce the Guidelines of the 14th National Congress of the Communist Party of China, Accelerate the Pace of Reform and Opening-up, and Strive for the Historic Progress of Railways* at the meeting, emphasizing that the national economy had reached a new level and a socialist market economic system was established, which brought severe challenges as well as opportunities to accelerate railway development. "To speed up the railway reform, we should closely

---

[7] Han Zhubin. (1994). Thoroughly implement the principle of the 14th National Congress of the Communist Party of China, accelerate the pace of reform and opening up, and strive for the historic development of railways: Speech at the National Conference of Leading Officers in the Railway Sector. In Wu Changyuan (ed.), *The important manuscripts on the reform and development of Chinese railways in 1993* (p. 57). Beijing: China Railway Publishing House.

[8] Zhang Guorong, There will be a Major Breakthrough in Railway Reform in the Eighth Five-Year Plan: Establishing a Railway Construction Fund, Charging Adjusted Prices for New Railway, and Vigorously Developing Joint Venture Railways, *People's Daily*, July 17, 1992.

[9] (1993, January 9). The National Conference of Leading Officers in the Railway Sector was held in Beijing. *People's Railway Daily.*

focus on the overall strategic goal of developing the socialist market economy, emancipate our minds, take bold steps, go to the market, and establish a new railway system and operational mechanism that adapts to the socialist market economy, embarking on a path of railway progress with Chinese characteristics".[10]

At this meeting, according to the guidelines of the 14th National Congress of the Communist Party of China, MOR summarized the practical experience of railway reform of the last decade. In consideration of the status quo of railway departments, the basic principles of a market-oriented railway were proposed at the meeting: taking the development of socialist market economy as the guide, taking "three benefits" as the standard, changing ideas, functions and mechanisms of the railways, separating the functions of government and enterprises, macro-controlling, diversifying operation, developing in group enterprises, enriching the people with high-quality railway services, and establishing a new management system and operation mechanism of railway with Chinese characteristics.[11] The core of these basic principles is to transform the railway from the traditional planned economic system into a new market economy system. Thus, it can become an organic part of the socialist market economic system.

The aforementioned reform principles explicitly show that the railways should adopt a market-oriented approach, establishing the central status of railway enterprises in the market and creating a new management system and operational mechanism with Chinese characteristics. This marks a theoretical breakthrough in railway reform. Since the late 1970s and early 1980s, the railway reform has been discussed repeatedly. However, due to the constraints of the system, the ideology and the concepts, the discussion on railway reform was always limited to the scope of planned economic system before 1992. After Deng Xiaoping's South Tour Speeches, and the 14th National Congress of the Communist Party of China, as the entire national economy gradually transitioned to a socialist market economy, the ideological constraints were progressively lifted. This accelerated the reform of the existing operation and management systems under the planned economy, with a consensus emerging across different social sectors, including the railway, to expedite the adoption of market-oriented approach. The basic idea of railway entering the market put forward by MOR at the National Conference of Railway Leading Officers in January, 1993 was a response to this call, which also determined the basic direction of Chinese railways reform during this period.

The railway sector had determined the basic direction of reform, but how to carry out the reform specifically? In January 1994, MOR convened a national conference for railway leading officers, reiterating that adhering to the market-oriented approach and deepening railway reform were the fundamental solutions to addressing the

[10] Han Zhubin. (1994). Thoroughly implement the principle of the 14th National Congress of the Communist Party of China, accelerate the pace of reform and opening up, and strive for the historic development of railways: Speech at the National Conference of Leading Officers in the Railway Sector. In Wu Changyuan (ed.), *The important manuscripts on the reform and development of Chinese railways in 1993* (pp. 57–78). Beijing: China Railway Publishing House.

[11] Ibid.: 61.

deep-seated contradictions and issues within the railway sector.[12] To strengthen railway reform, it is essential to use the *Decision of the Central Committee of the Communist Party of China on Some Issues Concerning the Establishment of the Socialist Market Economy* as the guideline. Therefore, at the National Conference of Leading Officers in the Railway Sector, MOR discussed and approved the *Opinions on Some Issues Concerning the Implementation of the Decision of the Third Plenary Session of the 14th CPC Central Committee and Deepening the Reform of Railways* (commonly referred to as the *Thirty Articles on Railway Reform*). This was submitted to the State Council on April 22, 1994, and subsequently distributed throughout the whole railway sector.

The *Thirty Articles on Railway Reform* highlights that strengthening reform is essential for achieving historic development in the railway sector. The overall goal of the reform is: "Under the macro-control of the nation, to leverage the foundational role of the market in resource allocation for railway capacity, establish a railway management system and operational mechanisms suited to a socialist market economy, accelerate railway construction, expand transportation capacity, improve management practices, enhance economic efficiency, and better serve national economic and social development".[13]

This encompasses several key areas: transforming the operating mechanism of railway enterprises to establish a modern corporate system; shifting government functions to improve the macro-control system; reforming the investment framework to gather funds through multiple channels to accelerate railway development; actively promoting diversified operation to optimize the allocation of railway resources; deepening reforms in railway technology and education systems; strengthening ideological and political work while enhancing railway culture. The *Thirty Articles of Railway Reform* further elaborates on the basic ideas and steps for market-oriented railway development. It emphasizes that strengthening railway reform should integrate the unique characteristics of the railway industry with the common requirements of national reform. It advocates for a market-oriented approach and stresses the importance of properly balancing centralized transportation command dispatch with enterprise autonomy, as well as social welfare with business interests. As the railways moves towards a market economy, they recognize that various sectors—such as transportation, infrastructure, industry, and enterprises of diversified business—have their own characteristics, requiring tailored measures to advance reform. Additionally, it stesses the need for unified planning,

[12] Han Zhubin. (1995). Implementing the "Decision", seizing opportunities, and achieving new breakthroughs in railway reform and development: Speech at the National Conference of Leading Officers in the Railway Sector. In Wu Changyuan (ed.), *The important manuscripts on the reform and development of Chinese railways in 1994* (pp. 45–64). Beijing: China Railway Publishing House.

[13] (1995). Opinions on some issues concerning the implementation of the Decision of the Third Plenary Session of the 14th CPC Central Committee and deepening the reform of railways. In Wu Changyuan (ed.), *The important manuscripts on the reform and development of Chinese railways in 1994* (p. 381). Beijing: China Railway Publishing House.

classified guidance, coordinated implementation, and orderly progress in enhancing railway reforms.

It can be said that *Thirty Articles of Railway Reform* was the elucidation and improvement of the basic ideas of railway reform put forward by MOR in January 1993. It further clarifies the market orientation of railway reform, establishes the overall goal and basic policy of railway reform, and outlines key tasks such as establishing the market's dominant position of railway enterprises.[14] This framework served as the foundation for deepening railway reform in the subsequent period. The reform tasks outlined in the document was advanced to varying degrees. Among them, two significant measures were the promotion of Separation of Railway Network and Transportation and Separating Railway Sideline Businesses from Core Businesses.

4.1.1.1 Attempt of "Separation of Railway Network and Transportation"

Transforming government functions and separating government and business functions are the basic requirements for establishing a socialist market economic system. Chinese railways have long been under a unified management system of "integration of government and business functions" and "integration of railway network and transportation", which has blurred the lines between public welfare and profit-driven enterprise, making it difficult for railway enterprises to truly engage with the market. Therefore, with the establishment and development of socialist market economic system, China's railway transportation management system should be reformed. One important aspect of the reform is the restructuring of railway transportation enterprises, enabling their organizational structures to better align with market-oriented operations and more effectively meet the demands for passenger and freight transport.

The restructuring of railway transportation enterprises generally shows two trends. One is the trend of disintegration of the existing state-owned railways, which manifests in various forms. For example, Japanese railways' disintegration involves the separation of the state-owned railways into six passenger companies, each operating within its own region. In Germany and the UK, there is a specialized one, where the state-owned railway is divided into several passenger companies, freight companies, parcel companies, and others. Sweden's demonstrates an "vertical separation" model. Swedish railway separates its infrastructure and services. The other trend is the concentration of non-state-owned railways, such as the expansion of major railway companies in the United States through market mergers and acquisitions. These two restructuring trends are mutually inverse in form, but their essence is to enhance the market adaptability and competitiveness of railways.[15]

---

[14] Ibid.: 381–392.

[15] Song Qiangtai, & Yang Yuefang. The disadvantages of the current system of railway transport and the choice of mode of separation of railway network and transportation. In the Railway Transport Committee of China Railway Society (Ed.), *Collection of academic papers on the*

How to restructure railway transportation enterprises and which method to adopt had become a critical decision for China's railways on the path of reform. Ultimately, in order to overcome various challenges in railway reform and establish a railway transportation management system that adapts to the socialist market economy, people both inside and outside the railway sector, after studying global railway reforms and drawing on international experience, proposed the idea of promoting railway reform in China by adopting the Separation of Railway Network and Transportation model. In other words, after repeated in-depth research and serious comparison of various plans, China adopted the approach of "vertical separation"—separating the management of the national railway network from passenger and freight transportation–marking a reform of the institutional structure.

Specifically, the model of Separation of Railway Network and Transportation reflects the natural monopoly and public service of the railway industry through the "network", while separating the "transportation" part to establish an independent market entity and legal entity, thus promoting the establishment of modern corporate systems and achieving the separation of government and business functions.[16] By adopting the the basic model of Separation of Railway Network and Transportation for the reform of China's railway transportation system, the state-owned railway infrastructure, which was naturally monopolistic, was separated from the competitive passenger and freight transportation operations. This would involve creating a unified national railway infrastructure company along with several robust companies for passenger and freight transportation, each managed separately. Such measures can systematically and reasonably resolve the long-standing institutional issues and structural contradictions that have hindered railway progress. It would enable Chinese railways to truly take the lead, enhancing the vitality of the national economy and facilitating strategic adjustments and reorganizations within the national economy.[17]

To this end, the Tenth Five-Year Plan explicitly mandated the Separation of Railway Network and Transportation within the railway sector. It also set forth the overarching strategy for strengthening transportation system reform as the "separation of government and business functions, the separation of railway network from transportation, the introduction of competition, and the enhancement of supervision".[18] The starting point for this reform was the establishment of passenger transport companies. On August 20, 1999, MOR issued *the Suggestions on Railway*

---

reform and development of railway transport enterprises (pp. 16–17). Beijing: China Railway Publishing House.

[16]Wu Jianzhong. (2001). *New polyconomy theory: Strategies and practices of reform in railway transportation enterprises* (p. 97). Beijing: China Railway Publishing House.

[17]General Office of MOR. 2000. The "Separation of Railway Network and Transportation": The basic idea behind the reform of China's railway transportation management system. *Railway Economic Research*, (2), 2–5.

[18]The Tenth Five-Year Plan of railways (Excerpts). (2001). In the Editorial Board of the China Railway Yearbook (Ed.), *China Railway Yearbook (2004)* (p.27). Archives and History Center of MOR.

Administrations Directly in Charge of Railway Stations Establishing Internal Passenger Transport Companies, outlining the guiding principles and framework requirements for their setup. In September 1999, the Hohhot Railway Administration and the Nanchang Railway Administration announced the establishment of their internal passenger transport companies. Following this, on October 20, 1999, the Kunming Railway Administration established its own internal passenger transport company, the China Railway Kunming Group Co., Ltd. On November 8, 1999, an internal passenger transport company of the Liuzhou Railway Administration was established, integrating marketing planning, ticket management, passenger carriage allocation, and passenger services.

The pilot programs in Hohhot, Nanchang, Liuzhou, and Kunming—each directly administrated by their respective railway administrations—primarily focused on the combination of the passenger train segments, passenger carriage segments, and the mixed cargo-passenger carriage segments. These companies operated under internal independent accounting and self-management systems, simulating the operations of a corporate entity. The business scope of the passenger companies included conducting market research and marketing planning for passenger transport, formulating the operation plans for passenger coaches, train marshalling plans, and ticket allocation schemes. They were also responsible for passenger services and baggage transportation on trains, as well as the dispatching, operation, and maintenance of passenger carriages. Additional responsibilities encompassed management and sales organization of tickets, as well as catering and advertising services on trains.[19] At that time, Wang Zhaocheng, then Vice Minister of Railways, pointed out, "The implementation of independent accounting by passenger companies marks a milestone in China's railway reform".[20]

Reform is a systematic project that cannot be accomplished overnight, and the reform of passenger transport companies is no exception; it should proceed steadily according to the objective laws of development. On April 18, 2000, MOR issued *Suggestions on Deepening Pilot Reforms of Passenger Transport Companies*, which required that liberating thoughts and updating concepts should be prioritized throughout the entire reform process. At the same time, the four pilot administrations were tasked with effectively performing and fulfilling the duties and responsibilities of the passenger transport companies, striving to achieve centralized management at one tier. This entails concentrating comprehensive management of human resources, finance, and planning within the company headquarters. The company itself is to serve as the profit-generation center and should improve financial settlement methods. Furthermore, it should also enhance reforms in personnel,

[19] Fu Shizhong (2000, November 7). An important chess piece on the board of reform: An analysis of the establishment and development of railway passenger transport companies. *People's Railway Daily*.

[20] Wang Zhaocheng (2002). Recognizing the situation, unifying thoughts, and resolutely achieving the goal of "Three Divisions": Speech at the National Conference on Diversified Operations in the Railway Sector. In Peng Kaizhou (Ed.), *The important manuscripts on the reform and development of Chinese railways in 2001* (p. 286). Beijing: China Railway Publishing House.

employment, and distribution systems, strengthen the integrated management of train crew members (including train attendants, train inspectors, and train police), and enhance supervision over the passenger transport companies.[21]

Under the guidance of the *Suggestions*, in May 2000, China Railway Kunming Group Co., Ltd. abolished its six subordinate branches, consolidating comprehensive management of its human resources, finance, and planning at its headquarters. This marked the first instance of direct management over six workshops and six teams, reducing the number of functional departments from 27 to 12 and workshops from 9 to 6. In August 2000, the Hohhot Railway Administration eliminated the interim management branches, resulting in a 22% reduction in administrative staff, laying off 116 staff members. It established a task force and dispatch office in Baotou, far from the headquarters, to enable direct management. On September 5, 2000, China Railway Liuzhou Group Co., Ltd. disbanded three branches, streamlining the number of officers from 592 to 298. It also relocated five of its ten functional departments to operate on the production front lines for direct management. As a result of these changes, "one-tier" management", as a deeper aspect of the passenger transport company's reform, had streamlined management levels, reduced administrative personnel, and cut management costs, thereby leading to a more optimized organizational structure, more streamlined and efficient staffing and enhancing efficiency to respond more swiftly to market changes.

On December 2, 2000, China Railway Guangzhou Group Co., Ltd. was officially inaugurated in Guangzhou. This marked the first pilot reform initiative to establish a passenger transport company under the framework of a railway administration with subordinate divisions. Fu Zhihuan, then Minister of Railways, commented on the establishment of the company, "This is a breakthrough in the reform of China's railway transportation system and a new starting point for establishing competitive entities in railway passenger transport".[22] The establishment of China Railway Guangzhou Group Co., Ltd. marked the beginning of the Separation of Railway Network and Transportation reform in the railway sector.

In March and August 2001, MOR successively held two meetings in Kunming and Hohhot, choosing the Shanghai, Jinan, and Zhengzhou Railway Administrations as pilot organizations for the passenger company reform. On this basis, MOR promoted the establishment of passenger transport companies across all railway administrations. By September 28, 2002, with the founding of China Railway Beijing Group Co., Ltd., all 14 railway administrations had completed the internal establishment of passenger transport companies.[23]

---

[21] Suggestions on deepening pilot reforms of passenger transport companies (TZF [2000] No. 151). In Peng Kaizhou (Ed.), *The important manuscripts on the reform and development of Chinese railways in 2001* (p. 586). Beijing: China Railway Publishing House.

[22] Liu Aiping (2000, December 5). Launch Ceremony of China Railway Guangzhou Group Co., Ltd. *People's Railway Daily.*

[23] The Editorial Board of *Chinese Railways at the Turn of the New Century* (2002). *Chinese railways at the turn of the new century* (pp. 77–78). Beijing: China Railway Publishing House.

Guiding by the concept of Separation of Railway Network and Transportation, after the reorganization of both passenger and freight companies, a national railway network corporation would be established to operate under a single legal entity, with subsidiaries formed through the appropriate reorganization of railway administrations after the separation of passenger and freight operations. The national railway network corporation would manage assets such as tracks, bridges, tunnels, signaling equipment, power supply facilities, and stations, while shouldering responsibility for their operation.[24] However, the separation model faced many challenges, particularly regarding the railway network itself. Questions arose about the natural monopoly of the railway network, the feasibility of its division, the necessity for unified scheduling and command, the construction of the infrastructure, and its management. Without addressing these issues, advancing railway reform and restructuring would be difficult. Therefore, the reform of the separation model remained contentious. Some scholars argued that the form of the separation model practiced in China was less efficient and merely provided alternative options for corporate reorganization in the railway sector.[25] Some scholars even claimed that the concept lacked an economic theoretical basis and successful practical examples, suggesting it might not break the monopoly of the railway network.[26] Despite the controversies surrounding the reform of the Separation of Railway Network and Transportation, it served as an integral part of China's railway system reform and represented an exploration towards modernizing the country's railways.

In addition to the attempt at Separation of Railway Network and Transportation, MOR also made efforts in reforming other aspects of the transportation management system. For one thing, it implemented a new system where railway administrations could directly manage stations. In 1994, the Ministry conducted research on the reform of the railway transportation management system, recognizing that the existing three-tiered legal entity and four-tiered management structure among the Ministry, administrations, sub-administrations, and stations led to issues such as a lack of distinction between government and enterprise functions, unclear authority and responsibilities, overlapping functions, and overstaffing. To address these issues, starting in August 1996, the Ministry began implementing a new railway transportation management model in certain railway administrations, allowing them to directly manage the grassroots stations, forming a three-tiered management system of the Ministry, administration, and station. On August 8, 1996, the Nanchang Railway Sub-administration was separated from the Shanghai Railway Administration to establish the Nanchang Railway Administration, as part of an initiative to reduce management levels, adjust the span of management, and

[24] Department of Science, Technology, and Education of MOR. (2001). *Fundamentals of economic management in railway transportation enterprises* (p. 240). Beijing: China Railway Publishing House.

[25] Rong Chaohe On the relationship between the "Separation of Railway Network and Transportation" and railway reorganization. *Journal of Northern Jiaotong University*, (2), 35–40.

[26] Zhao Jian 2016. *The reform and reorganization of Chinese railway and research on high-speed railway* (pp. 237-242). Beijing: Economic Press China.

establish a new transportation management system. Subsequently, on October 22, 1996, the Hohhot Railway Administration disbanded its Jining and Baotou Sub-administrations, thereby enabling the administration to directly manage the grass-roots stations. The practical experiences of the Nanchang Railway Administration and the Hohhot Railway Administration proved that a moderate span of management under the railway administrations, with direct leadership and command over stations, was a successful exploration, yielding high efficiency, agile command, and rapid response. Following this, the new model of railway administrations directly managing stations had gradually been promoted and applied more widely.

On the other hand, the implementation of the management and responsibility system of railway assets marked both a significant change in railway management methods and paved the way for railway transport enterprises to establish modern corporate systems. In 1998, MOR launched pilot projects for the asset management responsibility system in the Hohhot, Nanchang, Liuzhou, and Kunming Railway Administrations, China Railway Guangzhou Group Co., Ltd., and non-transportation enterprises affiliated with the Ministry. At the end of 1998, MOR formally issued the *Railway Administrations' Measures for the Implementation of the Management and Responsibility System of Assets*, deciding to implement the system for all transportation enterprises starting from 1999.[27] On April 5, 1999, Fu Zhihuan, then Minister of Railways, signed two-year asset management responsibility agreements (1999–2000) with the directors and party secretaries of 14 railway administrations across the country.[28] On June 4, 1999, MOR entered into asset management responsibility agreements with six state-owned enterprises directly under the Ministry, including China Civil Engineering Construction Corporation, China Railway International Service Corporation, China Railway International Group Corporation, Huayun Culture Tourism Group, China Railway Container Transport Center, and China Railway Specific Cargo Transport Center, covering the period from 1999 to 2000. With these agreements, the enterprises directly under MOR became competitive entities responsible for their own operations, risk-bearing decision-makers, investment bodies for self-development, and entities accountable for their own profits and losses.[29]

[27] Lin Liquan (1999). Overall implementation of the asset management responsibility system. In Yearbook of China Integrated Transport Press. *Yearbook of China transportation & communications* (p. 105). Beijing: Yearbook of China Integrated Transport Press.

[28] Li Xiaohua (1999, April 6). Signature Ceremony for Asset Management Responsibility Agreements held at Railway Administration. *People's Railway Daily*.

[29] Zhao Zhongyong (1999, June 8). Implementation of Asset Management Responsibility System for directly affiliated enterprises by MOR. *People's Railway Daily*.

4.1.2 The Beginning of Separating Railway Sideline Businesses from Core Businesses

Within the railway system, the direct provision of passenger and freight transport services, along with the maintenance and management of major railway equipment, are considered the main business of the railway transport sector. All other production and business activities fall under the category of sideline businesses. The separation of sideline businesses from core businesses primarily refers to removing entities under the jurisdiction of MOR that are less relevant to railway operations. For example, auxiliary departments within the railway system, such as construction sections, living management sections, and material supply sections, which are not independently accountable, should be spun off the main business. Moreover, diversified sections within the core business, such as locomotive depots, train depots, carriage depots and rolling stock maintenance division, should also be separated to ensure the smooth operation of the core business. Additionally, engineering group corporations and design institutes responsible for railway construction, as well as factories that manufacture railway locomotives and rolling stocks, should be reassigned to the State-owned Assets Supervision and Administration Commission (SASAC) or local government. Service-oriented enterprises associated with railway transportation, such as handling companies, travel agencies, and tobacco companies, should also be spun off and operate under independent accounting systems, facilitating the reform of the transportation sector.[30]

In the early years after the founding of PRC, social infrastructure was relatively underdeveloped. Alongside the construction of railways, various service departments were established to support the production and daily lives of railway workers. Over time, this led to a "comprehensive and self-sufficient" service system. Before 1998, the railway sector employed approximately 3.1 million people, peaking at 3.42 million.[31] This led to a humorous characterization of it as "a system having everything, including crematoria". From public security organs to procuratorates and courts, the railway system featured independent structures across various sectors, encompassing educational institutions from kindergartens, primary schools, and middle schools, to universities, as well as amenities such as canteens, guesthouses, hospitals and epidemic prevention stations. Moreover, the design and manufacture of locomotives and rolling stock, as well as the construction of roads, bridges, and tunnels, were entirely managed by the Railway Man, creating a miniature society that could function almost entirely self-sufficiently. Although this system played a positive role for a considerable period, it did not align with the requirements of a socialist market economy. It conflicted with the central

[30] Liang Meina (2004, March 22). Separation of Sideline Businesses from Core Business: The primary model of railway reform—An interview with Sun Yongfu, Deputy Minister of Railways. *China Business Journal.*

[31] Mai De (2004). The path of the Separation of Sideline Businesses from Core Businesses in Chinese railways. *China Post News*, (5), 58–59.

government's directives for separating government functions from business functions, community functions from enterprise ones, and public services from commercial activities. Furthermore, it was incompatible with the establishment of modern corporate systems.[32] Over time, this type of management system resulted in an excessively large workforce with low productivity, imposing heavy social burdens on the railway sector and severely constraining its reform and progress. Without separating the sideline businesses, it would be challenging to transform the railway management system, and the railway sector would struggle to achieve autonomous development.

Although the railway sector advocated to separate sideline businesses from core ones during the reform of MOR's organ in 1994, there was no substantial progress due to the complex and far-reaching influence of such a reform. However, a turning point occurred in 2000 when MOR formally spun off five major companies, including those involved in engineering and construction, along with ten higher education institutions, involving a total of more than 714,000 personnel. This reduced the total number of railway employees nationwide from 3.2 million to 2.488 million. By the end of 2002, the number of employees working in the "core business" sectors, such as passenger and freight transportation, had reached 1.4876 million, accounting for 59.8% of the total railway workforce. Meanwhile, the number of employees in "sideline business" sectors, including education, public security, procuratorial, judicial, and health services within the railway system, decreased to 1.001 million, representing 40.2% of the total railway workforce. In terms of assets, out of the national railway's total assets of 680 billion yuan, 86% were allocated to the "core business", while 14% were assigned to the "sideline business".[33]

In 2003, MOR proposed the leapfrog development of the railway sector as its strategic goal and emphasized accelerating the Separation of Sideline Businesses from Core Businesses, as well as the restructuring of auxiliary operations. According to Document No. 117 issued by MOR, entitled *Guidelines to Promote Separation of Sideline Businesses from Core Businesses and Reform of Sideline Businesses and Reemployment*, the Ministry required that starting from July 2003, efforts should be made to complete the transfer of educational and healthcare institutions of the railways within two years, and to substantially complete the separation of main and sideline businesses and the reform of sideline businesses within five years.[34]

On June 28, 2003, during a symposium on the leapfrog development of railways, the overall goals for the separation of main and sideline businesses, as well as reform of sideline businesses were established. These included: returning the governmental functions carried out by enterprises back to the government, in line with

[32] Cai Tiejun (2003, September 11). Lightening the load for better strides: Part Five of an in-depth report on the Leapfrog Development of Chinese railways. *People's Railway Daily*.

[33] Ibid.

[34] Guidelines to promote the Separation of Sideline Businesses from Core Businesses and reform of sideline businesses and reemployment (TB [2003] No. 117). In Wang Zhiguo (Ed.), *China railway yearbook (2004)* (p. 600), Archives and History Center of MOR.

the requirement of separating the functions of government and enterprises; returning the social functions performed by enterprises, in line with the principle of separating social functions from business functions; and implementing the reform of sideline businesses centered on property rights reform, ensuring a complete separation between railway sideline businesses and the core businesses.

The key approaches to advancing the separation of main and sideline businesses in the railway sector were outlined as follows: (1) Transferring public security, procuratorial, judicial, and health and quarantine departments of the railway system to the government; (2) Implementing localized management for railway schools and hospitals; (3) Restructuring and redirecting non-transportation enterprises.[35] On July 31, 2003, MOR convened a national symposium to promote the separation of core and sideline businesses and reform of sideline businesses. From August 9 to 10, 2003, the Ministry held another symposium on specialized railway reform. During this meeting, preliminary plans for three major reforms were approved: refining the internal passenger transportation companies of railway administrations, establishing specialized transportation companies, and spinning non-transportation enterprises off the core business.[36] These initiatives demonstrated the commitment to the separation reform.

After the decoupling of the China Railway Engineering Corporation, China Railway Construction Corporation, China National Railway Locomotive & Rolling Stock Industry Corporation, China Railway Signal & Communication Corporation, and China Civil Engineering Corporation from MOR, four survey and design institutes under the Ministry, 38 design and construction enterprises affiliated with the railway administrations, as well as China Railway Materials Group Corporation and China Tietong Telecom were transferred to the State-owned Assets Supervision and Administration Commission (SASAC). Additionally, all secondary schools and hospitals under the railway system were handed over to local governments.[37] As of the end of December 2003, a total of 225 primary and secondary schools, with 15,780 teaching staff, had been transferred, accounting for 25.7% and 27.6% of the total in the education system, respectively. Additionally, 50 kindergartens with 1,433 staff were transferred, representing 11.7% and 10.98% of the total. Thirteen hospitals, with 4,153 medical personnel, were also handed over, making up 6.25% and 5.32%, respectively. At the same time, agreements for the transfer of another 94 primary and secondary schools with 6,233 teaching staff, along with 39 hospitals

[35] Leadership Group of the Communist Party of MOR. Implementing the requirements of the "Three Represents" to seize new historical opportunities and strive for leapfrog development of Chinese Railways: Speech at the Seminar on Leapfrog Development of Railways. In *The important manuscripts on the reform and development of Chinese railways in 2003* (p. 69). Beijing: China Railway Publishing House.

[36] 2005, June 28). Significant achievements in the Separation of Sideline Businesses from Core Businesses in railways: Event Six of the Top Ten Events Displaying Two Years of implementation of the Leapfrog Development Strategy for Railways. *People's Railway Daily.*

[37] (2005, April 3). Separation of Railway Sideline Businesses from Core Businesses paves the way for reform of the railway system. *People's Railway Daily.*

and 14,847 staff, had been signed. Among them, all primary and secondary schools under the Shanghai Railway Administration were fully transferred to the local governments, as were all primary and secondary schools, vocational schools, and kindergartens under the Kunming and Hohhot Railway Administrations. Primary and secondary schools under the Qingdao Branch of the Jinan Railway Administration, and the Xining and Yinchuan Branches of the Lanzhou Railway Administration had also been completely transferred to local governments. The transfer process at other railway administrations was fully underway, with several schools being handed over to local governments.[38]

Alongside the separation of main and sideline businesses, the development of diversified operations was also a strategic choice for railway enterprises to survive and thrive in the face of intense market competition. Following Deng Xiaoping's South Tour Speeches in 1992, diversified operations in the railway sector witnessed rapid progress. In 1994, MOR established the "Collective Economic Development Center for Diversified Railway Operations" to conduct macro-level planning, policy guidance, and management for diversified operations across the railway system. Corresponding management bodies were also established within various institutions under MOR. The Ministry subsequently issued a series of documents and plans, including *The Development Plan of the Railway Diversified Operation during the Eighth Five-year Plan Period and 2000, The Development Plan of the Railway Diversified Operation during the Ninth Five-year Plan Period,* and *Provisions on Speeding up the Development of Railway Diversified Operation*, and *Suggestions on the Promotion of the Reform and Development of the Diversified Operation of Railway Transport Enterprises*. These documents offered guidelines for the reform and development of diversified operations throughout the railway system. In 1994, 1996, and 1999, the Ministry held three national conferences on diversified operations to discuss and formulate development strategies and policy measures.

By the end of the 1990s, diversified operations of Chinese railways had seen some progress, yet several issues remained unresolved. These issues manifested primarily in three ways: Firstly, in terms of scale, the operations were "small, scattered, and sluggish". Secondly, in terms of industrial structure, a disproportionately large number of operational projects were overly reliant on the transportation sector, lacking robust pillar industries. Thirdly, in terms of operational management, there was a lack of strict implementation of independent accounting. Problems in the appropriation of assets, cost allocation, and profit transfers between the transportation sector and diversified enterprises were prevalent. Furthermore, cost and capital management, as well as internal distribution, were characterized by a high degree of

[38] Department of Labor and Hygiene of MOR. (2004). Overview of the Separation of Railway Sideline Businesses from Core Businesses and the Reform of Sideline Businesses in 2003. In Wang Zhiguo (Ed.), *China railway yearbook (2004)* (p. 43), Archives and History Center of MOR.

arbitrariness, with common short-term behaviors, resulting in insufficient momentum for the long-term development of diversified operations.[39]

Although there were many reasons for these issues, at the core, they stem from problems in the management system of diversified operations. To accelerate the entry of railway transportation enterprises into the market and to establish a management system and operational mechanism compatible with the socialist market economy, Fu Zhihuan, then Minister of Railways, emphasized at the National Conference on Diversified Operations in the Railway Sector, held in Beijing on November 20–21, 1999, that the railways, as enterprises, should engage in diversified operations to survive and develop, as relying solely on two steel tracks would lead nowhere.[40] Consequently, MOR made a strategic decision to establish a new pattern of diversified operations for railway transportation enterprises.

The establishment of a new pattern for diversified operations in the railway sector can be summarized as breaking away from the traditional management model of transportation enterprises by separating the transportation industry from ancillary industries such as railway-related industrial production and other productive and service-oriented operations, construction, and logistics and support services, and then restructuring assets in collaboration with various diversified enterprises and labor service companies. The aim is to form a diversified operational system for transportation enterprises, and promote corporate governance, scale up industrial development, and standardize management practices, thereby forming a new landscape where the railway transportation industry and diversified operations mutually support and develop in coordination.[41] This transformation not only adjusts management relationships but also plays a crucial role in promoting asset restructuring, adjusting industrial structures, and establishing new systems for diversified operations that are compatible with the socialist market economy.

Under the new pattern of diversified operations in the railway sector, during the Ninth Five-Year Plan period, the transportation system generated a cumulative operating revenue of 18.45 billion yuan and a profit of 13.59 billion yuan, which represented 1.93 times and 1.11 times that of the Eighth Five-Year Plan period, respectively. In 2000, the operating revenue reached 44.4 billion yuan, with profits reaching 2.663 billion yuan, up 25% and 15% year-on-year respectively. Compared to the railway transportation industry, the total operating revenue from diversified

[39] Fu Zhihuan. (2000). Speech at the National Conference on Diversified Operations in the Railway Sector. In Peng Kaizhou (Ed.), *The important manuscripts on the reform and development of Chinese railways in 1999* (p. 157). Beijing: China Railway Publishing House.

[40] Fu Zhihuan. (2004). Striving to establish a new pattern to accelerate the development of diversified operations: Speech at the National Conference on Diversified Operations in the Railway Sector, November 20, 1999. In Fu Zhihuan (Ed.), *Exploration and practice of railway reform and development in China* (p. 239). Beijing: China Railway Publishing House.

[41] Sheng Guangzu. (2000). Seriously implementing the principle of the Fourth Plenary Session of the 15th Central Committee to accelerate the establishment of a new pattern for railway diversified operations: Speech at the National Conference on Diversified Operations in the Railway Sector. In Peng Kaizhou (Ed.), *The important manuscripts on the reform and development of Chinese railways in 1999* (p. 375). Beijing: China Railway Publishing House.

operations during the Ninth Five-Year Plan period was 123.7% of passenger transport income and 75% of freight transport revenue.[42]

Based on the establishment of a new pattern for diversified operations, in 2001, MOR issued the *Notice on Issues Concerning Non-transport Enterprises Establishing a Modern Enterprise System* (TZF [2001] No. 22),[43] which required a "three divisions" approach, that is, the "division of entities, division of financial accounts, and division of personnel" between railway transportation enterprises and diversified businesses.[44] Specifically, it mandated that transportation enterprises and diversified enterprises maintain clear asset relationships, establish an investor-investee relationship, clarify respective rights and responsibilities, and develop standard profit distribution and investment return systems, thereby evolving from the previous administrative management models. Additionally, these enterprises are required to conduct strict independent accounting, adhere to the principle of equivalent exchange, and handle economic interactions based on laws, regulations, and contracts. Furthermore, personnel in these enterprises shall be distinctly divided according to their positions, ensuring consistency among personnel, positions, and labor contracts, with standard labor and wage management. Through the "three divisions" approach, the operational scope of diversified enterprises and the transportation industry was clarified, which facilitated the clarification of the property rights of diversified enterprises and standardized financial and wage management practices. This approach provided institutional support for establishing a new pattern for diversified operations and promoting the healthy development of diversified enterprises.

4.2 New Step in the Opening Up of Railways

4.2.1 Exploration of a Diversified Financing System

In the early 1990s, the progress of Chinese railways lagged behind, primarily due to the country's chronic underinvestment in railway construction, which was compounded by the limited economic strength and tight fiscal revenues. In addition to this, during the 1980s, the government converted infrastructure construction investment allocations into loans (referred to as the policy of Conversion of Capital Allocations to Loans). However, under the planned economy, longstanding high

42 Wang Zhaocheng. (2001) A page of history, a grand chapter: A review of diversified operations in railways during the "Ninth Five-Year" Plan period. In Peng Kaizhou (Ed.), *China railway yearbook (2001)* (p. 94). Archives and History Center of MOR.

43 Wang Zhaocheng. (2002). Recognizing the situation, unifying thoughts, and resolutely achieving the goal of 'Three Divisions': Speech at the National Conference on Diversified Operations in the Railway Sector. In Peng Kaizhou (Ed.), *The important manuscripts on the reform and development of Chinese railways in 2001* (p. 296). Beijing: China Railway Publishing House.

44 Ibid.: 284.

taxes, low depreciation, and low freight rates remained unchanged, making it difficult for the railway sector to accumulate sufficient funds on its own to offset the shortfall in government investment. In this context, the Contract Responsibility System was unable to provide the substantial funding needed for railway construction.

According to the requirements of the national Eighth Five-Year Plan, the railway sector should increase its transportation capacity by over 50 million tons annually during this period. This objective not only required the reinforcement of existing lines but also the construction of a significant number of new lines. Based on 1990 prices, total investment would be projected to exceed 100 billion yuan, but existing financing methods and channels were inadequate to raise such a substantial amount of construction funds, resulting in a severe funding shortage. MOR estimated that the funding gap may reach 40%, approximately 60 billion yuan.[45] During the Ninth Five-Year Plan period, however, the increased construction mileage and the peak period for loan repayments on previous railway projects exacerbated the funding crisis. It was estimated that MOR would need to repay 137 billion yuan in principal and interest over five years (1996–2000), with the funding gap nearing the total investment required for railway construction during the Eighth Five-Year Plan.[46]

The shortage of funds for railway construction was the fundamental reason for the lag in railway development. To accelerate its progress, the most fundamental and effective solution was to increase investment in the sector to expedite construction. Since the 1980s, the government has implemented a series of measures to address the funding issue, including increasing loans, introducing contract management, and adjusting freight rates. However, due to the low-profit margins in the railway industry, the burden of repaying loans had become increasingly heavier. In addition, the "fixed freight rates" have failed to keep pace with "flexible costs", and the Contract Responsibility System could not meet the needs for scaling production in the railway sector. Even if freight rates were raised, the surplus generated from simple scaling production had to cover taxes, loan repayments, and profit margins for enterprises. With the continual rise in transportation costs, it was difficult to ensure sufficient funds for scaling production.

Given the insufficient national investment for railway construction, the most practical solution to meet investment needs is to transform the previous single-investor model, which relied solely on state funding, into a diversified and

---

[45] Railway Construction Fund Research Group. (1992, August 21). Establishing a railway development fund to accelerate railway construction. *People's Daily*; Han Zhubin. (1994). Thoroughly implement the principle of the 14th National Congress of the Communist Party of China, accelerate the pace of reform and opening up, and strive for the historic development of railways: Speech at the National Conference of Leading Officers in the Railway Sector. In Wu Changyuan (ed.), *The important manuscripts on the reform and development of Chinese railways in 1993* (pp. 56–57). Beijing: China Railway Publishing House.

[46] Fu Zhihuan, Urgent mobilization, organization for fighting and quickly setting off a new climax of railway construction: Speech at the Mobilization Conference for Accelerating Railway Construction. In Huang Sichuan (Ed.), *The important manuscripts on the reform and development of Chinese railways in 1998* (p. 51). Beijing: China Railway Publishing House.

multi-channel investment system. This requires reforming the existing investment system. Therefore, in discussions on how to enhance railway reform to address the issue of sluggish railway development, many proposed the need for reforming the investment system. Regarding how to reform the railway investment system, there was a general consensus from top to bottom, including both government and private sectors, that the current model of sole state investment should be changed. A diversified investment system involving various parties, including the state, railways, local governments, and enterprises, should be gradually established. This system would raise funds through various channels, such as levying railway construction funds, expanding domestic and international borrowing, issuing bonds, and adopting joint ventures and shareholding systems.[47]

At the National Conference of Leading Officers in the Railway Sector held in July 1992, MOR took a significant step toward enhancing investment system reform. Based on the government approval for establishing a railway construction fund and adjusted prices for new railways, the Ministry promoted the development of joint-venture railways, aiming to form a new mechanism for railway construction investment, characterized by the state, local governments, and enterprise's participation as the main investment entities, alongside various financing methods such as fundraising, loans, and bond issuance as the primary means of raising capital.[48] At another National Conference of Leading Officers in the Railway Sector in January 1993, the Ministry reiterated the importance of utilizing market mechanisms to attract and raise funds through multiple channels to accelerate railway construction.[49]

Firstly, the railway construction fund. The establishment of a fund primarily involves pooling a certain amount of capital through administrative and economic measures, dedicated to the construction and development of a specific socio-economic project. Since the reform and opening up, to raise capital through the establishment of funds for construction has proven effective in enhancing key projects in postal services, electricity, raw materials, energy, and highways. The railway sector has raised its construction fund through additional charges on freight rates, allowing it to allocate a certain amount of capital each year for scaling production, regardless of whether there was a deficit or surplus in simple scaling production. Following approval from the State Council, a policy was introduced to collect a railway construction fund from cargo owners, which stipulated that from March 1,

[47]Zhu Rongji. (2011). Easing the tension of railway transportation. In Zhu Rongji (Ed.), *Records of Zhu Rongji's Talks* (Volume 1) (pp. 188–189).Beijing: People's Publishing House.

Wu Shaoli, & Yu Chuan. (1993, April 20). Establishing a diversified railway freight rates system. *People's Daily.*

[48]Zhang Guorong. (1992, July 17). Major breakthroughs in railway reform during the "Eighth Five-Year" Plan period: Establishing a railway construction fund, charging adjusted prices for new railways, and actively developing joint-venture railways. *People's Daily.*

[49]Han Zhubin. (1994). Thoroughly implement the principle of the 14th National Congress of the Communist Party of China, accelerate the pace of reform and opening up, and strive for the historic development of railways: Speech at the National Conference of Leading Officers in the Railway Sector. In Wu Changyuan (ed.), *The important manuscripts on the reform and development of Chinese railways in 1993* (pp. 56–57). Beijing: China Railway Publishing House.

1991, a surcharge of 0.002 yuan per ton-kilometer would be levied on freight rates. Basic consumer goods and agricultural production materials related to people's livelihoods, such as salt, phosphate ore, grain, fertilizers, and edible vegetable oils, were exempted from this charge, as were containers and less-than-carload freight.[50] From July 1, 1992, an additional charge of 0.01 yuan per ton-kilometer was levied on top of the existing 0.002 yuan already collected in 1991. In 1993, another 0.015 yuan was added, bringing the total to 0.027 yuan per ton-kilometer. The price increase applied to all freight, including lines temporarily managed by railway transport enterprises, with the exception of fertilizers and pesticides. In 1996, while adjusting freight rates, MOR simplified the fund collection process by adding 0.001 yuan per ton-kilometer, raising the total to 0.028 yuan per ton-kilometer. In 1998, another increase of 0.005 yuan was levied, bringing it to 0.033 yuan per ton-kilometer. Statistics show that from 1991 to 1997, a total of 172.4 billion yuan was raised for the railway construction fund, all of which was invested in railway development.[51] The establishment of the railway construction fund has played a significant role in the construction and development of railways. In the absence of adequate government investment and social financing, the fund served as a relatively stable funding source for railway construction.

Secondly, the railway construction loans. Railway construction loans are primarily issued with MOR as the debtor and government credit as the support to debt agreement.[52] These loans are obtained by paying certain fees to banks. Although the policy of Conversion of Capital Allocations to Loans in the 1980s almost left the railway sector in dire straits, the situation began to improve in the 1990s with the introduction of railway construction fund policies. This marked a change in the funding channels for railway projects, evolving from reliance on a state policy bank to a dual support system involving both state policy banks and commercial banks, effectively alleviating the financial pressure on railway construction. To meet the needs of the national economy and railway development, the scale of bank loans borrowed by Chinese railways had continually expanded since the Seventh Five-Year Plan period. In the Eighth Five-Year Plan period, the total amount of bank loans used by the railway sector had reached 57.1 billion yuan, averaging over 10 billion yuan annually, thus making it an important source of funding for railway development.[53] In the first half of 1993, China Construction Bank raised funds through various means and issued loans totaling 3.71 billion yuan for railway construction, effectively supporting the funding of projects such as the Beijing-Kowloon Railway, the Houma–Yueshan Railway, the Baoji-Zhongwei Railway,

[50] Lin Xiaoyan. (2001). *Course in investment and financing management* (p. 94). Beijing: Economy & Management Publishing House.

[51] Ji Ling. (2012). *Railway transportation policies and regulations* (p. 53). Beijing: China Railway Publishing House.

[52] Ibid.

[53] Wang Kuizhong. (1996). Financial Management. In the Editorial Committee of the *Yearbook of China Transportation & Communications*, *Yearbook of China transportation & communications (1996)* (p. 90). Beijing: China Transportation Yearbook Publishing House.

the Nanning-Kunming Railway, and the Beijing West Railway Station.[54] With the establishment of the China Development Bank (CDB) in 1994, the railway sector became a key focus of its support. From 1994 to 1996, CDB provided a total of 32.82 billion yuan in loans for 49 railway projects, with loans for medium and large projects amounting to 32.68 billion yuan, accounting for 99.6% of the total. Loans for railway construction in the central and western regions amounted to 26.24 billion yuan, making up 80.0% of the total loans.[55] In 1998, CDB actively supported MOR in achieving its strategic objectives of "winning the decisive battle in southwest areas, mainly settling coal transportation problems, expanding roads network, and surpassing 70,000 kilometers of railway operating mileage" by signing the *Loan Agreement on Speeding up Railway Construction.* In that year alone, CDB issued 44.4 billion yuan in loans to the railway sector, a 70% increase compared to 1997.[56] Consequently, while the scale of bank loans for the railway sector was relatively small in the first three years of the Eighth Five-Year Plan period, it became a common practice for the institutions affiliated to MOR to take bank loans in the remaining years of the period.

Thirdly, the issuance of bonds. As the socialist market economy gained traction, China's financial markets began to gradually flourish in the early 1990s, making it possible to raise funds through the issuance of bonds for railway construction and investment. In 1992, with the approval of the State Economic and Trade Commission and the People's Bank of China, MOR issued Bonds for China Railway Investment for the first time, totaling 2 billion yuan, with a term of 3 years, an annual interest rate of 9.5%, and the principal and interest were fully repaid as scheduled in August 1995.[57] In December 1995, MOR issued the "1995 China Railway Construction Bonds" with a total amount of 1.53 billion yuan with a term of 3 years and an annual interest rate of 15%. These bonds were primarily issued to domestic enterprises, institutions, and individuals in China, and the funds raised were used to cover funding shortages for construction projects such as the Beijing-Kowloon Railway Line.[58] In 1997, with the approval of the State Development Planning Commission and the People's Bank of China, MOR issued 2.13237 billion yuan worth of railway construction bonds. These bonds were issued in book-entry form and divided into two types: 532.37 million yuan for 3-year bonds with an annual interest rate of 8% (no compound interest, and the principal and interest were paid in full upon maturity); and 1.6 billion yuan for 5-year bonds with an annual interest rate of 8.6% (no

[54] (1993, July 28). Securing funds for key projects: 3.7 billion yuan issued by China Construction Bank to railway construction in the first half of the year. *People's Daily.*

[55] Wang Zhizhong, & Wei Liying. (2001). *Research on China's railway reform and development (1978–1998)* (p.264). Beijing: Contemporary China Publishing House.

[56] Editorial Committee of the *History of China Development Bank.* (2013). *History of China Development Bank (1994–2012)* (p. 169). Beijing: China Financial Publishing House.

[57] Xin Zhen. (1992, August 29). Starting to issue 20 billion yuan of railway investment bonds. *People's Daily.*

[58] Qu Hong. (1997). Multiple sources of funds for railway development: Chinese railway financing strategies. *Chinese Railways*, (9), 10–12.

compound interest, and the principal and interest were paid in full upon maturity). The funds raised were primarily used for railway construction projects such as the Southern Xinjiang Railway, the Xi'an-Ankang Railway, the Baoji-Chengdu Railway, and the Chengdu-Kunming Electric Railway.[59] Statistics show that MOR issued a total of six bonds with a total amount of 10.765 billion yuan from 1992 to 1998, playing a vital role in accelerating railway construction.[60]

Fourthly, utilizing international loans. Utilizing international loans is also one of the important ways for China to actively expand diversified financing channels for railway construction. As China made its major decision to implement reform and opening-up, MOR began to make use of Japanese government loans in 1979 and World Bank loans in 1984; later in 1994, it began utilizing loans from the Asian Development Bank and bilateral government loans. By the end of 1995, the Chinese railways had secured three batches of loans from the Japanese government and seven batches from the World Bank, amounting to around 2.268 billion dollars and 2.03 billion dollars respectively. Moreover, there were also loans of 200 million dollars from the Asian Development Bank, 100 million dollars from the German government, and 15 million dollars from the Australian government. In total, these loans amounted to 4.613 billion dollars.[61]

While attracting international investment, MOR also paid keen attention to the management of international capital. On March 1, 1987, the Office of MOR for Using International Capital and Introducing Technology (formerly known as the Office of MOR for Loans, set up on June 20, 1981) was established. Its primary responsibilities included formulating plans, project negotiations, and final evaluations, as well as other duties related to the utilization of international capital and the introduction of technology throughout the railway system. On August 27, 1990, this Office was renamed the Center for Using International Capital and Introducing Technology of MOR. Since its inception, the Center has made numerous efforts to make effective use of and properly manage international capital.[62] At the 1996 National Railway Work Conference on Cooperation, Han Zhubin, then Minister of Railways, further emphasized, "It is necessary to establish and improve a comprehensive system covering the establishment, implementation, and post-evaluation of loan programs based on the national policies and principles for utilizing international capital".[63]

---

[59] (1998, June 10). Railway construction bonds issued publicly from today. *People's Daily*.

[60] Xu Zenglin. (1999). *Fifty years of the railroads of the People's Republic of China (1949–1999)* (p. 302). Beijing: China Railway Publishing House.

[61] Yang Ronglin. (1996). To accelerate railway development by expanding the use of foreign capital. In the Editorial Committee of the *Yearbook of China Transportation & Communications*, *Yearbook of China transportation & communications (1996)* (p. 131). Beijing: China Transportation Yearbook Publishing House.

[62] The Editorial Center of Chinese Railway History. (1996). *Memorabilia of Chinese railway (1876–1995)* (p. 410). Beijing: China Railway Publishing House.

[63] Han Zhubin. (1997). Seize the opportunity, actively explore, and form a new pattern of all-round expansion of opening-up for the railway industry: Speech at the National Railway Work

Fifthly, joint-venture railway lines. Joint-venture railway lines refer to significant main lines and branches that are jointly invested, constructed, and operated by MOR, local governments, enterprises, and other investment entities. They represent an important approach to leveraging the initiative of both the central and local governments.[64] In August 1992, the State Council issued the GF [1992] No. 44 document, endorsing the *Opinions of the State Development Planning Commission and MOR on Construction of Railways with Joint Investment from Local and Central Governments*. Accordingly, the State Development Planning Commission and MOR issued the *Measures for Building Railways Jointly Invested by Local and Central Governments* in November 1993, further promoting the construction of joint-venture railways. During the Eighth Five-Year Plan period, several joint-venture railway projects were launched, including the Jining-Tongliao Railway, the Nanning-Kunming Railway, the Xi'an-Yan'an Railway, the Guangzhou-Meizhou-Shantou Railway, the Xiangtang-Ji'an Railway, the Hefei-Jiujiang Railway, the Xiaoyi-Liulin Railway, the Baoji-Zhongwei Railway, and the Chengdu-Dazhou Railway.[65]

The principle that "benefits are determined by the amount of investment" applied to joint-venture railways is conducive to mobilizing the initiative of all parties involved in the construction and management of railways. Therefore, joint-venture railways often feature low investment, short durations, flexible operations, and promising benefits. For example, the North Xinjiang Railway was constructed through national and local fundraising, with local entities becoming the main contributors and beneficiaries. This resulted in a cost reduction of 200 million yuan and incalculable benefits. Similarly, the Lanzhou-Xinjiang Railway was a joint venture between the national and local governments. Normally, it would have taken 8 to 10 years to complete, but due to local preferential policies, sufficient capital investment, and ample raw materials, it was put into operation within three years.[66] The Guangdong Provincial Government once sought loans from the World Bank and the Asian Development Bank to build the Sanshui-Maoming Railway. However, due to concerns about Western sanctions, these two Banks once suspended funding. To ensure the continuity of construction, the Guangdong Provincial Government resolutely used administrative funds to cover the shortfall. At the same time, the government requested the Guangdong Development Bank to issue bonds and promote them to various cities and counties, realizing liquidity through the issuance of loans. Without the decisive measures taken by the Guangdong Provincial Government to

---

Cooperation on Conference. In Huang Sichuan (Ed.), *The important manuscripts on the reform and development of Chinese railways in 1996* (p. 71). Beijing: China Railway Publishing House.

[64] The Legislative Affairs Office of the State Council. (1993). *Compilation of laws and regulations of People's Republic of China from Janurary to December in 1992* (p. 790). Beijing: China Legal Publishing House.

[65] Xu Yaozhong, & Wang Yanrong. (1992, June 21). Establishing a number of joint-venture railways during the Eighth Five-Year Plan period. *People's Daily*.

[66] Wang Jinhu, & Wu Yirong. (1993, March 28). Necessary reform of railway itself: An interview with NPC Deputy Wu Changyuan. *People's Daily*.

pursue the joint-venture construction, the Sanshui-Maoming Railway Project might have been terminated.[67] By the end of 1998, there were a total of 29 joint-venture railway projects completed or under construction nationwide, distributed across 21 provinces, autonomous regions, and municipalities, with a total length of 8,400 kilometers, of which 4,897 kilometers were operational.[68] Joint-venture railways have become a crucial component of the Chinese railway network.

In short, the implementation of the railway construction fund policy in the country, the vigorous development of joint-venture railways, and the gradual establishment of a new investment mechanism for railway construction with investments led by the state, local governments, and enterprises through various financial instruments such as loans and bonds have greatly alleviated the shortage of railway construction funds caused by insufficient state investment, which is an important reason for the great achievements of Chinese railway construction during this period.

4.2.2 Gradual Opening up for Railway Industry

The first step is opening up to the markets beyond the railway industry.

The traditional railway system is like a "large and comprehensive" closed society, often featured with "highly centralized dispatching command, coordinative actions among transportation departments, semi-military approach to order execution".[69] Such features were gradually formed and proposed in the practice of transportation command in the 1950s, summarizing the productive nature of railway transportation. It is undeniable that this traditional railway system played a crucial role in the era of planned economy, enabling railway transportation to achieve higher efficiency and ensuring the completion of national transportation plans and key material transportation despite insufficient transportation capacity. However, when these features, originally formed in transportation organization, were extended to railway operation and management, it led to a highly centralized management style, further promoting the integration of government and enterprise functions, and resulted in blurred responsibilities between government and enterprise, exacerbating negative effects.

To escape the predicament caused by centralized operation and management, Chinese railway reform has made explorations and efforts in various aspects. Among them, opening the "iron gate of railway industry" to the market marks a significant step in the reform of the railway sector. Especially since 1992, the pace of railway opening-up has accelerated. The opening up of the railway industry can

[67] Yang Peixin. (1993, March 7). Calling for preferential policies first in railway. *People's Daily.*

[68] Xu Zenglin. (1999). *Fifty years of the railroads of the People's Republic of China (1949–1999)* (p. 304). Beijing: China Railway Publishing House.

[69] Lü Yinhua. (1994). Re-evaluating the concept of "Highly Centralized, Dispatching Command, Coordinative Actions among Transportation Departments, Semi-Military Approach to order of Order". In Zhao Guoqing (Ed.), *Challenges and Solutions* (p. 193). Beijing: China Railway Publishing House.

be understood from two dimensions. First, it is the opening-up to the external markets, which means transforming from the traditional system of state monopoly, vertical leadership, and semi-militarized management to a socialist market economy. Second, it is the opening-up to international markets, which involves, while carrying out domestic reforms, facing the global market, beginning to utilize international capital, conducting international cooperation, introducing advanced technology and equipment, developing international economic and trade business, and engaging in international transportation.

The opening up of the railway industry first involves the participation of local governments in railway construction and management. China's traditional approach to railway construction involved exclusive investment by the state and exclusive construction by MOR, with a single source of funding. However, in the actual process of railway construction, many lines cross multiple provinces and cities, leading to conflicts. For example, the construction of a railway from Suixi to Hai'an, though administratively part of Guangdong Province, falls under the jurisdiction of Liuzhou Railway Administration in the railway system, which is mainly located within Guangxi Zhuang Autonomous Region. This situation has dampened Guangdong Province's initiative for railway construction.[70] The emergence of joint-venture railways has effectively alleviated such conflicts. As early as the construction of the Nanning-Fangkonggang Railway in the early 1980s, Guangxi Zhuang Autonomous Region and MOR jointly explored possible avenues for cooperation, marking the emergence of the prototype of joint-venture railways.[71] The Sanshui-Maoming Railway in Guangdong, which was completed and opened to traffic in 1991, was also the first joint-venture railway between the central and local governments in China, with funding from both the Guangdong Provincial Government and MOR. The success of this initial exploration led to the central government's affirmation of the joint-venture railway model. In 1992, the State Council, addressing the construction of joint-venture railways, put forward that construction should achieve the goals of "overall planning, integration of departments and regions, hierarchical responsibility, and joint construction". It also endorsed the *Opinions of the State Development Planning Commission and MOR on Construction of Railways with Joint Investment from Local and Central Governments,* noting that joint-venture railway construction by the central and local governments is a new phenomenon emerging from reform and opening up. It is a way to mobilize the initiative of both the central and local governments, widely raise construction funds, and accelerate the development of Chinese railways.[72]

---

[70] Zhang Chengyao. (1994). Where the key point of railway reform. In Zhao Guoqing (Ed.), *Challenges and Solutions* (p. 176). Beijing: China Railway Publishing House.

[71] China Railway Society, & Statistics Center of MOR. (2003). *China Railway 1949–2001* (p. 127). Beijing: China Railway Publishing House.

[72] (2013). Opinions of the National Development Planning Commission and MOR on construction of railways with joint investment from local and central governments. In the Policy and Regulation Department of MOR (Ed.), *Compilation of China's railways laws, regulations and model documents* (pp. 15–16). Beijing: China Railway Publishing House.

The Beijing-Kowloon Railway, constructed during the Eighth Five-Year Plan period, represents Chinese joint-venture railways. The construction of the Beijing-Kowloon Railway commenced on on April 20, 1993 and became fully operational on September 1, 1996. It spans a total length of 2553 kilometers, making it China's third north-south railway after the Beijing-Guangzhou Railway and Beijing-Shanghai Railway. For this railway, MOR adopted a model of central investment as the primary source, supplemented by local fundraising and international loans, with joint efforts by both the central and local governments. In addition to China's preferential policies and the collection of railway construction funds, relevant provinces and cities contributed to the partial investment within their respective territories and allocated funds annually, thereby fully leveraging the initiative of both the central and local governments and ensuring the funding for the construction of the Beijing-Kowloon Railway.[73]

To facilitate the development of joint-venture railways, in November 1993, the State Planning Commission and MOR issued *the Measures for Building Railways Jointly Invested by Local and Central Governments*, clarifying the nature of joint-venture railways, organizational form, investment methods, rights and obligations of all investment parties, principles of construction management, operation and management methods, and preferential policies enjoyed by joint-venture railway companies. During this period, 13 joint-venture railways, including the Dazhou-Chengdu Railway, the Guangtong-Dali Railway, the Guangzhou-Meizhou-Shantou Railway, the Hefei-Jiujiang Railway, the Shilong-Changsha Railway, the Hengfeng-Nanping Railway, and others, were successively initiated and constructed, with the longest joint-venture railway, the Jining-Tongliao Railway, also completed.[74] By the end of 1996, the cumulative investment in joint-venture railways nationwide reached 12.46 billion yuan, with nine operational joint-venture railways totaling 3,316 kilometers and eight under construction totaling 1,966 kilometers.[75] By the end of 2001, the operational mileage of joint-venture railways in China reached 6,162 kilometers, forging an essential component of Chinese railway network.[76]

The construction of joint-venture railways has also brought about changes in the pattern of railway construction and operation. On December 31, 1993, Hunan Shimen-Changsha Railway Corporation was founded, and it is China's first joint-venture railway company established in accordance with modern enterprise systems. This Company is responsible for the construction, operation, and management of

---

[73] Sun Yongfu. (1996). A feat in the history of China's railway development: Reflections on the construction of Beijing-Kowloon Railway. In MOR of the People's Republic of China and Xinhua News Agency (Ed.), *Key projects of China's railway construction in Eighth Five-year Plan period* (p. 24). Economic Daily Press.

[74] China Railway Society, & Statistics Center of MOR. (2003). *China Railway 1949–2001* (p. 127). Beijing: China Railway Publishing House.

[75] Yearbook of China Integrated Transport Press. (1997). *Yearbook of China transportation & communications (1997)* (p. 83). Beijing: Yearbook of China Integrated Transport Press.

[76] The Editorial Board of *Chinese Railways at the Turn of the Century*. (2002) *Chinese railways at the turn of the century* (p. 63). Beijing: China Railway Publishing House.

the Shimen-Changsha Railway, featuring independent operation, separate accounting, self-development, and self-financing.[77] In 2004, following the introduction of the *Mid- and Long-Term Railway Network Plan*, railway construction reached a new climax, and the enthusiasm of local governments to invest in railways soared unprecedentedly. MOR successively signed agreements on joint railway construction with 31 provinces, autonomous regions, and municipalities, ushering in an unprecedented boom in the development of joint-venture railways.[78] The development of local railways represents another manifestation of local governments' participation in railway construction and management.

Local railways are those constructed with investment exclusively from local entities or in partnership with other railways, primarily tasked with short-distance transportation of public passengers and goods within local regions.[79] Compared with national railways, the management authority of local railways is shifted to local governments, representing the economic interests of their respective regions.

According to the Statistics Center of MOR, the first recorded investment statistics for local railways in Chinese railway infrastructure investment appeared in 1971, with an investment amount of 116 million yuan. After entering the 1980s, investment in local railways continued to increase, albeit on a relatively small scale, reaching 466 million yuan by the end of the Seventh Five-Year Plan period. In the 1990s, especially after 1992, the investment in local railways increased significantly, from 558 million yuan in 1992 to 1.861 billion yuan in 2003, marking a 2.3-fold increase. During the Eighth Five-Year Plan period, the total investment amounted to 4.582 billion yuan, and during the Ninth Five-Year Plan period, it totaled 5.719 billion yuan.[80] By the end of the Ninth Five-Year Plan period, a total of 12,000 kilometers of local railways had been constructed nationwide, of which more than 3,100 kilometers were subsequently transferred to MOR. Additionally, over 3,500 kilometers of railways, initially planned, designed, and constructed as local railways, were later converted into joint-venture railways.[81]

To guide and promote the reform of the local railway investment system and improve investment efficiency, on December 8, 2000, MOR issued the *Measures for the Management of Funds Invested by MOR in Local Railway Construction*, authorizing the China Railway Construction Development Center as the representative of MOR for investment management, and standardizing the use of

---

[77] Editorial Department of *Changsha Yearbook*. (1995). *Changsha Yearbook 1995* (p. 177). Shenzhen: Shenzhen Duolunmen Printing Company.

[78] Ji Ling. (2012). *Railway transportation policies and regulations* (p. 58). Beijing: China Railway Publishing House.

[79] Ibid.

[80] Statistics Center of MOR. (2008). *Collection of the historical statistics data of Chinese railways (1949–2006)* (internal resources) (pp. 229–230). Statistics Center of MOR.

[81] The Editorial Board of *Chinese Railways at the Turn of the Century*. (2002) *Chinese railways at the turn of the century* (p. 63). Beijing: China Railway Publishing House.

funds invested by MOR in local railway construction.[82] Subsequently, breakthrough progress in local railway reform has been attained because the Center, leveraging funds as a bridge, successfully restructured 13 projects in 2001, including the Huizhou-Aotou Railway in Guangdong Province, the Suiyang-Dongning Railway in Heilongjiang Province, the Linru-Dengfeng Railway in Henan Province, the Jinshawan-Junlian Railway in Sichuan Province, the Qinyuan-Qinxian Railway in Shanxi Province, and the Taocun-Weihai Railway in Shandong Province. MOR also increased its investment in local railway construction, with a total investment of 318.5 million yuan in 2001, almost equivalent to the total investment of 321 million yuan made by MOR in local railways during the entire Ninth Five-Year Plan period (1996–2000).[83]

The second step is opening up to the global market.

In addition to opening to the market beyond the railway industry, Chinese railways were also steadily opening towards the global market. At the Meeting of Leading Officers of the Railway System in 1993, Han Zhubin, then Minister of Railways, emphasized that railways "should actively participate in the competition in both domestic and international markets, making good use of both domestic and international resources".[84] In other words, the domestic reform of railways and its opening should be advanced simultaneously, which aligns with the trajectory of China's economic development after the Third Plenary Session of the 11th Central Committee. Economist Wu Jinglian pointed out, "A significant characteristic of economic development after China's reforms are market-oriented reforms and opening to the global market carried out simultaneously and reinforcing each other".[85] On this basis, in April 1996, MOR issued *the Railway Ninth Five-Year Plan to Expand Opening-up* which clarified the goals, main tasks, and requirements for strengthening macro-management, specialized management, and team building for the railway's opening to the outside world during the Ninth Five-Year Plan period.[86]

During this period, the openness of Chinese railways is mainly manifested in the following aspects.

First, the railways introduced and utilized international investment and advanced technology and equipment. Since 1993, there had been an investment boom in

[82] The Policy and Regulation Department of MOR. (2003). *Compilation of China's railways laws, regulations and model documents* (pp. 194–195). Beijing: China Railway Publishing House.

[83] Yearbook of China Integrated Transport Press. (2002). *Yearbook of China transportation & communications (2002)* (p. 77). Beijing: Yearbook of China Integrated Transport Press.

[84] Han Zhubin. (1994). Thoroughly implement the principle of the 14th National Congress of the Communist Party of China, accelerate the pace of reform and opening up, and strive for the historic development of railways: Speech at the National Conference of Leading Officers in the Railway Sector. In Wu Changyuan (ed.), *The important manuscripts on the reform and development of Chinese railways in 1993* (p. 64). Beijing: China Railway Publishing House.

[85] Wu Jinglian. (2018). *China's economy reform process* (p. 155). Beijing: China Encyclopedia Publishing House.

[86] (1997). *Railway Ninth Five-Year Plan to expand opening up* (TW [1996] No. 34). In Huang Sichuan (Ed.), The important manuscripts on the reform and development of Chinese railways in 1996 (pp. 539–546). Beijing: China Railway Publishing House.

China's railway construction, with 23.55 billion yuan invested in railway construction in 1993, which is 4.2 times the average annual investment of the Seventh Five-Year Plan period, and 32.2 billion yuan in 1994. However, there remained a significant funding gap due to the scale of railway construction during the Seventh Five-Year Plan period. The introduction and utilization of international investment effectively addressed this issue and promoted the advancement of China's railway openness. Japan has provided yen loans to China since 1979, mainly for infrastructure projects including railways, water conservancy, and ports, as well as projects in agriculture, forestry, urban construction, and environmental protection.[87] Since the Eighth Five-Year Plan period, Chinese railways had utilized a total of 132 billion yen in Japanese government loans to construct four new lines including the Baoji-Zhongwei Railway, the Hengshui-Shangqiu Railway, the Nanning-Kunming Railway, and the Jiaozhou-Huangdao Railway, totalling 1,828 kilometers. Additionally, 19.7 billion yen from three batches of yen loans were used for the construction preparation of the new electrified Xian-Ankang Railway, including the purchase of materials and the introduction of tunnel boring machines.[88] From 1989 to 1995, Chinese railways also utilized World Bank loans in a total of four batches (namely, the fourth, fifth, sixth, and seventh batches), amounting to a cumulative total of 2.02 billion dollars for projects including the renovation of the Sifang Locomotive and Rolling Stock Plant, the construction of the double-track line of the Zhejiang-Jiangxi Railway, the renovation of the Xuzhou railway hub, the electrification of the Chengdu-Kunming Railway and the Wuhan-Guangzhou Railway, and the construction of the railway communication network, among others.[89] Furthermore, the channels for railway loans in China have been continuously broadened. Starting in 1994, the Asian Development Bank loans totaling 200 million dollars were utilized for the purchase of communication equipment, signaling systems, track materials, and maintenance machinery for the Beijing-Kowloon Railway. A total of 15 million dollars in Australian loans were utilized for the communication project of the Lanzhou-Xinjiang Railway from 1995. By the end of 1998, the total amount of international investment loan agreements for railway development in China reached 6.55 billion dollars.[90]

While introducing and utilizing international investment, Chinese railways also emphasized the simultaneous introduction of advanced international technologies and equipment. This included diesel-electric locomotives and their manufacturing technologies, railway informatization technologies, complete equipment and technologies for electrified railways, large track maintenance machinery and related

---

[87] Xu Xianfen. (2011). *Japan's ODA diplomacy towards China: The dynamics of interests, power, and values* (pp. 67–70). Tokyo: Keiso Shobo.

[88] Xu Zenglin. (1999). *Fifty years of the railroads of the People's Republic of China (1949–1999)* (p. 314). Beijing: China Railway Publishing House.

[89] Lin Xiaoyan. (2001). *Course in investment and financing management* (p. 94). Beijing: Economy & Management Publishing House.

[90] Xu Zenglin. (1999). *Fifty years of the railroads of the People's Republic of China (1949–1999)* (p. 315). Beijing: China Railway Publishing House.

technologies and software, as well as key processing equipment and testing instruments. Regarding the introduction of railway trains and manufacturing technology, as early as May 1986, Chinese railways utilized a USD 10 million loan from the World Bank. It then signed a contract with a British railway engineering company about technical consultation for the modernization of the Changchun Railway Trains Factory and the introduction of new passenger train technology through international bidding. This increased the number of trains manufactured from 750–800 to 1,500, with excellent technical performance and a speed of 140 km/h. In 1994, a $30 million surplus from a yen loan was expended towards the Hengyang-Guangzhou Railway to jointly manufacture 30 stainless steel passenger coaches with a speed of 200 km/h in cooperation with a South Korean company. Additionally, a loan from Germany was used to import 96 subway cars from a German company.[91] These introductions of key railway technologies had played a positive role in the development of railway technology in China.

In the early period of the Eighth Five-Year Plan, MOR particularly emphasized the need to increase the technological investment and elevate the benchmark for imported technology when introducing international equipment. At the same time, it was important to accelerate the process of domesticating the introduced technology, during which the practice of Kunming Machinery Plant of China Railway Construction Corporation served as a typical example. The Plant, leveraging a loan from the World Bank and cooperating with the trade department, imported large-scale tamping machines, ballast cleaners, and manufacturing technology from Plasser & Theurer of Austria. By learning these new technologies, it developed dynamic stabilization vehicles and ballast distributing vehicles. These large-scale railway maintenance machines were on a par with their international counterparts from the 1990s and could be used as substitutes for imports. In 1995, when the Asian Development Bank granted loans for the procurement of 15 railway maintenance machines, the Kunming Machinery Plant won the bid for 13 machines, generating foreign exchange earnings of 12.65 million USD.[92]

Second, Sino-foreign joint venture is crucial for Chinese railways to further expand their opening up to the outside world. Han Zhumin, then Minister of Railways, pointed out that "utilizing existing enterprise assets to attract international investment is an effective way to revitalize those assets, adjust industrial structure, transform enterprise management mechanisms, and expand its opening-up. Establishing joint ventures and engaging in cooperation can not only address funding issues, but more importantly, introduce the critical technologies we

[91] Wang Zhizhong, & Wei Liying. (2001). *Research on China's railway reform and development (1978–1998)* (p. 437). Beijing: Contemporary China Publishing House.

[92] Han Zhubin. (1997). Seize the opportunity, actively explore, and form a new pattern of all-round expansion of opening-up for the railway industry: Speech at the National Railway Work Cooperation on Conference. In Huang Sichuan (Ed.), *The important manuscripts on the reform and development of Chinese railways in 1996* (p. 73). Beijing: China Railway Publishing House.

urgently need, upgrade products, enhance production levels and improve product quality".[93]

The earliest foreign joint ventures were established by the Guangzhou Railway Administration. As early as in the early 1980s, the Guangzhou Railway Administration actively introduced international investment. By the end of 1989, there were seven joint ventures and cooperative enterprises, including Shenzhen Station Travel Service Co., Ltd., Guangzhou Auto Show Travel Service Co., Ltd., Guangzhou Radio Wakie-Talkie Repair Center, Shenzhen Railway Plastic Tableware Product Co., Ltd., and Tongda Bamboo Leaf Co., Ltd. The annual operating revenue was 60.42 million yuan, with tax contributions of 2.88 million yuan, and the after-tax profit share for the Chinese side was 3.13 million yuan.[94] Among them, the earliest joint venture was Shenzhen Station Travel Service Co., Ltd., which was operated in collaboration with the Philadelphia International Trade Company of the United States and Chui Lee Fat Tobacco Co., Ltd. Shenzhen Station provided the operating site, and the Hong Kong investors invested 2.4 million Hong Kong dollars for renovation and equipment purchases, jointly operating businesses such as restaurants, bars, and shopping malls.[95] CASCO (CASCO Signal Co., Ltd.) was the first Sino-foreign joint venture in the Chinese railway system, established through a partnership between China Railway Communication Signal Company and an American railway signaling company, with a total investment of 4.8 million US dollars. The company was headquartered in Shanghai and began operations on March 14, 1986. It primarily introduced and adopted advanced technologies, manufacturing processes, and management methods from the United States in the 1980s to produce railway communication signal equipment and other products suitable for both Chinese and international markets, including technical services and maintenance. About 50% of its products were exported to countries and regions such as the United States, Singapore, and South Korea.[96] By the end of 1997, China Railway Communication Signal Company had established 15 Sino-foreign joint ventures, attracting international investment of 10.7607 million US dollars. Among these, 14 joint ventures manufactured railway products such as digital SPC exchange, overspeed protection devices, axle counting devices, and microcomputer interlocking systems. The introduction of technology for products like S12 SPC exchange, S700K electric spindle machine, and external locking devices played an important role in the construction of the Beijing-Kowloon Railway and in raising railway speed.[97]

[93] Ibid.: 72.

[94] Wang Zhizhong, & Wei Liying. (2001). *Research on China's railway reform and development (1978–1998)* (p.438). Beijing: Contemporary China Publishing House.

[95] Editorial Committee of the Yearbook of China *transportation & communications* (1989). *Yearbook of China transportation & communications (1989)* (p. 142). Beijing: Yearbook of China Integrated Transport Press.

[96] Li Chuanqing. (1991). *Quality management experiences of Shanghai's foreign-funded enterprises* (p. 83). Shanghai: Shanghai Scientific and Technical Publishers.

[97] Editorial Committee of the Yearbook of China *transportation & communications*. (1998). *Yearbook of China transportation & communications (1998)* (p. 121). Beijing: Yearbook of China

According to statistics, by the end of 1998, among the 218 registered foreign-invested enterprises, 123 enterprises had an investment scale of less than 1 million USD, accounting for 56.4% of the total; 75 had an investment scale between 1 million and 10 million USD, accounting for 34.4%; and 20 had an investment scale of over 10 million USD, accounting for 9.2%. Therefore, there was still a long way to go for Sino-foreign joint ventures at that time because they were generally small in scale.

Third, international transportation was expanded. Intermodal freight transportation is an important aspect of the railway's openness to the outside world. Intermodal freight transportation began in the early 1950s in China, with the first intermodal service being the China-Soviet Railway intermodal freight transportation in April 1951. Since then, China has established intermodal freight transportation with countries such as North Korea, Vietnam, Mongolia, Romania, Bulgaria, and Hungary, signing corresponding agreements on passenger and freight transport. With the continuous expansion of reform and opening up, the intermodal freight transportation of China had shown a positive development trend. In 1991, the import and export freight volume of China's intermodal freight transportation reached 6.9831 million tons, and the international trains and through trains carried a passenger volume of 165,000 person/trip.[98]

After 1992, intermodal freight transportation in China, particularly to Hong Kong, developed rapidly. The opening-up of rail ports had been expanded, with the addition of ports such as Alataw Pass. The number of ports for intermodal freight transportation reached nine, which connects the railways of Russia, Kazakhstan, Mongolia, North Korea, and Vietnam, respectively. These ports are dedicated to cargo handovers and intermodal transportation services for goods entering and exiting (or transiting) through the railways.[99] At the same time, the number of international passenger coaches and through trains operated by China Railways also gradually increased. For example, on June 20, 1992, the first international train from Almaty to Urumqi was launched; in December of the same year, the China-Kazakhstan intermodal freight transportation was initiated; and on January 8, 1993, a through train from Foshan to Kowloon began operation. At the end of October 1994, the ministers of railways (or transportation) of the seven countries held a meeting to promote further cooperation on intermodal freight transportation. In 1995, Alashan Port handled a total import and export freight volume of 1.116 million tons, which was 2.3 times that of 1994.[100] The China-Vietnam intermodal

Integrated Transport Press.

[98] Wang Zhizhong, & Wei Liying. (2001). *Research on China's railway reform and development (1978–1998)* (p.429). Beijing: Contemporary China Publishing House.

[99] Xu Zenglin. (1999). *Fifty years of the railroads of the People's Republic of China (1949–1999)* (p. 318). Beijing: China Railway Publishing House.

[100] Han Zhubin. (1997). Seize the opportunity, actively explore, and form a new pattern of all-round expansion of opening-up for the railway industry: Speech at the National Railway Work Cooperation on Conference. In Huang Sichuan (Ed.), *The important manuscripts on the reform and development of Chinese railways in 1996* (p. 73). Beijing: China Railway Publishing House.

freight transportation, which had been suspended since 1978, resumed operation on February 12, 1996 with a through train from Beijing to Hanoi. Additionally, international trains or through trains were also established from Urumqi to Almaty, Hohhot to Ulaanbaatar, Harbin to Vladivostok, Shenyang to Moscow, and Kunming North to Hanoi.

Fourth, international trade and contracting were developed. Before the reform and opening up, Chinese railways had already achieved significant accomplishments in international economic and technical assistance. From the 1950s to the early 1980s, recipient countries totaling 16 nations included North Korea, Vietnam, Tanzania, and Zambia, with total international assistance amounting to 2.693 billion yuan and 87,000 personnel dispatched abroad (excluding assistance to North Korea).[101] After the Third Plenary Session of the 11th Central Committee, MOR established China Civil Engineering Construction Corporation in 1979 to undertake international contracting projects and labor cooperation. By the end of 1991, this company had signed and implemented over 840 contracts, with a total contract value of 930 million USD and a turnover of 690 million USD, generating substantial economic and social benefits.[102] In 1980, China Railway Foreign Service Corporation was officially established, and its core business activities included undertaking commercial advertising, organizing tourism, facilitating commodity exhibitions on trains, and holding technical seminars.[103] By 1991, the China Railway Foreign Service Company had set up 14 branches and offices in places such as Hong Kong and Thailand, establishing business connections with 200 enterprises in over 30 countries and regions, including the United States, the United Kingdom, and Hong Kong. From 1981 to 1991, the company generated a profit of 30.57 million yuan and earned foreign exchange totaling 52.5 million USD.[104]

Entering the 1990s, China's railway industry experienced rapid growth in international trade and economic cooperation. The number of institutions authorized to engage in international trade and contracting increased swiftly and reached 13 by 1994. During the Eighth Five-Year Plan period, new contracts signed for international economic and technical cooperation in areas such as railway construction, labor cooperation, and design consulting amounted to a total of 1.85 billion USD, with a cumulative turnover of 740 million USD.[105] Among them, China Civil Engineering Construction Corporation signed the *Contract on the Nigerian Railway Rehabilitation and Reconstruction Project* with the Nigerian Ministry of Transport

[101] Li Jixiang et al. (1990). *Railway construction cause in contemporary China* (Volume II) (pp. 421–423). Beijing: China Social Sciences Press.

[102] Editorial Committee of the Yearbook of China *transportation & communications.* (1992). *Yearbook of China transportation & communications (1992)* (p. 121). Beijing: Yearbook of China Integrated Transport Press.

[103] (1980, June 20). The establishing of China Railway Foreign Service Corporation, *People's Daily.*

[104] (1991, December 14). China Railway Foreign Service Corporation earns over 50 million US dollars, *People's Daily.*

[105] Xu Zenglin. (1999). *Fifty years of the railroads of the People's Republic of China (1949–1999)* (p. 317). Beijing: China Railway Publishing House.

at the end of 1995, with a value of 530 million USD. This was the largest international economic cooperation project in the history of China's railway industry at that time.[106]

During the Ninth Five-Year Plan period, China's railway industry experienced a surge in international trade and economic cooperation. China Civil Engineering Construction Corporation, China Railway Construction Corporation Co., Ltd., and China Railway Harbin Group Co., Ltd. successively expanded their business scope into international trade, and became international trade companies with comprehensive qualifications for external operations. In the early stage of the Ninth Five-Year Plan period, MOR had increased the number of such enterprises and institutions to 45. The scale of railway international trade and economic cooperation continued to expand. The scope of business mainly included international contracting, labor cooperation, design consulting, and import-export trade. According to statistics, from 1991 to 1998, the total contract amount for labor cooperation, international contracting projects, and design consulting across the railway system reached 2.978 billion USD, with a turnover of 1.904 billion USD.[107] Meanwhile, the export of railway products grew year by year, with locomotives and rolling stock being the major commodities. In 1995 alone, the export turnover of locomotive and rolling stock industrial products reached 210 million USD, surpassing the total turnover of the previous four years.[108]

Fifth, China's accession to the World Trade Organization in 2001 brought new opportunities and challenges for the reform and development of Chinese railways. On December 10, 2001, China officially joined the WTO after 15 years of arduous negotiations, entering a new phase in China's opening up to the outside world. This allows China to participate in international economic competition and cooperation on a broader scale, in wider fields, and at higher levels, and better to adapt to the economic globalization.

Railway transportation falls under the category of service trade. According to *Annex 9: Schedule of Specific Commitments on Services of the Compilation of the Legal Instruments on China's Accession to the World Trade Organization*, the Chinese government is committed to a phased opening of the railway freight market, as well as related warehousing and freight forwarding services. In the first phase, starting from the date of accession to the WTO, international investment is permitted to hold shares in Sino-foreign joint venture railway freight companies,

---

[106] Han Zhubin. (1997). Seize the opportunity, actively explore, and form a new pattern of all-round expansion of opening-up for the railway industry: Speech at the National Railway Work Cooperation on Conference. In Huang Sichuan (Ed.), *The important manuscripts on the reform and development of Chinese railways in 1996* (p. 75). Beijing: China Railway Publishing House.

[107] According to the calculations based on the railway international trade and economic data from the *Yearbook of China Transportation & Communications (1991–1998)*.

[108] Han Zhubin. (1997). Seize the opportunity, actively explore, and form a new pattern of all-round expansion of opening-up for the railway industry: Speech at the National Railway Work Cooperation on Conference. In Huang Sichuan (Ed.), *The important manuscripts on the reform and development of Chinese railways in 1996* (p. 74). Beijing: China Railway Publishing House.

with an international ownership limit of no more than 49%. In the second phase, within three years after joining the WTO, international investors will be permitted to have controlling stakes in railway freight companies. In the third phase, within six years after joining the WTO, international investors will be allowed to establish wholly foreign-owned railway freight companies.[109]

Joining the WTO to open the railway transportation market represents both a unique opportunity and a significant challenge for Chinese railways. According to a study by Sun Lin and Xu Minglu from the China Railway Economic and Planning Research Institute in 2002, such an act will bring several opportunities for the reform and development of Chinese railways. First, it will accelerate the reform of railway transportation and economic systems, particularly for state-owned railway enterprises, injecting new momentum for significant breakthroughs. Second, it will help attract international investment, accelerate technological advancements, and improve management standards in the railway sector. Third, it will promote the formation of a competitive railway transportation market and enhance the efficiency of railway companies. Fourth, it will facilitate the expansion into new markets and the development of new business opportunities. Finally, it will ensure that Chinese enterprises enjoy fair treatment in the international market, allowing them to actively participate in multinational competition and cooperation.[110]

At the same time, with the gradual opening of the domestic market, the railway transportation industry also faces significant challenges. According to the research by Wu Jianhong and Xu Minglu, these challenges are primarily reflected in the following aspects: the existing railway management system, organizational structure, and operational management methods do not meet the demands of a market economy; there is a lack of laws and regulations for opening up the railway transport sector, which may contradict the transparency principles required by the WTO; international parties may adopt a "cherry-picking" strategy, increasingly entering high-profit, high-potential, and emerging specialized transport markets, posing a threat to the new growth points of China's railway freight; this exacerbates the capacity strain on busy railway lines during peak demand periods; foreign-funded enterprises engaged in road transport in China create substitute competition for railways; in conjunction with their freight operations, international companies may engage in corresponding warehousing and freight forwarding businesses, significantly impacting the related business of Chinese railways, especially in international freight forwarding; entry of international entrepreneurs will intensify the competition for high-quality technical and managerial talent, and skilled workers

---

[109] (2003). Opportunities, challenges and countermeasures for the railway transportation industry after China's accession to the WTO (TZF No.70 [2002]). In Xie Gaochao (Ed.), *The important manuscripts on the reform and development of Chinese railways in 2002* (p. 609). Beijing: China Railway Publishing House.

[110] Sun Lin, & Xu Minglu. (2002). Challenges and opportunities: Influences on China's railway transportation industry and related industries after accession to the WTO. In the Secretariat of Chinese Railway Society (Ed.), *China's railway reform, development and scientific innovation after accession to the WTO* (p. 89). Beijing: China Railway Publishing House.

in Chinese railways, further leading to talent loss in the railway sector.[111] Therefore, as pointed out by Fu Zhihuan, then Minister of Railways, in the face of both opportunities and challenges, China's railway sector shall address the following: first, the government's role and approach in managing railways should be transformed; second, the reform of the railway transport system should be accelerated; third, it is essential to meet challenges in talent cultivation, utilization, and motivation.[112]

4.3 Solid Foundation for the Removal of the "Bottleneck"

4.3.1 A Great Opportunity for Railway Construction

"With the national economy reaching a new level, what should be done about the railways?"[113] The Central Committee of the Communist Party, the State Council, and MOR, as the national railway authority, placed great importance on how to address the serious lag in railway development. From January 8 to 11, 1993, MOR held a conference in Beijing for railway leading officers, focusing on the central theme of "studying and implementing the principle of the 14th National Congress of the Communist Party, and discussing the basic ideas for railway reform under the new conditions of establishing a socialist market economy".[114] Since railway work was a top priority at the time, Zhu Rongji, then vice prime minister of State Council, made an exception to attend the conference and delivered a speech entitled *Accelerating the Pace of Railway Construction and Reform*. He stated at the meeting that the difficulties faced by the railways had significantly constrained the national economy. It's essential to prioritize railway construction, as focusing on this aspect can drive overall development.[115] On April 21, 1993, during his inspection of rural and railway construction in Hunan, Zhu Rongji reiterated, "Among the key industries that need strengthening—such as energy, transportation (including roads, ports, and civil aviation), and communications—the most important is still railways; railways are the top priority… Railway construction should be viewed as

---

[111] Wu Jianhong, & Xu Minglu. (2000, June 14). Is China's railway ready for China's accession to the WTO? *People's Railway Daily.*

[112] Fu Zhihuan. (2002). Opportunities, challenges and countermeasures for the railway transportation industry after China's accession to the WTO. In the Research Office of the State Council (Ed.), *China's accession to the WTO: opportunities· challenges· countermeasures* (pp. 257–258). Beijing: China Yanshi Press.

[113] Zhu Rongji. (2011). Easing the tension of railway transportation. In Zhu Rongji (Ed.), *Records of Zhu Rongji's Talks* (Volume 1) (pp 183).Beijing: People's Publishing House.

[114] (1993, January 9). National Conference of Leading Officers in the Railway Sector was held in Beijing, *People's Railway.*

[115] Wu Changyuan. (1994). *The important manuscripts on the reform and development of Chinese railways in 1993* (p. 9). Beijing: China Railway Publishing House.

a strategic issue in economic development. The Beijing-Kowloon Railway is a representative example of this strategic step".[116] Zhu Rongji's speech highlighted the central leadership's significant concern and urgent desire to rectify the serious lag in China's railway development.

Indeed, following the rapid economic progress after 1992, the severe lag in railway development increasingly revealed its "bottleneck" effect on the national economy. As described in a report by the *People's Daily*, it reached a critical point where railway construction required the same urgency as the efforts in developing atomic and hydrogen bombs.[117] Due to the serious lag affecting the overall progress of China's economy and society, this issue has garnered significant attention from the central leadership and MOR, becoming a key concern for various sectors and the general public nationwide.

Beginning on February 22, 1993, the *People's Daily* launched a series of reports on the serious lag in transportation, starting with *When Will Chinese Railways Break Free from the "Bottleneck"?* This initiative sparked strong reactions across various social sectors.[118] *People's Daily* reporters visited relevant national ministries and various enterprises and institutions in Beijing, finding that "many ministers, department heads, and general managers had copies of the article or its clippings on their desks. Some had even marked them with red and blue lines. This indicated that the leadership was paying close attention to the issue, and discussions were lively". Zhang Pan, then Deputy Director of the Development Research Center of the State Council, and researcher Gao Bo said in an interview that the article highlighted a crucial strategic issue for revitalizing China's economy. Now that the national economy has progressed significantly, railways should also follow. It reflects both economic demands and the voices of millions. Li Duanshen, then director of the Transportation Division of the National Development Planning Commission, said that the article reflected the serious lag in China's railway development. He noted that it helped to further enhance everyone's sense of responsibility and urgency regarding the acceleration of railway development. Cao Rujia, then deputy general manager of the National Transportation Investment Company, stated that the article profoundly reflected the severe stagnation in railway transportation, presenting clear viewpoints that provoked deep thought. He emphasized that, as China gradually establishes a socialist market economy system, addressing this issue has become urgent and cannot be delayed.[119]

---

[116] Zhu Rongji. (1994). Viewing railway construction as a strategic issue. In Wu Changyuan (Ed.), *The important manuscripts on the reform and development of Chinese railways in 1993* (p. 9). Beijing: China Railway Publishing House.

[117] Zhang Guorong. (1993, February 22). When will Chinese railways out of "bottleneck" (Second Part). *People's Daily.*

[118] Zhang Guorong. (1993, February 22). When will Chinese railways out of "bottleneck" (First Part). *People's Daily.*

[119] Lu Mu. (1993, March 1). The leaders attach importance, the public concerns and understands, supports, "When will Chinese railways out of "bottleneck" receives a strong response. *People's Daily.*

The series of reports in the *People's Daily* received the strongest response from MOR. People eagerly circulated the newspapers, and when there weren't enough copies, they resorted to photocopies. There was a widespread consensus that the article comprehensively and objectively depicted the severe stagnation of the railway system and its underlying causes, emphasizing that the railway issues had reached a point where they should be addressed. Shi Xiyu, the then Vice Minister of Railways, said, "After reading this article, I felt a strong sense of urgency, realizing that the railway's long struggle is now gaining social understanding". Zhang Youmin, the then Chief Economist of MOR, stated, "This article expresses the genuine concerns of 3.4 million Chinese railway employees, presenting the significant issue of railway transport—currently the most prominent weakness in our country's economic progress—before the entire Party and the people. It encourages greater support for the large-scale development of railway construction. The fact that *the People's Daily* has published a lengthy report on railway issues in a prominent position on the front page indicates both the importance and urgency of this matter, and shows that the *People's Daily* is implementing the principle of the 14th Party Congress, focusing on economic development and addressing key issues in its coverage".[120]

The series of reports by the *People's Daily* had created a public opinion environment across society that fosters understanding, concern, and support for the railways. From December 1991 to January 1992, the China Railway Economic and Planning Research Institute conducted a sampling survey entitled "National Awareness of Railways and the Social Image of Railways". The aim was to assess the public's understanding of the nature and function of railways, their satisfaction with railway services, and their views on railway development and future reform directions. The respondents included officials and scholars engaged in macroeconomics and transportation management, as well as freight owners and passengers in Beijing. A total of 6,770 questionnaires were distributed, with a response rate of 70.49%. The survey results showed that 93.1% of respondents expressed concern or significant concern about the development of railways and their impact on society.[121]

At the 1993 sessions of the National People's Congress and the Chinese People's Political Consultative Conference, apart from agriculture, the topics that garnered the most discussion among representatives and committee members at the 1993 sessions of the National People's Congress and the Chinese People's Political Consultative Conference were railway. There were 32 proposals on railway development and construction, an increase of 50% over the previous year, of which 26 requested that railway construction be accelerated.[122] This clearly indicated that

---

[120] Ibid.

[121] Yu Yan, & Cui Wen. (1994). Society calls for reform: Survey on national railway awareness and social image of railways. In Zhao Guoqing (Ed.), *Challenges and Solutions* (pp. 84–84). Beijing: China Railway Publishing House.

[122] Zhang Guorong. (1993, June 15). Seizing opportunities and accelerating the construction of railways: Interview records of Han Zhubin, the Minister of MOR. *People's Daily.*

acceleration of railway construction and development had become a common aspiration of the entire nation.

A farmer named Chen Wuhua from Dafangping Village, Yuyangguan Town, Wufeng Tujia Autonomous County, Hubei Province, read the series of reports on *When Will Chinese Railways Break Free from the "Bottleneck"?* in the *People's Daily* on February 22 and 23, 1993. The striking words and shocking statistics in the articles constantly touched the heart of this farmer. He felt "deeply anxious about the severe lag in Chinese railways". On February 28, 1993, Chen Wuhua couldn't help but write a letter to Han Zhubin, then Minister of Railways, expressing in plain language the deep concern and support of an ordinary Chinese farmer for China's railway construction. "Railway construction is the responsibility of MOR, but it is also the responsibility of the entire Party and the people of the whole country. I believe that every citizen of the People's Republic of China should bear the pressure, take the responsibility, and strengthen the sense of urgency and mission for the development of Chinese railways… If everyone treats China's railway construction as they support and care about the Asian Games and the bid for the Olympic Games, I don't think it will be difficult for Chinese railways to change". Chen Wuhua also expressed his willingness to contribute 1,000 yuan each year from 1993 to 2000 to support the national railway construction. Along with the letter, he enclosed 1,000 yuan as a token of a farmer's heartfelt support for China's railway development.[123] This touching real-life story reflects, on the one hand, the deep concern and support of ordinary Chinese citizens for the development of Chinese railways. On the other hand, it highlights the severe shortage of railway capacity in the early 1990s, underscoring the urgent need to accelerate railway construction to meet the demands of the national economy.

Thanks to the significant expansion of railway construction during the Eighth Five-Year Plan period, China's railway transportation capacity saw considerable improvement, which, to some extent, alleviated the constraints on the growth of the national economy. However, during this period, the scale of China's railway network, its structure, technical standards, and transportation quality failed to meet the demands of national economic and social development. Therefore, after the 15th National Congress of the CPC, the Party Central Committee and the State Council made the important decision to continue accelerating railway construction. Particularly after the outbreak of the Asian financial crisis in 1997, the Party Central Committee and the State Council made the significant decision to intensify infrastructure construction in order to expand domestic demand and stimulate economic growth, with the aim of ensuring the sustained, rapid, and healthy development of the national economy. Consequently, this presented an unprecedented opportunity for China's railway construction.

[123] (1993, March 27). Chen Wuhua, a farmer, writing a letter to Han Zhubin, the Minister of MOR, noting everyone should contribute to the construction of railways. *People's Railway Daily*.

(1993, April 17). Causing a strong reaction due to our report on railways, farmar Chen Wuhua writing to the Minister of Railways. *People's Daily*.

Under these circumstances, MOR revised the overall railway construction plan for the Ninth Five-Year Plan period and made a strategy for railway construction in the next century. This strategy was designed to "Meet the tough challenges of railway construction in the Southwest, secure coal transport at any expense, construct high-speed railways, expand railway network, and surpass the goal of seventy thousand kilometers", aiming to concentrate efforts on building a number of high-capacity main lines that would have a significant impact on the national economy and serve as the backbone of the railway network. It also focused on projects that would improve the layout of the railway network and accelerate regional development.[124] In terms of construction scale, the initial plan was to build 5,340 kilometers of new railways over five years and 2,580 kilometers of double-track railways on existing railways, and electrify 4400 kilometers of old railways.[125]

On March 28, 1998, MOR held a special mobilization conference to accelerate railway construction. During the meeting, then Minister of Railways Fu Zhihuan delivered a speech entitled *An Emergency Mobilization to Organize Campaign, and We Need to Stir a New Wave of Railway Construction*. He urged all departments of railway construction, design, and supervision to adhere to the principles of "accelerated pace, systematic execution, and optimal efficiency". He called for a greater determination than that shown during the railway construction campaign of the Eighth Five-Year Plan, and emphasized the need for decisive and effective measures to achieve complete victory in this cross-century railway construction campaign.[126] This marked the beginning of an unprecedented cross-century railway construction campaign.

As we entered the twenty-first century, with the implementation of the Western Development Strategy, the focus of railway construction further shifted towards the western regions, adding new content and vitality into the acceleration of railway development in the west. In order to implement the Western Development Strategy, MOR focused on railway construction in the West in the Tenth Five-Year Plan for railways. It specifically issued *the Tenth Five-Year Plan of the Railway Construction in Western China*, which outlined the requirements for railway development in the West during this period: expanding internal and external channels, strengthening technological upgrades, enhancing transportation capacity, and adapting to development needs.[127] Against this backdrop, the pace of railway construction in the West

---

[124] (1998, March 29). Determined Five-Year Plan of China's railway construction: Winning the decisive battle in southwest areas, mainly settling coal transportation problems, building expressway and expanding roads network. *People's Daily.*

[125] Sun Yongfu. (1999). Speech at the Mobilization Meeting to Speed up Railway Construction: Seizing opportunities, aiming for new goals and creating a new brilliant performance in railway construction. In Huang Sichuan (Ed.), *The important manuscripts on the reform and development of Chinese railways in 1998* (p. 177). Beijing: China Railway Publishing House.

[126] Fu Zhihuan. (2004). Urgent mobilization, organization for fighting and quickly setting off a new climax of railway construction. In Fu Zhihuan (Ed.), *Exploration and practice of railway reform and development in China* (p. 406). Beijing: China Railway Publishing House.

[127] Fu Zhihuan. (2000, October 8). Implementation of strategy for large-scale development of West China to accelerate the construction of railways in western areas. *People's Daily.*

continued to accelerate. Starting from 2000, a total of 26 medium and large-scale infrastructure projects for western railways were launched.[128] During this period, construction of the first railway to Tibet—the Qinghai–Xizang Railway—also commenced, becoming a landmark project in China's implementation of the Western Development Strategy.

4.3.2 Unprecedented Investment in Railway Construction

One important reason for the slow development of railway transportation compared to the growth of the national economy is that the proportion of railway investment in the country's total investment has long been relatively low and has been on a downward trend. During the three-year economic recovery period from 1950 to 1952, railway investment accounted for 14.5% of the country's total investment. This proportion was 10.7% during the First Five-Year Plan, 9% during the Second Five-Year Plan, 12.3% during the Third Five-Year Plan, 10.5% during the Fourth Five-Year Plan, 6.4% during the Fifth Five-Year Plan, 7% during the Sixth Five-Year Plan, and 6.6% during the Seventh Five-Year Plan. By 1991, it had dropped to 5%, marking the lowest since the founding of PRC in 1949.[129] As the result of the long-term imbalance in the proportion of railway investment, railways have not only failed to play their role as a "forerunner" but also lagged significantly behind the progress of the national economy.

To remove the railway bottleneck hindering national economic progress, in the latter half of 1992, the Central Committee of the CPC and the State Council made a significant decision to accelerate railway construction, sparking a new wave of railway construction and investment in this sector. During the Eighth Five-Year Plan period (1991–1995), railway construction investment rapidly increased with the scale and the pace of construction continuously growing. Particularly during the last three years of the period, hundreds of thousands of railway construction workers were mobilized urgently to focus their efforts on "intensively building the Beijing-Kowloon and Lanzhou-Xinjiang lines, swiftly completing the Houma-Yueshan and Baoji-Zhongwei lines, further developing networks in East China and Southwest China, and improving the Datong-Qinhuangdao line", thereby launching a grand railway construction campaign.[130]

[128] The Editorial Board of *Chinese Railways at the Turn of the Century*. (2002) *Chinese railways at the turn of the century* (p. 51). Beijing: China Railway Publishing House.

[129] Zhang Shengshu, and Saeki Koji. (1998). *China modern logistics research* (pp. 69–70). Beijing: China Materials Press.

[130] Han Shubin. (1997). Seizing the historical opportunity to implement the "two transformations", and striving for the grand goals of railway reform and development during the Ninth Five-Year Plan: Speech at the National Conference of Leading Officers in the Railway Sector. In Huang Sichuan (Ed.), *The Important Manuscripts of China's Railway Reform and Development in 1996* (p. 32). Beijing: China Railway Publishing House.

After the 15th CPC National Congress in 1997, China entered a new stage of reform, opening up, and socialist modernization. Under these new circumstances, investment in railway construction reached a new level. Starting in the second half of 1997, the Southeast Asian financial crisis began to affect other regions of Asia and beyond. To effectively counteract the effects of the financial crisis and achieve a national economic growth target of 8%, the Central Committee of the CPC and the State Council took significant measures. One of the key measures was to accelerate infrastructure construction, particularly in railways. This measure aimed to further promote railway construction, develop and improve the railway network structure, advance the modernization of railways, and increase domestic demand to stimulate economic growth. The *Report on the Work of the Government* delivered at the First Session of the Ninth National People's Congress also prioritized infrastructure development, including transportation, as a focus of fixed asset investment.[131] China decided to invest 50 billion yuan annually in railway construction from 1998 to 2002—a scale of investment unprecedented in Chinese railway history, surpassing that in any previous Five-Year Plan.[132]

Following the year 2000, with the implementation of the Strategy for Large-Scale Development of Western China, funding for railway construction began to shift toward projects in the western regions. To implement the Strategy set forth by the Central Committee of the CPC and the State Council, MOR highlighted western railway construction in its Tenth Five-Year Plan. A dedicated plan, the *Tenth Five-Year Plan of the Railway Construction in Western China*, was formulated,[133] aiming to achieve an infrastructure construction investment scale of approximately 140 billion yuan in the western region during the 10th Five-Year Plan period, accounting for about 52% of the total planned investment in railway construction nationwide. The plan was targeted at the completion of 3,670 kilometers of new lines, 1,250 kilometers of double-track lines on existing lines, and 510 kilometers of electrification on existing lines. By the end of 2001, 45.86 billion yuan had been invested in western railway construction, accounting for 50.2% of the total investment in medium- and large-sized infrastructure projects during the same period. Out of this, 13.42 billion yuan came from national bonds, comprising 81.8% of MOR's national bond allocation for the same period. An additional 28.06 billion yuan was allocated from the railway construction fund, making up 47.2% of the total amount of the fund used in medium- and large-sized infrastructure projects during the same period.[134]

[131] Fu Zhihuan. (2004). Urgent mobilization, organization for fighting and quickly setting off a new climax of railway construction. In Fu Zhihuan (Ed.), *Exploration and practice of railway reform and development in China* (pp. 405–406). Beijing: China Railway Publishing House.

[132] Jiang Shijie (1998, June 10). Railways: the locomotive driving economic growth. *People's Daily*.

[133] The Editorial Board of *Chinese Railways at the Turn of the Century*. (2002) *Chinese railways at the turn of the century* (p. 51). Beijing: China Railway Publishing House.

[134] MOR of the People's Republic of China. (2003). *The great decision of the century* (p.141). Beijing: China Railway Publishing House.

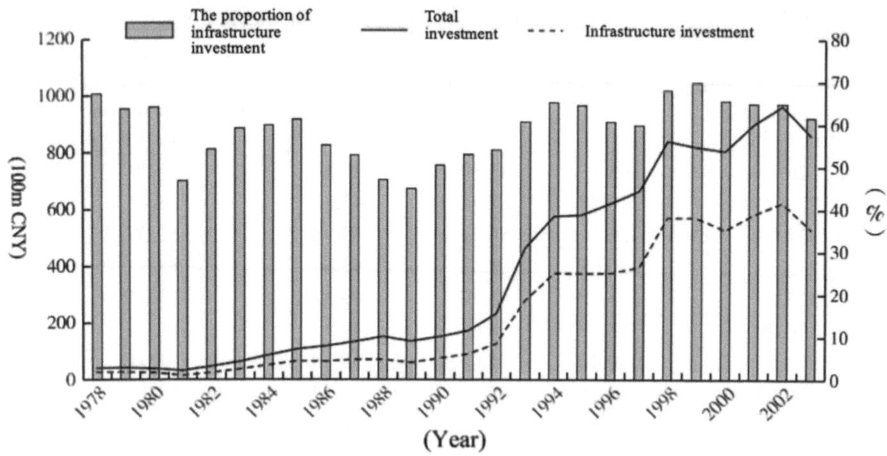

Fig. 4.1 The fixed asset investment in Chinese railways from 1978 to 2003 (*Source* Data derived from Tables 6-1 "National Railway Fixed Asset Investment" and 6-2 "National Railway Infrastructure Investment" in the *Compilations of the Historical Statistics Data of Chinese Railways* [1949–2006] [internal resources], compiled by the Statistics Center of MOR)

Notably, on June 29, 2001, construction of the Qinghai-Tibet Railway began, and 1.18 billion yuan was invested in the project that same year.[135]

As shown in Fig. 4.1, the fixed asset investment in Chinese railways increased year by year since 1978, with a noticeable upward trend after 1992. During the Seventh Five-Year Plan period (1986–1990), the railways' total fixed-asset investment was 69.219 billion yuan. This figure rose to 202.524 billion yuan during the Eighth Five-Year Plan period (1991–1995), and during the Ninth Five-Year Plan period (1996–2000), it had surged to 374.866 billion yuan, more than five times the amount during the Seventh Five-Year Plan period. Notably, in 1998 alone, the fixed-asset investment in railways reached 83.969 billion yuan, exceeding the entire fixed-asset investment in railways during the Seventh Five-Year Plan period.[136]

In the national railway fixed asset investment, railway infrastructure investment constitutes the largest portion. As shown in Fig. 4.1, from 1992 to 2003, the share of railway infrastructure investment in the total railway fixed-asset investment was generally above 50%, reaching close to 70% by 1999. From 1978 to 2003, the increasing trend of total basic railway construction investment was consistent with that of total railway fixed asset investment, with a noticeably sharper increase after 1992. In 1994, railway infrastructure investment increased from 12.613 billion yuan

[135] The Editorial Board of *China's Railway Construction across the Century.* (2003). *China's railway construction across the century (1998–2002)* (p. 54). Beijing: China Railway Publishing House.

[136] Statistics Center of MOR. (2008). *Collection of the historical statistics data of Chinese railways (1949–2006)* (internal resources) (pp. 224–225). Statistics Center of MOR.

Table 4.1 The sources of railway infrastructure investment (Unit: 100 m CNY, %)

Year	Total infrastructure development investment	Investment from MOR	Proportion of investment from MOR	Investment from local governments and enterprises	Proportion of investment from local governments and enterprises
1992	126.13	105.7	83.80	20.43	16.20
1993	279.69	236.38	84.51	43.31	15.49
1994	374.08	330.1	88.24	43.98	11.76
1995	373.74	330.08	88.32	43.66	11.68
1996	375.27	325.60	86.76	49.67	13.24
1997	396.09	348.72	88.04	47.37	11.96
1998	570.50	490.40	85.96	80.10	14.04
1999	570.17	489.05	85.77	81.12	14.23
2000	526.78	456.84	86.72	69.94	13.28
2001	581.39	519.84	89.41	61.55	10.59
2002	623.52	576.49	92.46	47.03	7.54
2003	528.62	490.46	92.78	38.16	7.22

Source Data derived from Table 6-2 of "National Railway Infrastructure Investment" in the *Compilations of the Historical Statistics Data of Chinese Railways* (1949–2006), an internal document, compiled by the Statistics Center of MOR.

in 1992 to 37.408 billion yuan, reached 57.050 billion yuan in 1998, and grew further to 62.352 billion yuan in 2002 (refer to Table 4.1).[137]

Categorized by types of construction, the railway infrastructure investment in new lines during the Eighth Five-Year Plan period (1991–1995) was 71.96 billion yuan, while investment in existing lines reached 43.814 billion yuan. During the Ninth Five-Year Plan period (1996–2000), the investment increased to 119.676 billion yuan for new lines and 109.63 billion yuan for existing lines, representing increases of 1.7 and 2.5 times over the Eighth Five-Year Plan period, respectively.[138] During the Eighth Five-Year Plan period, the construction of the Beijing-Kowloon Railway, with a total length of 2,536 kilometers, saw a total investment of nearly 40 billion yuan. The annual investment in the Railway project accounted for 40% of the total investment in all medium- and large-sized projects across the railway network. It was the largest-scale railway project at the time, with the highest investment and the longest mileage completed in a single phase.[139]

With regard to the sources of railway infrastructure investment (refer to Table 4.1), the investment is from MOR and from the local governments and enterprises. According to data from 1992 to 2003, the majority of the railway

[137] Ibid.: 230–231.

[138] Ibid.

[139] Sun Yongfu. (1996). A feat in the history of China's railway development: Reflections on the construction of Beijing-Kowloon Railway. In MOR of the People's Republic of China and Xinhua News Agency (Ed.), *Key projects of China's railway construction in Eighth Five-year Plan period* (p. 24). Beijing: Economic Daily Press.

infrastructure investment came from MOR, with contributions from local governments and enterprises playing a secondary role. However, after 1992, the investment from local governments and enterprises increased year by year, rising from 2.043 billion yuan in 1992 to 8.01 billion yuan in 1998, an increase of nearly fourfold. Although there was a decline afterward, the total investment remained higher than that in 1992.

As mentioned before, the Beijing-Kowloon Railway is a typical example of joint investments between the central government and local authorities. A key reason for the significant investment in the Beijing-Kowloon Railway in the 1990s was its departure from the traditional model, where MOR was the sole investor. Instead, the railway adopted a model where central government investment played a leading role, supported by local funds and complemented by international capital to fill the gap in construction financing. This approach mobilized the efforts of both the central government and local authorities in railway construction. Although the funds of local governments along the railway line constituted only a small fraction compared to central government investment, they helped resolve many issues and significantly alleviated the financial burden on the state. Similarly, the Lanzhou-Xinjiang Railway, whose construction began in 1992, adopted the same investment model. Of a total investment of 4.5 billion yuan, two-thirds was invested by MOR, and the remaining one-third by the Xinjiang Uygur Autonomous Region and Gansu Province. The Nanning-Kunming Railway, whose construction started in the same year, had a total investment of 6.5 billion yuan and was jointly funded and constructed by MOR and the provinces of Yunnan, Guangxi Zhuang Autonomous Region, and Guizhou Province.[140]

4.3.3 *Significant Achievements in Railway Construction*

From 1992 to 2003, with the continuous advancement of railway reforms, the management and operational mechanisms of the railway system, which was more adaptable to the socialist market economy, gradually broke free from the old planned economy model. Modern enterprise systems began to take shape among railway companies, and their status as market entities started to be established. This allowed for significant development in railway construction.

The first major development was the continuous increase in operational mileage. From 1992 to 2003, the total operational mileage of railways in China (refer to Table 4.2) increased from 58,076 kilometers to 73,002 kilometers. To be more specific, the mileage of state-owned railways increased from 53,565 kilometers to 60,446 kilometers. Joint venture railways saw a more notable increase, rising from 2,738 kilometers in 1995 to 7,738 kilometers in 2003, nearly 1.8 times the previous

[140] Wu Changyuan. (1994). *The important manuscripts on the reform and development of Chinese railways in 1993* (pp. 654–657). Beijing: China Railway Publishing House.

Table 4.2 National railway operational mileage (Unit: kilometer)

Year	State-owned railways	Joint venture railways	Local railways	Total
1992	53,565		4511	58,076
1993	53,802		4800	58,602
1994	53,992		5038	59,030
1995	54,616	2738	5034	62,388
1996	56,678	3044	5210	64,932
1997	57,566	3064	5339	65,969
1998	57,584	3918	4927	66,429
1999	57,923	4696	4776	67,395
2000	58,656	5181	4813	68,650

Source Data derived from Table 6-2 of "National Railway Infrastructure Investment" in the *Compilations of the Historical Statistics Data of Chinese Railways* (1949–2006), an internal document, compiled by the Statistics Center of MOR.

figure. Although the changes in local railways were relatively modest, their mileage increased from 4,511 kilometers in 1992 to 4,818 kilometers in 2003.

Specifically, during the last three years of the Eighth Five-Year Plan period, under the concerted efforts of hundreds of thousands of railway workers, the pace of railway construction was accelerated steadily, investment controls proved highly effective, and the overall quality of engineering works generally improved. On November 16, 1995, track for the Beijing-Kowloon Railway was fully laid; on June 1, 1995, the Baoji-Zhongwei Railway opened to traffic; on June 30, 1995, the Lanzhou-Xinjiang Railway commenced operations; by the end of 1995, the first phase of the Houma-Yueshan Railway and the Zhejiang-Jiangxi Railway went into operation; and on January 21, 1996, Beijing West Railway Station opened for service.

During the Eighth Five-Year Plan period, China railways' track laid totaled 11,181 kilometers, which was 2.2 times that of the Seventh Five-Year Plan period. Of this, 6011 kilometers were new railway tracks, and 5170 kilometers were for double-track line expansion, representing 2.5 times and 1.9 times the figures of the Seventh Five-Year Plan period, respectively. Additionally, 3719 kilometers of new lines, 3847 kilometers of double-track lines, and 2937 kilometers of electric lines were delivered into operation, which were 2.2 times, 1.6 times, and 1.1 times the respective figures of the Seventh Five-Year Plan period. By the end of the Eighth Five-Year Plan period, the operational mileage of the national railway network had exceeded 60,000 kilometers.[141]

During the Ninth Five-Year Plan period, 4,800 kilometers of new railway track were laid, 4,100 kilometers of double-track railways were constructed, and 4,300 kilometers of electric railways were built. By the end of 2000, the operational

[141] Han Shubin. (1997). Seizing the historical opportunity to implement the "two transformations", and striving for the grand goals of railway reform and development during the Ninth Five-Year Plan: Speech at the National Conference of Leading Officers in the Railway Sector. In Huang Sichuan (Ed.), *The Important Manuscripts of China's Railway Reform and Development in 1996* (p. 32). Beijing: China Railway Publishing House.

mileage of the national railway network had reached 68,650 kilometers, an increase of nearly 6,000 kilometers from the end of the Eighth Five-Year Plan period, reaching the highest level among all the five-year-plans in operational mileage since the start of the reform and opening up.[142] In 1998, the turn-of-the-century railway construction campaign began. The Beijing-Zhengzhou Electric Railway, the Jinhua-Wenzhou Railway, the Shimen-Changsha Railway, the Guangtong-Dali Railway, and the Hengfeng-Nanping Railway all commenced operations. Throughout the year, 1,092 kilometers of new railway track were laid, 1,208 kilometers of double-track railways were constructed, 1,105 kilometers of new railway lines, and 590 kilometers of double-track lines were put into operation, along with 988 kilometers of electric railways, achieving initial targets set for the campaign.[143]

In the following year, 1999, the Party Leadership Group of MOR identified it as a crucial year to ensure the smooth implementation of the turn-of-the-century railway construction campaign, maintaining high levels of investment, output, and speed. During the year, many projects, such as the completion of the Second Qiantang River Bridge, the Hangzhou Hub, and the electrification of the Chongqing-Dazhou Railway, were successfully concluded. On August 16, 1999, the Qinhuangdao-Shenyang High-speed Railway, the symbol of railway modernization in our country, officially commenced construction. On August 29, 1999, the Qinling Tunnel I of the Xi'an-Ankang Railway was successfully completed. On December 6, 1999, the Southern Xinjiang Railway was officially opened to traffic. Key projects including the Xiangtang-Longchuan section of the Beijing-Kowloon Railway, the Handan-Jinan Line, the Litang-Zhanjiang Railway, the Zhuzhou-Liupanshui Railway, the Baotou-Lanzhou Railway, and the Chengdu-Kunming Railway—comprising new lines, double-track sections, and electric railways—were also put into full or partial operation, promptly yielding the benefits of these investments.[144]

At the turn of the twenty-first century, in 2000, construction of the Xi'an-Nanjing Railway and the Chongqing-Huaihua Railway commenced. On June 29, 2001, construction of the Qinghai-Xizang Railway was launched, with an investment exceeding 1 billion yuan that year. Several railway projects in western China were successively completed, including the Xi'an-Ankang Railway, the Shenmu-Yan'an Railway, and the Zhuzhou-Liupanshui Railway, generating new operational capabilities. Tracks for the Dazhou-Wanzhou Railway, the Neijiang-Kunming Railway, and the Liupanshui-Baiguo Railway were all fully laid. By the end of 2001, the total operational mileage of Chinese railways exceeded 70,000 kilometers, reaching 70,058 kilometers. On May 12, 2002, the Neijiang-Kunming Railway was fully operational, and on November 1, the Shuozhou-Huanghua Railway, a major project

[142] Fu Zhihuan (2001). Creating further glory when marching in the new century. in Peng Kaizhou (Ed.), *The important manuscripts on the reform and development of Chinese railways in 2001* (p. 2). Beijing: China Railway Publishing House.

[143] The Editorial Board of *China's Railway Construction across the Century.* (2003). *China's railway construction across the century (1998–2002)* (p. 54). Beijing: China Railway Publishing House.

[144] Liu Jiangtao (2000, September 20). Century of masterstrokes: One of the overviews of railway achievements in Ninth Five-Year Plan period. *People's Railway Daily.*

at the turn of the century, was completed and opened to traffic a year ahead of schedule. Key projects of the Qinghai-Xizang Railway, including the Kunlun Mountain Tunnel and Fenghuoshan Tunnel, were completed, and the track-laying task for the Golmud-Wangkun section was finished ahead of schedule, with a cumulative investment of 6.5 billion yuan. Construction of the Wanzhou Yangtze River Bridge on the Yichang-Wanzhou Railway also commenced. The year of 2003 was particularly promising. On June 18, tracks for the Xi'an-Hefei section of the Xi'an South Suburban Railway were laid ahead of schedule. The 2nd line of the Baoji-Lanzhou Railway, which was the "bottleneck section" of the Lanzhou-Lianyungang Railway, was put into operation in June. Construction of the Wushaoling Tunnel commenced. It was the control engineering project of the Lanzhou-Wuwei Railway as well as the longest tunnel in China at 20.25 kilometers.[145]

Between 1992 and 2003, China's railway network not only expanded in scale, but also became more rationally organized. During this period, the construction of railways in western China saw rapid progress. By the end of 2001, seven projects had been completed and put into operation, including the Baoji-Chengdu Railway, the Baotou-Lanzhou Railway, the Yangpingguan-Ankang Line, the Litang-Zhanjiang Railway, the capacity expansion of the Qinghai-Xizang Railway from Xining to Golmud, the Southern Xinjiang Railway, and the Guangtong-Dali Railway. The total operational mileage in the 12 provinces, autonomous regions, and municipalities in western China reached 25,900 kilometers, accounting for 37% of the national total.[146] By the end of 2003, this figure in western China had risen to 26,660 kilometers.[147] In particular, commencement of the construction of the Qinghai-Xizang Railway ended the isolation of the snowy plateau from the railway network, playing a significant role in promoting ethnic unity, safeguarding national unity, and facilitating socio-economic progress in Xizang.

With the smooth advancement of large-scale railway construction, there has been a significant increase in railway operation mileage. In addition, widespread raising of train speeds had directly resulted in a substantial increase in rail transportation capacity. As a result, the "bottleneck" effect caused by the previous lag in railway development, which had hindered the national economy growth, was largely alleviated. China's railway have been shifted from a constrained model to one that actively supports and adapts to economic growth. Especially after the outbreak of the Asian financial crisis in 1997, the launch of a major railway construction campaign played a significant role in supporting the national economic development. As Fu Zhihuan, then Minister of Railways, pointed out, "accelerating railway construction was of special significance under the economic circumstances at the time". Specifically, "railway construction plays a vital role in promoting land development, ensuring

[145] Lei Fengxing (2004, March 11). Large-scale development of China's western region: railway construction inventory. *People's Railway Daily.*

[146] The Editorial Board of *Chinese Railways at the Turn of the Century*. (2002) *Chinese railways at the turn of the century* (pp. 77–78). Beijing: China Railway Publishing House.

[147] Lei Fengxing (2004, March 11). Large-scale development of China's western region: railway construction inventory. *People's Railway Daily.*

coordinated regional economic development, increasing employment, and fostering a unified national market. Additionally, railway construction involves substantial investment and long industrial chains, thus increasing investment in railway construction can drive the development of a large number of related industries such as steel, construction materials and machinery, electronic equipment, and energy sectors. Accelerating railway construction is crucial for maintaining current economic growth rates, mitigating the impact of the Southeast Asian financial crisis, stimulating domestic demand, boosting economic growth, and enhancing the sustainability of economic development".[148]

According to statistical analyses by relevant departments, from 1998 to 2001, railway construction consumed 24.44 million tons of steel, 29.71 million tons of cement, 11.45 million tons of coal, 4.22 million tons of diesel, 3.49 million tons of gasoline, 3.53 million cubic meters of timber, and 94.8 billion kilowatt-hours of electricity. During these four years, railway construction created an average of 600,000 jobs per year in the construction industry and 2.89 million jobs per year in other sectors. Such a vast demand for materials and labor played a significant role in driving national economic growth.[149]

[148] Fu Zhihuan. (2004). Urgent mobilization, organization for fighting and quickly setting off a new climax of railway construction. In Fu Zhihuan (Ed.), *Exploration and practice of railway reform and development in China* (p. 406). Beijing: China Railway Publishing House.

[149] Wang Zheng, Lei Fengxing, & Liu Jiangtao. (2002, June 13). Railway construction drives economic growth. *People's Daily*.

Chapter 5
Great Stride Forward: Deepening Reform Under Systematic Planning (2004–2008)

In the Rising Period of Industrialization from 1992 to 2003, China's GDP grew at an average annual rate of 9.5%. After the entry of WTO in 2001 in particular, the "troika" of export, investment, and consumption significantly boosted China's social productivity, and the process of industrialization assimilated into that of the world. At the turn of the century, China achieved the first two strategic goals for modernization, namely the all-round development of the economy and society and the rise of people's living standards to a moderately prosperous life on the whole. The momentum was carried on into the twenty-first century, and railway modernization in China also entered a new era with large-scale construction and market-oriented reform in full swing after 2004.

5.1 System Planning to Remove "Bottlenecks"

Faced with a barrage of pressures and demands in the last few decades, China's railways focused more on solving imminent problems and less on making systematic plans for the future development of the railway network. The first move to reshape the landscape is to formulate mid- and long-term plans for the national network.

"Developed to meet the demands of passenger and freight transport under certain historical conditions, the railway network is an interconnected system consisting of trunk railways, branch lines, connecting lines, railway stations, and railway junctions within a certain spatial range (in a nation or between regions or nations)".[1] "Planners generally determine the sites for stations first according to the transportation needs and natural conditions, and position them in transportation centers such as major industrial cities, political and economic centers, mineral resources bases, energy supply bases, as well as coastal ports. Railway lines come the next to connect the stations and later expand gradually to other regions. They are developed

[1] Qu Guangyi. (2011). *Environmental impact assessment of railway network planning: Methodology and its application* (p. 16). Doctoral dissertation, Harbin Institute of Technology, Harbin, Heilongjiang.

X. Zhang, *China's Railway Expansion in 40 Years of Economic Reform and Opening Up*, Understanding China,
https://doi.org/10.1007/978-981-95-0333-9_5

into passenger and freight transport channels in different regions to formulate a network eventually".[2]

A major contributor to the formation and development of railway networks in modern times is industrialization. With the national economic and social progress expressed in national or regional productivity elevation, industrial upgrading, trade acceleration, and improvement of people's living standards, the railway network has contributed its due and reasonable share together with other mode of transportation. In addition, "transportation is one of the key features of industrialization, as well as an economic process that occurs alongside it".[3] Modern Western industrialization is the cradle of railways, and railway networks exemplify a common trend across different countries.

Transportation is the very foundation of the national economic development. Throughout the world's industrialization process, all countries have taken the railway as the lead, putting prioritized investment in and great efforts to it, and most of the industrialized countries spent six or seven decades to build a railway network. This shows that industrialization after the railway network represents the only road to progress and prosperity.[4] Compared to these countries, China's railway network construction did not ever outpace its industrialization, and the railway was once the weak link or even bottleneck of China's industrialization. China's railway network has its distinct features during its formation: the construction and development of China's railway network since the reform and opening up is synchronized with the development of the national economy and society; the sufficient transportation demands for the modernization of the country is secured through such systemic reforms as tapping the potential and expanding capacity, scaling up, improving the structure, and optimizing the railway layout; great achievements for domestic economic reform and opening up to the outside world, such as the establishment of a comprehensive material production system, the rising international status, the full integration into the world economic system, the rapid and steady economic growth, and the significant improvement in people's living standards, have been a backbone continuously promoting the shaping of the railway network.

---

[2] Ibid.

[3] Rong Chaohe. (2001). From transport development to sustainable transport. *China Railway Science*, (3).

[4] Wang Zhuting, & You Juxiang. (1984). On the construction of China's railway network. *Development of Railway Science and Technology*, (7).

5.1.1 *The Introduction of* Mid- and Long-Term Railway Network Plan

China's railway network scale was relatively small, therefore promoting large-scale railway construction and building a well-developed modern railway network in an efficient and fast way has been the relentless pursuit of several generations of railway-related workers. Before the founding of PRC, less than 22,000 km of railways were open to traffic, and 60% of them were in northeastern and north China. Since the founding of PRC, especially after the reform and opening up, the layout of China's railway network has undergone great changes. After the construction of a batch of new railways, the framework of China's railway network was basically in place, addressing the previous issues of "limited mileage, imbalanced layout, and low density", and China's railway modernization was also groping forward in the spiraling trend.

During the period of the Seventh Five-Year Plan (1986–1990), the primary focus was on tapping the potential and expanding capacity, with the construction of new lines as the secondary task to break the emerging bottleneck against railway development, which was, in contrast, intensified after introducing the Contract Responsibility System. During the Eighth Five-Year Plan period (1991–1995), tables were turned through market-oriented railway reforms from which railway construction was flourishing. During the Ninth Five-Year Plan period (1996–2000), the bottlenecks in the way of railway development were overcome through five-year large-scale construction across the century, as a result, a catch-up development was achieved at the turn of the century. "From 1998 to 2001, an average amount of over 54 billion yuan was invested in railway construction per year. At the end of 2001, Chinese railway mileage in service, double-track railway mileage, and electrified railway mileage all ranked first in Asia, among which its over 70,000 kilometers of railway in operation ranked third in the world. With these advancements, the railway system's capacity increased significantly with more railways and upgraded facilities. Furthermore, the railway network was further expanded, its layout improved, constrained sections in transportation reduced, and the heavy burden on major transportation passages was relieved. The long-existing bottlenecks that had hindered the national economy were broken, thus improving the transportation capacity of railways to better align with the social and economic development".[5] Since then, the infrastructure of China's railway network has been improved. Through railway reforms to tackle bottlenecks, systematic plans targeting the upgrading of the Chinese railway network have been introduced, providing the essential foundations to win the battle down to the ground.

Unfortunately, "the growth of national economy always outpaces that of railways, so the construction of Chinese railways has always been driven by the

[5] Fu Zhihuan. (2002). The development of Chinese railways across the century. *Chinese Railways*, (11).

mushrooming demands of transportation".[6] "Since China's reform and opening up, its railway construction has always been haunted by continuous bottlenecks. Each time a bottleneck seemed to be easing, a new one would emerge before the previous one had been fully resolved".[7] In 2003 for instance, Chinese railway construction was poised for growth in that year but encountered some bottlenecks in different aspects once again. On the railway transport of passengers, the tickets were out-of-demand, especially on special holidays like the Spring Festival travel season and Golden Holiday Week in May. The demand for tickets was reduced by an outbreak of Severe Acute Respiratory Syndrome (SARS). "the number of departure passengers across China in Spring Festival travel rush still reached 1.3 billion people, with a year-on-year growth of about 1.66 million people. MOR formulated three-tiered transportation plans in response to the normal, peak, and emergency passenger flows respectively, arranging a total of 210 pairs of temporary through trains for passengers, an increase of 43 pairs year-on-year".[8]

On the other hand, trains for railway freight were also in short supply, leading to long-term overloading on busy railways. "During the second half of December 2003, an additional 10 million tons of coals were transported to different regions within 20 days, with about 50,000 carloads of coals every day, effectively alleviating the shortage of coal supply in China".[9] "Since the second half of 2003, power rationing became a special phenomenon occurring in some regions as basic energy supplies such as coal, electricity and petroleum ran short in many Chinese regions, representing both China's rapid economy and the need to improve the infrastructure supporting China's economic growth. Railway construction once again suffered from bottlenecks while China ran short of energy supplies".[10] "Despite steady and rapid economic growth, the outstanding problems in economic activities have yet to be solved from the root, the imbalance between the growing demand and undersupply of coal, electricity, and petroleum in particular. The railway is an important part of linking supply to demand, but its overall traffic capacity remains insufficient to meet the growing demands, and for some time to come, the traffic capacity of some railways for certain periods in some regions is getting worse. To lift the burden on transportation is of great importance to boost economic growth, especially to strengthen macro-control".[11]

---

[6] Yan Pengfei. (2004). Reform on China's railroads investment system: A late choice for China's railroads. *Fortune World*, (10).

[7] Ibid.

[8] Statistics Center of MOR. (2003). Railway statistics bulletin of MOR of the People's Republic of China in 2003.

[9] Ibid.

[10] Lu Cairong. (2004, March 5). The first adopted mid-and long-term railway network plan. *Guangming Daily.*

[11] Huang Ju. (2006). Seize the opportunity to accelerate the pace of construction and strive to achieve leap-forward development of railways. In Wang Zhiguo (Ed.), *The important manuscripts on the leaping development of China's railways in 2005* (p. 5). Beijing: China Railway Publishing House.

Based on the status quo in China's economic and social development, the railway sector, taking the relatively low density and inadequate coverage of the railway network into consideration, has managed to meet the actual demands in railway passenger and freight transportation by renovating existing railways to tap their potentials and improve their efficiency, and investing significant manpower and resources to improve the transport capacity. By doing so, Chinese railways, with less than one-tenth of the world's total rail mileage, have handled one-fourth of the world's traffic volume, becoming one of the busiest railway systems in the world. This is not a normal state of economic cooperation, but a result of the imbalance between the slow railway development and the fast economic growth, making railway development always lag behind economic growth.

For instance, "the Datong-Qinhuangdao Railway was open to traffic in 1992 and carried the designed freight volume of 100 million tons by 2002. Since 2004, continuous technical reforms to expand its capacity have been implemented, and its annual freight volume reached 400 million tons in 2010 and 450 million tons in 2014, 4.5 times the original design capacity. But at the same time, the over-loaded operation of the Railway was to blame for badly damaged facilities and personnel fatigue, especially among locomotive crews, who had been enduring excessive workloads for a long time. A large freight volume does not necessarily equate to high transport efficiency, and such overloaded operation cannot last long".[12]

To meet the growing demands of passenger and freight transportation after the entry into the twenty-first century, the railways had to step up the efforts to build a modern railway network, which should also align with the goals of socialist modernization and, to some extent, run ahead of current needs. Only by transforming the development of Chinese railways from a passive, burdened approach to a proactive one under "scientific", "systematic", and "advanced planning", can the recurring bottlenecks be eradicated. This transformation is key to ensuring that railway reforms in China continue to advance.

The Mid- and Long-Term Railway Network Plan (hereinafter referred to as the *Plan in 2004*) was approved in principle at the Executive Meeting of the State Council convened on 7 January 2004. "The meeting recognized that railways, with their large transportation capacity and low cost, play a major role in transportation. The formulation of such a *Plan* is of great importance to accelerate the development of the railway, boost the national economy, and build a moderately prosperous society in all respects".[13] This is the first *Mid- and Long-Term Railway Network Plan* in PRC's history, and also the first industry plan the State Council approved, marking the beginning of China's railway development.

---

[12] Chen Ronghu. Supply-side reform forces the transformation and development of railways. *People's Railway Network*. http://www.peoplerail.com/rail/show-466-258556-1.html.

[13] (2004, January 8). Premier Wen jiabao presides over an executive meeting of the state council to discuss and approve in principle the *Mid- and long-term railway network plan* and *Decision of the state council on further strengthening work safety. People's Daily.*

"Meanwhile, the mid-and long-term plans for such important industries as highway transport are also formulated".[14] At that time, China's transportation industry was going through a diversified development with the rise of other transportation networks like highway systems, civil aviation, water transportation, etc., breaking the monopoly of railway transportation. Under such circumstances, compared to other plans, the plan for railway network construction, the first one approved by the State Council, carried a special meaning. At the same time, it also underscores that the railway is still the most competitive leading industry in transportation.

Furthermore, the proposal for "leap-forward development of railways" is linked to the introduction of *the Mid- and Long-Term Railway Network Plan.*

The railway serves as a network-based organization (NBO) that plays a significant role in the economy, national defense, policy, and society across different countries and regions. Different from the surplus capacity of railway transportation in foreign countries, especially in developed countries, Chinese railways have long been haunted by the shortage of traffic capacity, which is a fundamental mismatch between the scale of its network and overall population, national-territorial area, and social and economic development. After China's official accession to the WTO on 11 December 2001, China entered a period of economic boom, featuring stable growth in passenger transportation and growing demands for freight transportation, leading to persistent supply shortages in both sectors. According to statistics, from 1995 to 2006, the volume of railway passenger traffic increased by 70% (in passenger-kilometer), and the railway freight volume by 60% (in tonne-kilometer). However, China's railway length accounted for only 6% of the world's total in 2006.

At the same time, at the beginning of the twenty-first century, the developed countries took the lead in railway progress, with continuous breakthroughs in high-speed railway passenger transportation, container transport, logistics flow, and information technology. Railways have increasingly played a backbone role in urban rail transit. However, China's railway construction showed clear gaps in comparison with global railways in terms of railway network density, railway quality, transport equipment, communication and signaling technologies, modernization in operations and management, as well as technology and equipment for safety. At that time, China didn't even have a railway connected to the airport and faced multiple bottlenecks in the development of seamless multimodal transportation.

The gaps between the supply and demand of railways at home, and between China's railway progress and that of the world all indicated that China was entering a new historical stage in building a moderately prosperous society in all respects, which requires a greater scale and higher quality of railway improvement in order to keep pace with the socialist modernization. The new overall plan for railway construction in the 21 century has been formulated in *the Outline of the Tenth Five-Year Plan for National Economic and Social Development of the People's Republic*

[14] Lu Cairong. (2004, March 5). The first adopted mid-and long-term railway network plan. *Guangming Daily.*

of China, which pointed out the need for a safe, convenient modern integrated transportation system.

"The railway departments will invest 270 billion yuan, an increase of 24 billion yuan than that in the *Ninth Five-Year Plan* period, to build a railway network centered on eight vertical and eight horizontal main transport corridors. By 2005, China's total railway operating mileage would reach about 75,000 kilometers, an increase of 7,000 kilometers over 2000, exceeding the original goal of 6,000 kilometers set in the *Ninth Five-Year Plan*".[15] Actually, after five rounds of speed increase of China's railways in 1997, 1998, 2000, 2001, and 2004, especially the fifth large-scale speed increase of China's railways and adjustment of the passenger train operation map at 0:00 on April 18, 2004, "the new map increased the railway passenger transport capacity by 18.5%, and the freight transport capacity by about 15%.

Despite these progresses, China's railway network scale could not match the demands of the country's population, national-territorial area, and economic development, and its technologies, equipment, and operating speed still lagged far behind the world-leading ones. The capacity constraints on the main railway lines also remained largely unresolved. As the railway departments pointed out, overcoming these bottlenecks requires a leap-forward development in railway construction".[16] To adapt to national economic development, MOR held a seminar on the leap-forward development of railways from June 28 to July 1, 2003, and officially launched the leap-forward development of railways strategy to rapidly expand transportation capacity and accelerate the upgrade of technologies and equipment to completely tackle the bottleneck in railway construction head-on, and take the construction of an advanced railway network as one of the core tasks.

The *Mid- and Long-Term Railway Network Plan* serves as a response to the "leap-forward development of railways strategy".[17] "The *Plan* is of great importance to the leap-forward development of China's railway, which aims at boosting the steady, fast growth of China's national economy and achieving the goal of building a moderately prosperous society in all respects".[18] To make the leap-forward development of China's railway into reality, the railways, based on China's reality, were committed to learning from other countries' successful experiences in railway construction, and finding its forward-looking path aligned with the cutting-edge trends to step up the construction of advanced railway network and support China's industrialization.

To satisfy the needs of mid- and long-term development of China's railway, the main battlefield for national economic growth, the systemic overall plans for

[15] Wen Jie. (2001, July 4). During the *Tenth Five-Year Plan* period, China will build an "Eight Vertical and Eight Horizontal" railway network. *Beijing Sci-Technology Report*.

[16] Wang Zheng. (2004, May 27). How does the Chinese railway realize leap-forward development? *People's Daily*.

[17] Lu Cairong. (2004, March 5). The first adopted mid-and long-term railway network plan. *Guangming Daily*.

[18] Ibid.

mid- and long-term railway construction and technological innovations were devised, showcasing that China would adopt more reasonable reforms to eradicate the "bottlenecks" in the way of railway development. This Plan pointed out a clear direction for railway development, including speeding up the construction of new railways at high standards, expanding the railway network scale, gradually realizing the synergistic effect of the scale, structure, and quality of the railway network, and drawing up an effective blueprint for the leapfrog development of China's railway. The *Mid- and Long-Term Railway Network Plan* is not only an important political document to break bottlenecks, further reforms, and champion comprehensive, coordinated, and sustainable economic and social development, but also a guideline for China to narrow the gap in railway development with developed countries and keep pace with the world's railway progress.

Furthermore, a new development philosophy that puts people first and aims at comprehensive, coordinated, and sustainable development was proposed in the *Decision of the Communist Party of China (CPC) Central Committee on Some Major Issues Concerning the Improvement of the System of Socialist Market Economy* adopted at the Third Plenary Session of the Sixteenth Central Committee of the CPC on October 14, 2003. In the Central Economic Work Conference held at the end of 2003, the need to establish and implement the Scientific Outlook on Development was reiterated.

As China's industrialization accelerates, the introduction and implementation of the Scientific Outlook on Development in Railways have become a vital guide to accelerate China's access to the rapid railway modernization period with "scientific", "systematic", and "advanced" planning. In comparison to the previous railway network planning largely driven by the leadership decisions, local initiatives, and the urgent development needs of specific regions, the current one better suits the demand of the mid- and long-term development of the national economy, as well as the railway development laws and characteristics. It provides a more systematic and scientific guide for the overall layout and implementation of the leapfrog development of railways in the new period more systematically and scientifically, which shows how the Scientific Outlook on Development is put into action in China's railway.

China's railway construction has entered a golden period under the guidance of the *Mid- and Long-Term Railway Network Plan.*

The *Mid- and Long-Term Railway Network Plan* is an overall plan to improve the scale, structure, and quality of China's railway network by 2020. Its goal is to optimize the railway transportation capacity to satisfy the demands of building a moderately prosperous society with an advanced, complete railway network. According to the Plan, the national railway operating mileage would reach 100,000 kilometers by 2020, half of which consists of double-track railways and electric railways. To be specific, the focuses of the *Plan* are as follows:

Firstly, the railway network will adopt dedicated lines for the passenger train and freight train to improve the overall efficiency. Before 2004, except for the Qinhuangdao-Shenyang Passenger Dedicated Railway, all railways in China were shared by both passenger and freight trains, making it difficult to balance the need for increasing the speed of passenger trains with the requirements for transporting heavier goods for freight trains. This arrangement failed to meet the growing transportation needs, especially for the passenger service. Three requirements were proposed in *the Mid- and Long-Term Railway Network Plan*, namely, arranging dedicated lines for passenger and freight trains, building the "four vertical and four horizontal" passenger dedicated lines with advanced technologies at high standards, and developing three intercity transit systems in Bohai, Yangtze River Delta, and Pearl River Delta respectively to support the intensive passenger traffic between developed cities such as provincial capitals as well as large and medium-sized cities. The total mileage of passenger dedicated lines would reach over 12,000 kilometers, with the speed reaching 200 kilometers per hour and above.

The mixed traffic railway line or passenger-freight shared line, refers to the railways that the passenger trains and freight trains share the line with a design speed of 200 kilometers per hour or less. In contrast, transporting passengers and goods in separate railways has been adopted as one of the multiple-track railway[19] transport modes. "Germany, France, the United Kingdom, and other countries have begun building double-track or multi-track railways in busy corridors as early as the middle and late 20th century to separate passenger and freight traffic",[20] with main corridors in many developed countries now having four or more tracks for separate operations. The separation of passenger and freight trains in China, to a certain extent, reflects that the scale of the railway network is growing and the transportation capacity is improving as well at the same time. This shows that the railway network is transforming into a three-dimensional one balancing transportation efficiency, profit, and quality.

The second is to construct the linking corridor and improve the layout of the railway network. In 2002, while the railway mileage reached 46,000 kilometers, and about 169 kilometers of railways operated per 10,000 square kilometers in eastern and middle China, those were only 26,000 kilometers and 38 kilometers in every 10,000 square kilometers respectively in western China. With the development of the western region in China, the problems regarding railway density, transportation capacity, and connectivity to the eastern and middle China were more obvious.

The *Mid and Long-term Railway Network Plan* proposed to build approximately 16,000 kilometers of new railways, including the construction of a batch of new railways in western China, to speed up the construction of the Qinghai-Xizang Railway construction, to accelerate the development of the western-eastern China corridors, and to create the railway network framework connecting northwest China to East China and North China, and southwest China to South Central China. Necessary connecting lines would be built in Eastern and Mid-China to better meet the transportation demands; and new and reconstructed international railway

[19] In order to meet the requirements of quality and quantity of transportation brought about by the increase of passenger and freight traffic, the railways with more than three lines built in the same direction of operation or route direction are called multi-line railroads.

[20] Xu Limin. (2012). Research on the development of railway passenger and freight line transportation in China. *Railway Transportation and Economy*, (11).

corridors would be established in the northwest and southwest China, to facilitate the land connections with the outside world. The shaping of the railway network includes both the transformation of the existing railways and the construction of new ones, and both should be taken into account and developed in an integrated and coordinated manner.

Huang Ju, then vice-premier of the State Council, emphasized at the national railway working conference held on January 12, 2005, that "Speeding up the construction of new lines is important to fundamentally alleviate the tension in the transportation of coal, electricity, and petroleum, and promote the development of the national economy".[21] An official of MOR also pointed out that after five consecutive rounds of major speed increases, the potential of China's existing railways had been fully tapped, and the railway system would soon face new "bottlenecks" if the operating mileage of railways could not be increased. At the beginning of the railway construction campaign during the *Eighth Five-Year Plan* Period, some experts had estimated that China would need at least 200,000 kilometers of railways to meet the demands of social and economic development".[22] By the end of 2003, although the operating mileage of China's railways had reached 73,002 kilometers,[23] it was far from the estimated requirement of 200,000.

The third is to enhance the transport capacity of existing railways while emphasizing coordinated development with other modes of transportation. The capacity of existing main trunk lines should be upgraded through technological transformation, expansion of transportation capacity, construction of double-track lines, railway electrification, etc. A total of 13,000 kilometers of double-track lines and 16,000 kilometers of electric railways were planned. Based on the practical coal transport routes, the capacity of coal transport channels shall reach 1.8 billion tons. To improve the coordinated development of hubs and railways, the railway hubs should be arranged systematically for better integration with other modes of transportation, and such transport support facilities as locomotive and vehicle maintenance bases should be strengthened accordingly.

The fourth is to promote technological innovation and improve the localization of railway equipment. At the beginning of the twenty-first century, the level of railway technology and equipment in China was generally equivalent to that of developed countries in the 1980s, especially China's high-speed electric multiple unit (EMU) technology was only in the research and development stage. In addition to its focus on increasing speed and enhancing heavy-haul railways, the *Mid- and Long-term Railway Network Plan* also highlighted that the improvement of the

---

[21] Huang Ju. (2006). Seize the opportunity to accelerate the pace of construction and strive to achieve leap-forward development of railways. In Wang Zhiguo (Ed.), *The important manuscripts on the leaping development of China's railways in 2005* (p. 5). Beijing: China Railway Publishing House.

[22] Lu Cairong. (2004, March 5). The first adopted mid-and long-term railway network plan. *Guangming Daily.*

[23] Statistics Center of MOR. (2003). Railway statistics bulletin of MOR of the People's Republic of China in 2003.

localization level of equipment should be taken as its main direction, specifying that locomotives, vehicles and engines with a speed of more than 200 kilometers per hour should be localized as soon as possible. Besides, the Plan also stipulated the need to improve the design and manufacturing capabilities for heavy-haul trains, as well as to advance the technologies in areas such as railways, bridges and tunnels, traction power supply, communication signals, and railway informatization. Thus, a new round of large-scale railway construction started.

"In 2005, over 500 billion yuan was invested in a large number of railway projects following the approval of feasibility studies. Among these, 23 key railway projects including the construction of the Dali-Lijiang Railway and the Anhui Section of the Tongling-Jiujiang Railway, started successively, and 13 key railway construction projects were completed, contributing to the growth of network with 1,085 kilometers of new railways, 662 kilometers of double-track lines and 562 kilometers of electric railways".[24] In 2007, "the national railway fixed asset investment in the whole year had reached 252.07 billion yuan, 43.23 billion yuan more than the previous year. The railway operating mileage had reached 78,000 kilometers, an increase of nearly 1,000 kilometers over 2006. The sixth round of widespread speed increase of the railways, marked by the operation of EMUs, was implemented, extending the total mileage of railway lines with a speed of 200 kilometers per hour and above to 6,227 kilometers, and those with a speed of 250 kilometers per hour reached 1,019 kilometers".[25]

Local governments at all levels also rendered their strong support to the Plan. For instance, "MOR has adopted the cooperation strategies of both going out and inviting in, and signed strategic cooperation agreements on railway construction with more than 20 provinces, municipalities and autonomous regions".[26] Many regions, including Guangxi Province, Zhejiang Province, and Northeast China, broke new ground in furthering railway construction through various cooperation modes with railway departments and local governments. In such cooperation modes, joint investments were made, with local governments assuming responsibility for land requisition and related costs, as well as promoting domestic and foreign investment. For instance, in the construction of the Harbin-Dalian Passenger Dedicated Line, more than 50% of the funds were invested by MOR and the provincial governments of the three northeastern provinces, of which three provinces became shareholders through land requisition and demolition, and the remaining funds were raised through the investment from domestic and foreign investors, leading to the establishment of the Harbin Railway Passenger Dedicated Line Co., Ltd.

The four years (2004–2008) witnessed the golden period of Chinese railway progress, during which the construction of high-speed railways, passenger

---

[24] (2005, January 8). One year after the implementation of the *Mid- and long-term railway network plan*, 23 key railway projects have been started. *PLA Daily*.

[25] Statistics Center of MOR. (2007). Railway statistics bulletin of MOR of the People's Republic of China in 2007.

[26] Wang Zhiguo. (2006). *The important manuscripts on the leaping development of China's railways in 2005* (p. 20). Beijing: China Railway Publishing House.

dedicated lines, intercity railways, coal transport corridors, interregional trunk railways, and the railways in western China gradually commenced. "A total of more than 2 trillion yuan was invested in the approved projects, among which about 809 billion yuan was invested in infrastructure development, three times that of the previous five years. The investment from the local governments and enterprises (140 billion yuan) was 5.2 times that of the previous five years; and the investment in the railways in 2008 alone amounted to over 337 billion yuan, 1.1 times the total in infrastructure construction during the *Tenth Five-Year Plan* period".[27] China's railway network extended rapidly, showcasing the flourishing development of China's economy and society to the world.

On July 1, 2006, the Qinghai-Xizang Railway, the world's longest railway in the plateau region with the highest altitude, was open to traffic. On December 27, 2007, the annual traffic volume of the Datong-Qinhuangdao Railway, China's first heavy-haul railway, exceeded 300 million tons, breaking the theoretical limit in this regard, that is, the traffic volume of a single heavy-haul railway in the world could handle no more than 200 million tons. On August 1, 2008, Beijing-Tianjin Intercity Railway, an important part of the intercity rail transit network in the Bohai Rim Region, was open to traffic with its maximum speed reaching 350 kilometers per hour before the Beijing 2008 Olympic Games. In addition, major projects such as the construction of the Beijing-Shanghai High-Speed Railway, the Beijing-Guangzhou Passenger Dedicated Line, the Lanzhou-Chongqing Railway, and the Guiyang-Guangzhou High-Speed Railway commenced. "By the end of 2008, the national railway mileage in operation reached 80,000 kilometers, with passenger volume exceeding 1.46 billion and freight volume surpassing 3.3 billion tons, an increase of 8.4% and 8.2% over 2003 respectively".[28]

It can be said that "Chinese railway construction embraced the golden age from 2004 to 2008, with the largest scale, the highest standards, and the fastest growth—a feat unparalleled in both history of Chinese railway construction and the world's railway development record".[29]

During this period, as the large-scale railway construction progressed, breakthroughs of the Chinese railway were made in many fields, including construction, operations, technology innovations, and management mechanisms. The independent innovation capabilities of the Chinese railways have been promoted through the "Chinese model" of original innovation, integrated innovation, and the introduction, assimilation, and re-innovation of foreign technologies.

---

[27] Transcript of press conference on the adjustment of *Mid- and long-term railway network plan*. www.gov.cn. http://www.gov.cn/wszb/zhibo286/wzsl.htm.

[28] Statistics Center of MOR. (2008). Railway statistics bulletin of MOR of the People's Republic of China in 2008.

[29] Gao Lipeng. (2008). The development of the railway construction during the four years after the implementation of the *Mid- and long-term railway network plan. Railway Freight Transport,* (2).

5.1.2 Adjustment of the Mid-and Long-Term Railway Plan

By 2006, the bottlenecks of the transport capacity and the quality of the services failed to be fully removed, and the discrepancy between supply and demand still hindered China's progress. Under such circumstances, the requirement for accelerating the railway development was proposed in the *Outline of the Eleventh Five-Year Plan for National Economic and Social Development of the People's Republic of China*, with the construction of passengers dedicated lines, intercity rail transit, and coal transport corridors as its focuses to boost the shaping of an express passenger transport network and a coal transport network. In the same year, MOR, in cooperation with the National Development and Reform Commission, started to adjust *the Mid- and Long-term Railway Plan* formulated in 2004 based on China's specific conditions after borrowing successful experiences of other countries and gathering opinions and suggestions from experts, scholars, relevant departments, and local governments.

In November 2007, the *Mid- and Long-Term Comprehensive Railway Network Plan* (FGJY [2007] No.3045) was approved by the State Council. According to it, by 2000 the national railway operational mileage would reach 120,000 kilometers, among which the double-track railways should account for 50%, and the electrified railways 60% respectively of the total. To achieve the goal, MOR expedited the process of research-feasibility study-demonstration-approval of the Plan. On October 31, 2008, the National Development and Reform Commission approved *the Adjustment of Mid- and Long-Term Railway Network Plan in 2008* (hereinafter referred to as *the Plan in 2008*).

The Scientific Outlook on Development calls for comprehensive, balanced, and sustainable development in economy, politics, and natural environment. Following such a guideline, the state implemented a strategy for coordinated regional development to boost industrialization, marketization, urbanization, and internationalization, and put forward new and higher requirements for the sustainable development of the key areas for the national economy. In this context, the *Plan in 2004* could not satisfy the new requirements for railway development anymore. Besides, the external environment and development conditions of the railway network had also turned rapidly and in a positive direction.

The first change is the major adjustments of the National Macroeconomic Policies. The construction of railway itself involves many industries, so it can drive the development of them. It can stimulate the progress of the national economy, for instance, on the one hand, the investment in the construction of the railway itself can increase China's GDP and drive economic development; on the other hand, the completion and operation of railways helped improve transport capacity and the transport conditions. This results in direct and visible economic benefits for railway users through reduced freight costs, increased efficiency, and fewer safety accidents. At the same time, the improvement of transportation channels and the surrounding comprehensive environment will also continually stimulate the development of the relevant regions and industries, which will play a more significant role in driving long-lasting economic progress.[30]

In 2008, the worldwide financial crisis slowed down the growth of the world economy significantly. China's real economy such as export trade, steel, real estate, and automobiles were greatly affected. To recover from the crisis, major macroeconomic policies were adjusted accordingly. For instance, China would continue to expand the domestic demand, stay committed to forceful and realistic steps targeting specific areas as required, and introduce 10 measures to accelerate the construction of such major infrastructures as railways, highways, and airports and the post-disaster reconstruction of earthquake-stricken areas, and to promote stable and rapid economic growth. By the end of 2010, about 4 trillion yuan would be invested in 10 major projects corresponding to the 10 major measures. On the one hand, by implementing these national policies, the adjustment brought a large amount of new investment to the railway, requiring the railway to speed up the reform, adjust the development plans in stages, and take the initiative to "digest" the new investment with new output. On the other hand, to give full play to the role of the industry in the context of China's macroeconomic policy adjustment, accelerating the railway construction and adjusting railway planning promptly would increase investment in fixed assets, expand domestic demand, and ensure stable and rapid economic growth, strongly backing the rapid development of China's economy and society against the impact of the crisis.

The second change is China's urbanization. Urbanization includes the transformation of people's occupations, industrial structures, and the change of land and geographical space. The driving forces for urbanization include abundant social service resources, convenient transportation, and cultural facilities, under which people are driven out of rural areas and move into the cities. In 2008, the urbanization in China reached 45.68%,[31] approaching 50%. The rapid urbanization process urgently requires convenient and popular transportation. Railways, especially intercity railways, have increasingly become a new "engine" driving urbanization. There is an urgent need for the construction and expansion of new railways for new urban rail transit, such as intercity railways that are highly dense and public transport-oriented, connecting large and medium-sized urban clusters.

The third change is the high enthusiasm for speeding up railway construction. In October 2006, the construction of a resource-saving and environment-friendly society was identified as a strategic task in the mid- and long-term planning of national economic and social development at the Fifth Plenary Session of the 16th Central Committee of the Communist Party of China. China has a vast territory. However, its population is too large, and China is bothered by resource shortages and an unbalanced energy mix. The railway has a large transport capacity, connects many regions, and serves different purposes. With such technical advantages, railway

[30] Transcript of press conference on the adjustment of *Mid- and long-term railway network plan*. www.gov.cn. http://www.gov.cn/wszb/zhibo286/wzsl.htm.

[31] (2013, March 1). China's Urbanization Rate over the Years. *Oriental Morning Post*.

transportation is a more ideal mode of transportation for building a resource-saving, environment-friendly society. As the constraints from development conditions were increasingly intensified, the society attached more importance to the role of railways in regional economic development. The local governments urgently wanted to enhance regional development capabilities through the construction of railway-based transportation hubs, some local projects for railway construction even exceeded the scope of *Plan in 2004*, and the sustainable development of the transportation industry required railways to play a greater role in it.

Taking Sichuan Province, located in western China, as an example, at the end of 2007, a major strategic decision to build a "Comprehensive Transportation Hub in western China" was outlined at the Fourth Plenary Session of the Ninth Plenary Session of the Sichuan Provincial CPC Committee. The Plan proposed that 18 railways, 21 highways, and two water transport channels would be gradually constructed to form a comprehensive transportation network accessible in all directions. During the *Twelfth Five-Year Plan* period, of the 700 billion yuan to be invested in Sichuan's transportation construction, the investment in railway investment was 280 billion.

The fourth change is that the demand for passenger and freight transportation has rapidly and steadily grown. Regarding the rate of macro-economic development, the overall economic landscape from 2002 to 2007 was favorable, with GDP increasing at the rate of more than 10% per year and all industries moving forward at a positive speed. In the railway industry, an annual increase of 8.5% in the turnover of passenger and freight was hit, greatly exceeding the 4.7% estimated by the experts of the industry during the same period. The railway mileage was expected to grow at an annual rate of about 2% when the *Plan in 2004* was unveiled. However, the plight of the railway networks was aggravated because the railway construction had long lagged while the demand for passenger and freight transportation had been steadily increasing within a certain period. Especially when the reform and opening up came to a new stage, since China's economic cooperation and exchanges with foreign countries had been expanded, the call for export-oriented transportation accesses was amplified accordingly, such as railways connecting sea and river ports and ports of entry.

According to adjustments made in the *Mid- and Long-Term Railway Network Plan*, the *Plan in 2008* was based on the *Mid-and Long-Term Development Plan of Multi-modal Transport Network* released in 2007, the *Plan in 2004,* and *the Plan of the Railway during the Eleventh Five-Year Plan*, with the new conditions in China's economic and social development being taken into consideration. The *Plan in 2008* coordinated the development of key regions, construction of major energy accesses, foreign economic and trade exchanges, the development and utilization of land and resources, and the demand for and trend of passengers and freight transportation. Apart from the above mentioned, the *Plan in 2008* also took account of other means of transportation and the major industries of China's economy, and ultimately adjusted the scale of railway construction and overall plan accordingly.

China was committed to delivering a railway network featuring a proper layout, well-organized structure, diversified functions, and proper connection. On the one hand, China's railway mileage in 2020 was expected to rise from 100,000 kilometers to more than 120,000 kilometers, and its high-speed railway from 12,000 kilometers to 16,000 kilometers; on the other, the rate of electrification was expected to move up from 50% to 60%. Meanwhile, based on the four vertical and four horizontal high-speed railway corridors, the passenger dedicated lines and intercity railways were extended to more areas; to be more specific, high-speed railways of more than 50,000 kilometers would be completed, and all provincial capitals and cities with a population of more than 500,000 connected, thus the time for traveling among the capitals would be saved by more than 50%. It was estimated that high-speed railways would be available to more than 90% of the Chinese people.

It has been specified that a high-speed railway network of more than 50,000 kilometers for passengers would be constructed. The release and implementation of the plans and adjustments are vital to a sustainable and comprehensive transportation network and can shed light on further railway construction. To fully act on the plans and adjustments, the domestic demand will be expanded, the employment increased, and the economy boosted; in the long run, compared with other means of transportation, the advantages of rail traffic can be brought into full play, which helps cut the total costs of transportation for the society at large, accomplishing the mission of energy conservation and emission reduction, and driving the formulation of a moderately prosperous society in all respects. In 2009 alone, 600 trillion yuan was invested in the railway infrastructure. According to the related analysis, 20 million tons of steel and 120 million tons of cement will be needed to turn the investment into reality, besides, 6 million jobs will be created in that course,[32] Lu Dongfu, then Vice Minister of MOR, pointed out. In fact, "China's fixed asset investment in railways (including investment in infrastructure development, and upgrading and purchase of rolling stocks) in 2009 was 701.321 trillion yuan, growing by 286.579 trillion yuan compared with that in 2008, marking a 69.1% increase, and it amounts to 80.3% of the total volume of investment during the *Ninth Five-Year Plan* and the *Tenth Five-Year Plan*, hitting a historic high".[33]

As scheduled in the *Plan in 2008*, of the 41,000 kilometers of railways planned for construction from 2008 to 2020, 16,000 kilometers were designed for high-speed operation at over 250 kilometers per hour. The planned mileage of China's high-speed railway would be three times the 6,350 kilometers that had been put into service across the world; investment in railways would be more than 3.5 trillion yuan, exceeding the total invested from 1949 to 2008, which necessitated the consumption of tens of millions of steel for infrastructure construction, seven million tons of heavy steel rails, hundreds of millions of cement, thousands of high-speed trains, millions of workers, and tens of thousands of construction machinery. Once the *Plan in 2008* was turned into reality, a framework of China's fundamental

---

[32] Transcript of press conference on the adjustment of *Mid- and long-term railway network plan*. www.gov.cn. http://www.gov.cn/wszb/zhibo286/wzsl.htm.

[33] Statistics Center of MOR. (2009). Railway statistics bulletin of MOR of the People's Republic of China in 2009.

railway network would be accessible, making the transportation of passengers and freight faster and easier.

With the *Plan in 2008* as the guideline, the Wuhan-Guangzhou High-speed Railway was completed and put into service in December 2009, which indicated that China was on its course to mastering the technology of high-speed railway comprehensively; the Beijing-Shanghai High-speed Railway, the longest high-speed railway with the biggest investment and the highest standard in China at that time, was put into operation in June 2011; the Harbin-Dalian High-speed Railway, the first high-speed line in the extremely cold region in the world, came into service in December 2012; by the end of 2015, China's railway in operation measured 121,000 kilometers and the high-speed railway 19,000 kilometers, the second and the first in the world respectively. During the period of the *Twelfth Five-Year Plan*, the maximum profits have been achieved through the investment, the number of new lines has hit the biggest, and construction has been completed with the highest quality, compared with the former plans. "Within the five years, fixed asset investment in railway has amounted to 3.58 trillion yuan and new lines in operation 30,500 kilometers, increasing by 47.3% and 109% respectively compared with those in the *Eleventh Five-Year Plan*. At that time, China excelled over the rest of the world in the total volume of passenger and freight transportation".[34]

5.1.3 Revision of Mid- and Long-Term Railway Network Plan

On June 29, 2016, the State Council's Executive Meeting passed the revised *Mid- and Long-Term Railway Network Plan*. As a result, the goals of railway development were also changed, and the railway's function, supporting China's economic development, was enhanced. Stimulating the upgrading and development of society through the construction of transportation arteries entailed the expansion of the railway network, that is, to build a comprehensive transportation system involving railways, roads, waterways, airlines, etc. According to the revised Plan, the mileage of China's railway network by 2020 would reach 150,000 kilometers and high-speed railway 30,000 kilometers, covering more than 80% of China's big cities. By 2025, the network would measure about 175,000 kilometers, and the high-speed railway about 38,000 kilometers. By 2030, such goals as connectivity between China and foreign countries, multiple railways connecting different regions, provincial capitals being interwoven by high-speed railways, cities being reached within a short time, and most counties being covered would be achieved.

Firstly, the revision of the Plan aimed at improving the plan of the Chinese railway network against the backdrop of the new normal of the economy.

[34] Qi Hui. (2016, June 30). The *Mid- and long-term railway network plan* was adopted in principle, opening up the main traffic artery. *Economic Daily.*

Under the guidance of the *Mid- and Long-Term Railway Network Plan* and *Adjustment of Mid- and Long-Term Railway Network Plan in 2008*, China saw a boom in its railway construction, with the part scheduled to be completed in 2020 being finished in around 2015. By the end of 2016, Chinese railways in operation amounted to 124,000 kilometers, and high-speed railways 22,000 kilometers (only 8,358 kilometers were in use by 2010). From the perspective of the relation between the development of railways and the supply and demand of railway transportation, China's insufficient capacity for railway transportation has been largely changed and the bottleneck constraining economic and social development has been broken. The improvement of the railway network has enhanced its ability to serve the public, thus those days in which tickets were difficult to buy and trains were hardly adequate have gone forever.

However, according to the statistics released by the National Development and Reform Commission, the total volume of railway freight transportation was 3.36 billion tons across the country in 2015, a year-on-year decrease of 11.9%; in the same year, Chinese railway freight turnover was 2.3754 trillion ton-kilometers, a year-on-year decrease of 13.7%; during that period, the total volume of freight transportation in the whole society was 45.02 billion tons, an increase of 4.4%, and the freight turnover amounted to 18.0583 trillion ton-kilometers, a decrease of 0.5%. Compared with other means, the railway saw a sharper decline in freight transportation. What was the reason for the unexpected decrease in the volume of freight transportation and railway freight turnover? Was there overcapacity in Chinese railways?

In November 2015, President Xi Jinping stressed at the 11th Meeting of the Central Leading Group for Financial and Economic Affairs that in the course of economic restructuring, five types of major policies were necessary, namely, stable macro-economic policies, targeted industrial policies, flexible micro-economic policies, effective policies for reforms, and social policies that ensure the satisfaction of basic needs; besides, he also pointed out that "while expanding the demand within an appropriate range, we should also improve the quality and efficiency of the supply-side, invigorate and sustain the economic development, and promote the overall development of China's productivity".[35] What President Xi had said was considered the "map" for China's supply-side reform, and the requisite for the country to achieve economic new normal and act on the development philosophy to pursue innovation, balance, eco-friendliness, openness, and inclusiveness. The major measures to further supply-side reform were to cut overcapacity, destock, and deleverage while the demand was expanded within an appropriate range, thus effective supply could be increased and the ineffective reduced.

In China, such industries with serious overcapacity as steel, coal, non-ferrous metals, petroleum, and petrochemistry were not only the key fields of supply-side structural reform but also the principal sources of traditional railway freight

[35] Literature Research Office of the CPC Central Committee. *Excerpts from Xi Jinping's discourses on the development of socialist economy*, (pp. 87-88, 87). Beijing: Central Party Literature Press.

transportation. It is easy to tell that supply-side structural reform was indeed the "invisible hand" leading Chinese railways into a predicament in which "no freight was in need of transportation". In 2015, China saw a decline in the volume of railway freight transportation, which truly reflected that the country had ushered in the new normal of economic development and economic transition. The supply-side structural reform serving as the impetus to further the reform after Chinese railways were free from the "bottleneck", and it was also a brand-new mission in front of Chinese railways, which implied that in the framework of China's socialist market economy with Chinese characteristics consisting of an indivisible whole in which the market and the government had their roles to play, it was inevitable that the "Big Brother of Railways" (MOR) would be turned into a modern enterprise.

One thing worthy of noting is that the decline in the volume of railway freight transportation does not imply that China will cease its railway construction. Compared with foreign countries, China's density of railways was much lower than that of developed countries, for instance, China had a population that was four times as large as that of the US, but the mileage of its railways was less than one-third of that in the US. Within China, railways were disproportionately distributed, and the areas covered by railways should be expanded; the railways in central and western China were particularly underdeveloped. The railway transport was inefficient, with the railroads connected but still largely obstructed and the transportation capacity of the railways being constrained; the connection among various means of transportation needed enhancement because the structural problems within the transportation system were prominent, the layout of the network remained complicated, and the intercity transportation was in slow growth. The support for China's economic growth from the development of railways was increasingly prominent, especially against the specific landscape of the new normal. Railways, the artery of the Chinese economy, had their unique advantages in such respects as ensuring stable economic growth, restructuring, attracting investment, and promoting consumption. Therefore, furthering supply-side structural reform of railways was a profound move, bringing multiple benefits.

On the one hand, the major missions specified in the *Mid- and Long-Term Railway Network Plan* were accomplished in advance, and railway construction stepped onto a more reasonable course, hence the railway network should be re-planned on the basis of China's realities at that time and demand for going further in 2025 and 2030. On the other, new conditions of China's economic and social development required Chinese railways to meet new requirements and fulfill new goals accordingly; in a new phase in which the focus of Chinese railways was shifted from pursuing hard power to balancing soft and hard power, the planning of railway network should be upgraded; meanwhile, different from the previous railway construction, increasing the mileage to break the "bottleneck", quality and efficiency would be more stressed in the future to meet the needs of high-quality development. The above is not only the historical background of the revision of the *Mid- and Long-Term Railway Network Plan* in 2016, but also the general idea of the adjustments of the *Plan*.

Secondly, the revision also indicated that the reform of Chinese railways would enter a golden age of transformation and development.

President Xi pointed out that "the *Thirteenth Five-Year Plan* period is a golden age for the infrastructure of transportation to be ungraded, transformed, and developed, and its services to be improved".[36] This age was not just an extension of the period in which the scale and speed of development were the focuses, but a brand-new starting point of the period in which quality and efficiency would be the subject matter; meanwhile, it was also a golden age to accelerate the formation of a network with integrated transportation infrastructure, transformation and upgrading of the transportation sector, and building of the modern integrated transportation system.

The requirement that more than 800 billion yuan should be invested into railways in 2016 was made in the Report on the Work of the Government of 2016. *The Outline of the Thirteenth Five-Year Plan for National Economic and Social Development of the People's Republic of China*, released on March 17, 2016, outlined an integrated transportation network that would be constructed through network layout, intelligent management, integrated services, and eco-friendly development. The focus was on four respects, namely transportation accesses connecting China and foreign countries, modern efficient means of intercity transportation, integrated transportation hubs, and transportation services with intelligence, safety, and low carbon emission. China's philosophy of transportation, "Great Transportation" and "Big Arteries", highlighted the connectivity of "railways, roads, and airports", featuring integration, eco-friendliness, intelligence, and safety. The philosophy manifested itself in such aspects as speeding up the pace of weaving a high-speed railway network, accelerating railway construction in central and western China, promoting intercity and commuter rails in urban areas, improving integrated transportation hubs at the national and regional levels and in different areas, and enhancing the capabilities of the systems of intermodality, intelligent management, and public information. According to the Plan, the priority of railway construction, high-speed railways in particular, during the *Thirteenth Five-Year Plan* period would be shifted to central and western China. At the same time, many provinces in succession published their plans of investment in railway construction during the *Thirteenth Five-Year Plan* period, for instance, Sichuan planned to invest more than 200 billion yuan in railway construction and Fujian was about to spend 750 billion yuan in optimizing the modern transportation system.

Regarding the relation between the *Mid- and Long-Term Railway Network Plan* and *the Outline of the Thirteenth Five-Year Plan for National Economic and Social Development of the People's Republic of China*, the revision of the former was made for the implementation of the latter. According to the revision, the goal of "four vertical and four horizontal high-speed railway networks" would be converted into "eight vertical and eight horizontal high-speed railway networks"; the idea of the "high-speed railway network" was put forward for the first time, in addition,

[36] People's Daily editorial: Seize the golden age, be a pioneer. *People's Daily Online*. http://opinion.peple.com.cn/n1/2017/0103/c1003-28993160.html.

such network would be integrated with "railway network" to form an "Internet". With optimization of the layout of various transportation modes, sharing of the accesses and promotion of intermodality within the network, effective connection and integration of different transportation modes could be formulated.

The planned supportive measures were aimed at the core of the supply-side structural reform of railways, and a modern enterprise system was required to be introduced into MOR to activate the available assets at that time. In addition, diversified and market-oriented approaches should be introduced into financing, and market access broadened. At the same time, local governments and the general public where private and foreign capitals existed should be encouraged to participate in the investment in railway construction.

The most distinctive feature of the revision was that the idea of developing a high-speed railway economy was proposed for the first time. Fostering new forms of the economy through high-speed railways became one of the focuses of economic development. On the basis of high-speed railway lines and stations, communication, cooperation, and distribution of resources within the areas along railway lines could be promoted and optimized. The manufacturing industry and China's economy at large could be upgraded by speeding up the process of industrial gradient transfer. The above-mentioned indicated that the grand transportation network with connectivity and overall planning at different levels, featuring "eight vertical and eight horizontal high-speed railway networks", has been put in place, the era in which China's economy would be driven by railways has come, and the integration of Chinese railways and social and economic development would be furthered.

As China had entered a new era of socialism with Chinese characteristics, a two-step approach with 2035 and 2050 as the cut-off points was proposed in the Report at the 19th CPC National Congress; in addition, it was also clearly stated that China would become a country with strong transportation network. Trying to answer a series of calls in the new times, the central and local governments revised the former *Mid- and Long-Term Railway Network Plan* again, aiming to achieve the goal by 2035. Some railway experts argued that in order to fulfill the requirements of high-quality development, such revision might introduce an increase to the mileage of railways, but compared with the revision of the *Mid- and Long-Term Railway Network Plan* in 2016, the planned mileage would not rise by tens of thousands of kilometers. On November 20, 2017, multiple Ministries and Commissions jointly issued *The Development Plan of the Railway during The Thirteenth Five-Year Plan*, which indicated that the layout of the railway network would be optimized by 2020. By that year, the mileage of Chinese railways would reach 150,000 kilometers, high-speed railways accounting for one-fifth, and the proportions of double-track and electrified railways would be around 60% and 70% respectively, thus a railway network with proper layout, widely-covered areas, clearly-structured framework, safety, and efficiency would gradually take shape.

In central and western China, the mileage of railways amounted to around 90,000 kilometers; the intercity and commuter railways measured around 2000 kilometers. "Zero-distance transfer" for passengers and "seamless" connectivity of freight transportation were largely available; within two to eight hours, passengers and freight could arrive at most provincial capitals from Beijing, and neighboring large and medium-sized cities could be mutually reachable within one to four hours, and within major city clusters, time for commuting would be half an hour to two hours. To further the construction of the "eight vertical and eight horizontal high-speed railway networks" as the main access, high-speed railway projects with favorable conditions to attract passengers and meet the needs of development would be implemented.

In respect of the expected indicators, the rate of the average annual increase of the mileage of high-speed railways in operation was 11.6%, far beyond 4.8%, the overall rate of the increase of Chinese railways. *The Development Plan of the Railway during the Thirteenth Five-Year Plan* represented the *Third Five-Year Plan* for Chinese railways since 2004 (The former ones were *The Plan of the Railway during the Eleventh Five-Year Plan* released in 2009 and *The Development Plan of the Railway during the Twelfth Five-Year Plan* in 2012—Author's note), which was also the action plan for the implementation of *The Outline of the Thirteenth Five-Year Plan for National Economic and Social Development of the People's Republic of China, The Development Plan of Modern Integrated Transportation System during the Thirteenth Five-Year Plan Period*, and the *Mid- and Long-Term Railway Network Plan* for development in a specific stage. Under the guidance of the Plans, China was on its way to become a country with strong transportation network from a country with vast transport infrastructure.

5.2 Investment and Financing Oriented Railway Reform and Opening UP

In the twenty-first century, systems of railway administration, investment and financing, enterprise property rights, operation, etc. have been unsuitable for the development of the market economy in a new stage; what's worse, many obstacles caused by those systems held back the high-quality and rapid development of Chinese railways, hence an increasing number of Chinese people called for accelerating the reform of the railway systems.

Investment and financing in railways, which has long been haunting the development of railways in countries all over the world, became a hot-button issue of the reform. In China, "one of the reasons why the growth of Chinese railways lagged behind foreign countries is the lack of investment, thus the reform of Chinese railways shall be furthered".[37] Especially since the implementation of the *Mid- and Long-Term Railway Network Plan* in 2004, Chinese railways have seen leapfrog development. The standard of railway construction became extremely high, and the number of new lines was exceedingly large, which was unprecedented. The surge in

[37] Lu Dongfu. (2005, September 22). Lu Dongpu's Opinions on the reform of investment and financing of railways. *People's Railway Daily.*

investment in railway construction became the driving force to speed up the reform of investment and financing. "Railway construction has such industrial features as long construction period, high sunk cost, and huge investment".[38] Regarding the investors, the central government (MOR) was the major one to finance railway construction. With respect of the proportion of the investment, the funds for railway construction (the most important and stable), depreciation, financial budget from the central government, retained profits of Chinese railways, bonds, and bank loans were the principal channels of investment and financing of railway construction.

"In 2004, the funds for railway construction and loans from domestic banks accounted for three-quarters of the total amount financed".[39] As the construction of the railway network was at a fast pace, the planned network was also being expanded, becoming more well-shaped; consequently, "raising funds for railway construction via increasing the charges for transportation and policy-based loans provided by China Development Bank has now been inadequate to meet the investment demands".[40] Since the railway construction on large scale had brought many benefits, the economic elements played an increasingly important role in the process.

"By the end of 2003, the long-term debt balance amounted to 200 billion yuan, but principal and interest were more than 40 billion yuan while the income of the funds of railway construction was only 40.99 billion yuan, less than the amount of the principal and interest"[41]; "it is impractical to rely excessively on bank loans for financing railway construction".[42] Shortage of capital led to great cuts in the speed and efficiency of railway construction. Only by accelerating the reform of the investment and financing system, innovating the methods of investment and financing, and opening up new financing channels, can sufficient and sustained capital for the construction of the railway network be ensured.

Some scholars believe that at the beginning of the twenty-first century, investment in projects of Chinese railways went through four stages of development, featuring "steadiness, rapidness, high speed, and adjustments" respectively.[43] Before 2004, the railways were in steady development, but since the release of the *Mid- and Long-Term Railway Network Plan* in 2004, Chinese railways have seen rapid growth; in 2008, due to the financial crisis, the Chinese government carried out incentive plans of four trillion yuan to stimulate economic growth, representing a more powerful driving force for railway construction, thus Chinese railways moved

[38] Kou Linlin. (2013). Exploration on the China's railway financing model in the new era. *Construction Economy*, (5).

[39] Huang Ming. (2005). The direction of investment and financing reform for Chinese railways. *China Transportation Review*, (10).

[40] Wang Zhaocheng. (2006). Thoughts on the reform of investment and financing system of Chinese railways. In Wang Zhiguo (Ed.), *The important manuscripts on the leaping development of China's railways in 2005* (p. 267). Beijing: China Railway Publishing House.

[41] Ibid.

[42] Ibid.

[43] Ren Xin, & Zhou Yu. (2016). Investment and Financing of Chinese Railways: An analysis of the status quo and the problems, and the orientation of the reform policies. *Macroeconomics*, (5).

forward at a high-speed pace; in 2011, swung by national macro-economic policies and the Wenzhou 7/23 rear-end collision accident, railway construction suffered from a shortage of capital, thus investment and financing should be adjusted accordingly. On the one hand, as a result of the elements within and outside railways, the process of reform of the investment and financing system, corresponding to the growth of the amount of investment in railways, presented such features as solving problems of sources of capital through multiple methods, optimizing the proportion of investment and financing, and improving the efficiency of investment and financing. On the other hand, breaking free from the constraints of systems and mechanisms to create better domestic and foreign conditions became the major approach to further the reform of the investment and financing system during that period.

5.2.1 Groundbreaking Progress of Reform of Railway Investment and Financing System During the Periods of Tenth Five-Year Plan and Eleventh Five-Year Plan

The top-down reform of the railway investment and financing system was a compulsory institutional transformation. Prior to the Reform and Opening Up, the railway investment and financing were mostly planned, and the government directly invested in and financed railways. Since the founding of PRC, the central government has attached great importance to railway transportation, and most of the capital used to improve transportation has been transferred to railway construction, thus Chinese railways saw rapid growth; "railway construction has been halted for more than ten years during the 'Cultural Revolution', so the mileage of railways was only around 50,000 kilometers".[44] However, owing to the changes in the economic system since the Reform and Opening-Up and the shortage of financial support from the central government, sustaining investments in railway construction became increasingly difficult.

In the 1980s, MOR started to shoulder the responsibilities for the investment and financing of railway construction. During the Seventh Five-Year Plan period, the Contract Responsibility System was introduced into the railways, which was an economic responsibility contract system in the railways, featuring a "self-sustaining approach to building new railways and maintaining the existing ones", besides, bank loans began to play a role in the railways. The system was considered the practice of "letting Chinese railways make their own breaks" by Deng Xiaoping. Since the Eighth Five-Year Plan period, the State Council began to raise funds with the help of the Ministry of Finance. As the socialist market economy was established, the model of planned economy featuring railways' over-reliance on financial support from the central government was replaced by a market mechanism.

[44]Yan Pengfei. (2004). Reform on China's railroads investment system: A late choice for China's railroads. *Fortune World*, (10).

Nevertheless, the reform did not succeed in addressing the real issues, and raising funds at a large scale from the market was restricted by the railway system, hence the development of Chinese railways greatly lagged behind that of other industries.

"During the five years from 1997 to 2002, investment in railways was less than 300 billion yuan while that in highways was more than 1.4 trillion yuan; the amount of investment in highways in 2002 was more than the total in railways during the five years".[45] Behind such a huge contrast was the outdated railway investment and financing system. Back in the 1980s, a novel system of investment and financing, known as "building roads by loans and repaying loans by tolls", was established in highways, however, the railways, acting as a natural monopoly, were still on the outdated tracks, relying on administrative measures to allocate the financial resources.

A deeper reason was the barriers to investment and financing caused by the slow reform progress of the railway management system. Besides, characteristics of railways, such as pursuing profits and public welfare, indicated the complexity and difficulty of the reform. The first round of reform of the railway system, known as "a great attempt and two breakthroughs",[46] was phased out in 1992 after the Contract Responsibility System having being implemented for six years during which a succession of safety incidents had happened. Starting from 1993, the government has been trying to start a pilot initiative of a modern enterprise system in the railways, introducing structural separation, but the fundamental problems of the system were not solved. In 2000, the second round of reform with "the separation of transport network and transportation" as a focus was initiated. During the reform, MOR unveiled the plan of the restructuring of the railway industry for the first time. However, owing to the disorder in the reform, the plan was rejected by the State Council, thus those companies of passenger transportation founded after 2002 were gradually deregistered. "The arrangement for unemployed workers and the sources of capital" were the major bottlenecks in the process of the reform. "Investors dare not to act blindly because the railway system has yet to be reformed, the government, public institution and enterprise functions have not been separated, and the

[45] Ibid.

[46] "A great attempt" refers to bringing the "Contract Responsibility System" (1986-1990) into the railways. During the period, MOR delegated its powers in such aspects as planning, finance, labor and capital, goods and materials, and human resources to its subordinate departments, and started to put a Contract Responsibility System of economic responsibility into trial use within the industry. One of the "two breakthroughs" is that from 1986 to 1992, MOR, on the basis of the "Contract Responsibility System", brought the economic responsibility contract system in the railways featuring a "self-sustaining approach to building new railways", "raising funds for railway maintenance via tolls", and input and output with five years as a cycle. The other is that in 1993, the Ministry made an attempt to start a pilot initiative of modern enterprise system in the railways. By 1998, five companies, China Railway Engineering Corporation, China Railway Construction Company, China National Railway Locomotive & Rolling Stock Industry Corporation, China Railway Materials Commercial Corporate, and China National Railway Signal & Communication Company, and MOR agreed on the asset management responsibility system and structural separation.

transport network and transportation have also remained combined".[47] The above plights were the principal problems haunting the reform of railway investment and financing system.

In April 2002, the Yantai-Dalian Railway Ferry Project, connecting Yantai, Shandong Province, and Dalian, Liaoning Province, became open to foreign investors for the first time, which was an important move in the reform of the railway investment and financing system in the twenty-first century. At the end of 2003, MOR started to find a place for transportation enterprises in the market by establishing three companies: China Railway Container Transport Corp., Ltd., China Railway Special Cargo Logistics Co., Ltd., and China Railway Parcel Express Co., Ltd. What's more, MOR also ordered further reform of the three companies, trying to bring in a shareholding system in accordance with the system of modern enterprises. In addition, it also revealed that a shareholding system would be introduced for qualified railway companies running diversified businesses based on the resources of existing listed companies, including those of Guangshen Railway Co., Ltd. and China Railway Tielong Container Logistics Co., Ltd. These moves signified railway departments' reform of investment and financing system based on the leapfrog development.

Around 2004, guiding policies at the national level and a series of regulations for market-oriented financing and social capital invested in railway construction were released. A sequence of vital decisions for "transition" and "opening-up", including further reform of the national investment system and broadened access for railways, created unprecedentedly favorable conditions for the reform of railway investment and financing system. On October 14, 2003, *The Decision of the CPC Central Committee on Major Issues Concerning the Improvement of Socialist Market Economic System* was passed at the Third Plenary Session of the Sixteenth Central Committee, putting forward a large number of novel ideas, such as driving the growth of the mixed-ownership economy, diversifying the investment channels, and making the shareholding system as the principal mode to achieve state ownership. In addition, it was also specified that a modern market system shall be put in place to convert the planned economy into a market-oriented economy.

In January 2004, "Nine Articles of the Country", aiming to promote the capital market, were unveiled in *Several Opinions of the State Council on Promoting the Reform and Opening Up and Stable Development of the Capital Market*. It was required in the *Decision of the State Council on Reform of the Investment System* released in August 2004, that non-government-funded projects should be authorized and recorded according to the specific circumstances; a large number of supporting authorization measures for companies' projects and overseas investments were carried out in a short period. Besides, in light of China's commitment to taking "a three-step approach"[48] to open the railway freight transportation market, made in

[47] Yan Pengfei. (2004). Reform on China's railroads investment system: A late choice for China's railroads. *Fortune World*, (10).

[48] Since the accession to WTO, the only foreign investors allowed to enter Chinese market were joint ventures in which the foreign capital should account for less than 49% of the company's

2001 following China's access to the World Trade Organization (WTO), none of the areas in the railway sector was prohibited from investment according to the revised *Catalogue for the Guidance of Foreign Investment Industries* in 2004. In 2005, policies and measures including those in *Several Opinions of the State Council on Encouraging, Supporting and Guiding the Development of Individual and Private Economy and Other Non-Public Sectors of the Economy*, were revealed to encourage and guide private capital to enter industries and areas that had been made inaccessible to investors. With the help of the country's macro measures, the railway market, which was relatively closed in the past, saw an entirely new opportunity which allowed it to open up to both domestic and international markets.

During this period, funds for special purposes were also permitted to be invested in railways. For instance, it was clearly stated in *Interim Provisions on the Administration of Insurance Assets Management Companies*, coming into effect on June 1, 2004, that the management and utilization of insurance funds should be confined within such fields as bank balance, purchase and sale of government bonds, and financial bonds, and other areas specified by the State Council. According to *Measures for the Management of Pilot Project of Infrastructure Construction Indirectly Invested by Insurance Funds* released in March 2006, insurance funds could be entrusted to the trustee and invested indirectly on the basis of fiduciary relationships in national key projects of infrastructure construction, including transportation, communication, energy, municipal utilities, and environmental protection.

At that time, the railway administration was aware of the prominent problems in the reform of the railway investment and financing system, to be specific, inadequate diversity of investors, channels for raising funds, manners of financing, and the dated management of investment. For example, "owing to some historical factors, the lack of diversity also exists in Chinese railways' ownership structure and property rights, and the state ownership almost completely dominates the railway sector. Even in jointly-funded and local railways, the state-owned capital plays a dominant role, too, some owning more than 95% of the shares of the railways".[49] An initiative for leapfrog development of railways was brought forward in 2003 when the railway administration mapped out the reform of railway investment and financing, known for "three 'diversifications', two 'classifications', and three 'combinations'": "three 'diversifications' referred to the diversification of investors and channels for raising funds; two 'classifications' were short for the classification of investment projects and decision-makers"; three 'combinations' represented the combination of investment and financing system with the modern enterprise system, innovation of construction management, and acceleration of upgrading of railway

shares; three years later, the foreign capital would be permitted to own the majority of the shares; foreign capital would be authorized to set up wholly-owned subsidiaries after six years since the accession.

[49] Zhou Xinjun. (2012). Reform of Chinese railways: Particularities, plights, and basic ideas. *Economic Affairs*, (4).

technologies and equipment".[50] Huang Ju, then Vice Premier of the State Council, repeated the above ideas of reform at two nationwide railway working conferences. Since then, those ideas have served as the guiding philosophy for many concrete measures to further the reform of the investment and financing system.

On July 16, 2004, MOR published *MOR's Notice on Forwarding the Decisions of the State Council to Reform the Investment System* (TZF [2004] No. 98), which illustrated the overall philosophy of reform of investment and financing system, known as "government's guidance, diversified investment, and market-based operation". It was put forward that apart from pursuing diversification of the investors, encouraging the general public to play a role in investment in railways, and searching for various ways of market-based financing, the government should also promote railway construction with the help of non-governmental organizations and institutions through such channels as capital infusion, investment subsidies, loan transfer, and subsidized loans.

"The idea is suitable for China's reality and Chinese railways, so it receives a warm welcome from governments at all levels, enterprises, and all sectors of the society".[51] In July 2005, MOR released *Opinions on Encouraging, Supporting and Guiding Private Economy to Participate in Railway Construction and Operation*, making investment in areas, including railway construction, passenger and freight transportation, manufacture of equipment, and diversification of businesses, available for the social capital which was encouraged to be invested in major construction projects, such as railway passenger and freight transportation, energy transport corridors, and restructuring of the state-owned enterprises in a variety of ways, for example, jointly funded railways.

Thanks to the above two documents, railways became more attractive to social capital. In September 2005, to garner more recognition and insightful ideas, MOR held an international forum for the first time in Beijing International Hotel, China Railway Investment and Financing Reform Forum, regarding the reform of Chinese railways investment and financing system. Three issues, namely the government's role, social capital's investment in railways, and manners of financing in railways, were intensely discussed. Meanwhile, during the *Fifteenth Five-Year Plan* period, the government demanded that "we should not only enhance the role of railway companies, entities of the market, or optimize the responsibility system of assets operation, but also separate government, supply and marketing cooperatives, and government functions from business ones, downsize staff to improve efficiency, and form companies for passenger and freight transportation".[52] Significant reforms,

[50] Xiong Xuejun. (2006). Issues related to establishing railway joint ventures. *Railway Economics Research*, (3).

[51] Lu Dongfu. (2005, September 22). Lu Dongpu's Opinions on the reform of investment and financing of railways. *People's Railway Daily.*

[52] National Development and Reform Commission. (2001). National Development and Reform Commission's notice on printing and distributing special major plans for the development of comprehensive transportation system during the Thirteenth Five-Year Plan period (JGH [2001] No. 710).

including those of the separation of sideline businesses from core businesses in railway transportation, the system of railway stations and sections under the direct management of railway corporations, and the performance evaluation system for transportation companies, were successively carried out, speeding up the process of reform of railway investment and financing system.

Increasing the mileage of jointly-funded railways and improving the market-oriented financing platform were the principal missions of reform of railway investment and financing system during the Tenth Five-Year Plan period. Among the measures to further the reform, constructing jointly funded railways was the most powerful and effective one. Since the mid-1980s, the central government has allowed local governments and enterprises to introduce the shareholding system into the railway investment to address the problem, that is, the lack of diversity in the sources of investment. However, the capital invested in accordance with the shareholding system from the above institutions was limited to a low proportion. Nevertheless, since the reform of the railway investment and financing system was carried out during the Tenth Five-Year Plan period, the number of railway construction projects under agreements signed by and between MOR and the local governments amounted to 158 in 2004 and 2005 alone. In April 2005, MOR managed to diversify the sources of financing in the 43 projects that were open to the general public; by June 2006, more than 70 non-governmental institutions, private companies, and state-owned enterprises that were not in the railway industry reached agreements with MOR to implement more than 20 projects related to corridors of coal transportation and passenger dedicated lines. In that process, the total amount of equity investment from the society was 44 billion yuan, among which 2.5 billion was from the private sector.

On October 28, 2004, China Railway Investment Co., Ltd. was incorporated, representing the establishment of a major platform of market-oriented financing for railways, which completely changed the existing system under which the government had direct and unlimited liability for funding and managing the projects, thus the barriers for the market mechanism to be introduced into the railway industry were removed. China Railway Investment Co., Ltd. soon played its role in the construction and property rights transfer for projects such as the Wuhan-Guangzhou Passenger Dedicated Line and the Sanshui-Maoming Railway.

As the process of bringing the shareholding system to state-owned enterprises was being accelerated, companies in different industries, including power, telecommunication, highway, and civil aviation, were restructured to get listed in the stock markets. In 2004, under the guidance of "generating added value on existing properties", MOR, committed to breaking new ground for and promoting equity financing, decided to launch five pilot projects of equity reform in the Datong-Qinhuangdao Railway, Guangshen Railway Co., Ltd., China Railway Tielong Container Logistics Co., Ltd., China Railway Express Co., Ltd., and the Qingdao-Jinan-Lancun-Yantai Railway. Those experimental and pioneering measures to further the reform of the railway investment and financing system during the Tenth Five-Year Plan period introduced a "brand-new mechanism" for large-scale railway construction in the future. But we also found that although the channels for fund raising were increasing, the social capital did not flow into railways on a large scale regarding its total amount; even though the mileage of railways jointly funded by MOR and local governments was improved, the Ministry in the equity holding was in the dominant status, indicating that the previous situation, the central government as the major investor, has yet to be completely rectified.

In the first year of Eleventh Five-Year Plan, MOR issued *Plan for the Reform of Railway Investment and Financing System during the Eleventh Five-Year Plan*, representing further reform in the railway investment and financing system. The Plan complies with the basic philosophy of "government leading, diversified investment, and market-oriented operation", aiming to attract various investors both at home and abroad, and local governments to directly invest in railway construction. Many new terms concerning the investment and financing, such as "strategic investor", "to issue bonds based on a fixed period", "industrial investment fund" and "financial leasing" are included, and the goal of "establishing a new mechanism for railway investment to meet the requirements of the development of the socialist market economy" is set out in the plan.

In terms of reform practice, the reform of the railway investment and financing system, which started in the Tenth Five-Year Plan period in the twenty-first century, was intensified during the Eleventh Five-Year Plan period, thus establishing a diversified framework for railway investment. On August 1, 2006, Daqin Railway Co., Ltd. successfully launched its IPO in the A-shares market and raised 15 billion yuan. It was the first time that railway enterprises had been listed in the domestic capital market, marking the breakthrough in the reform of railway stock rights. On December 22th, 2006, Guangshen Railway Company also successfully listed its A shares and raised 10.3 billion yuan. These breakthroughs have not only strengthened the commitment to the reform of the railway system but also provided valuable experience for more social capital to share in the interests of the railway market.

On 7 September 2006, MOR and China Development Bank signed *Summaries of Talks over Cooperation on Development Financing during the Eleventh Five-Year Plan*, China Development Bank provided 250 billion yuan as the policy-based lending for railway during the Eleventh Five-Year Plan Period, to meet the funding needs of the construction of mid- and long-term railway network. On 27 December 2007, the Beijing-Shanghai High-Speed Railway Co, Ltd. was established in Beijing, representing the implementation of the plan for market-based financing of the Beijing-Shanghai high speed railway. In 2008, MOR began to issue medium-term notes to raise funds. Meanwhile, the Central Government continued to support the railway construction with government bonds and financing so that the funds for railway construction could stably grow annually. Local government and other investors also had an active role in railway construction by means of direct investment or bearing the cost of land requisition and resettlement, and funds provided by local governments for some projects were up from 30% to 50%. In addition, China Railways also raised construction funds through various channels such as issuing railway enterprise bonds and lending from home and abroad. According to statistics, enterprises from all sectors invested more than 60 billion yuan in railway construction and operation from 2002 to 2008. In addition to the social security fund, railway construction was financed by over 150 well-known enterprises at home and abroad such as China National Petroleum Corporation and Deutsche Bahn.

In the later years of Eleventh Five-Year Plan Period, railway investment and financing had gradually changed its focus. In 2007, with the support of banks, MOR increased its annual investment in railway construction from an average of 50 to 60 billion yuan during the Tenth Five-Year Plan Period to nearly 200 billion yuan. With a 4 trillion-yuan investment plan from the Central Government after the financial crisis in 2008 as well as the adjustment of *the Plan in 2008*, funds for railway construction reached nearly 700 billion yuan in 2009.

Investment has witnessed dramatic growth. At that time, railway construction was stably funded by the state at about 50 billion yuan annually, and the rest was raised by MOR itself. In the case that social capital, accessible to railway sectors in a short period, had no scale effect and that the source of equity financing was limited, MOR gradually shifted its investment and financing focus to debt financing like lending to support the large-scale and rapid development of railway construction. During the whole Eleventh Five-Year Plan Period, due to the rapid growth in railway construction, and the wide application of various financing modes like railway bonds, commercial loans, and foreign investments, railway construction funds accounted for 13% of total construction funds, down from 45% from Tenth Five-Year Plan Period, and the proportion of debt financing increased from 48% in 2005 to 64.26% in 2009.

From the Tenth Five-Year Plan to the Eleventh Five-Year Plan, MOR made great progress in the reform of the investment and financing system, such as speeding up the railway construction invested by a joint venture, promoting the listing of high-quality railway assets, and improving the issue of railway construction bonds. However, the scale and strength of the reform were still limited due to both internal and external conditions, such as the imperfect national public financing system, the backward reform of railway management, and railway freight control, as well as the gap between large-scale railway construction and inadequate ability of railway investment and financing.

5.2.2 *Evolution and Further Reform in Railway Investment and Financing System During the Twelfth Five-Year Plan Period*

1.97 trillion yuan was invested in railway infrastructure development during the Eleventh Five-Year Plan, which was 6.3 times the total investment during the Tenth Five-Year Plan Period, and a large amount of railway projects had been launched. Confronted with the fund shortage, MOR mainly depended on the bank loans to fill the investment gap, which accounted for more than 50% of the total railway construction funds. Because of railway's insufficient repayment, inadequate operation and low profitability, MOR's indebtedness continually increased, up to 1.89 trillion yuan at the end of 2010. However, the total income of railway transportation was 450 billion yuan at that time, while the net profit was only 15 million yuan.

2011 is the first year of the Twelfth Five-Year Plan and also a special one for China's railway. At the beginning of the year, MOR announced that the investment in railway construction would be the same as that in 2010, around 700 billion yuan, however, the actual investment to infrastructure development by the end of the year was around 350 billion yuan. Meanwhile, the asset-liability ratio increased from 58.5% in the first half of the year to 80% at the end of 2011. Due to the tight state macro policies, rectification of fraudulent conduct initiated by MOR, and the high-speed railway accident on 23 July, railway investment and financing witnessed a major decline, railway payment capability was weakened, and predicament in sourcing railway construction funds was encountered.

In April 2011, Hu Jintao inspected railway work in Hainan and made significant instructions. "Scientific development and accelerating the transformation of development patterns should be integrated into all aspects of railway construction and operation, especially structure optimization and improvement in transportation service to continually upgrade the quality and efficiency of railway development and provide better service for economic society development and people's well-being".[53] The Central Committee of the CPC attached great importance to railwaydevelopment and encouraged railway workers to strive forward. Confronted with problems of railway investment and financing, the government also initiated control in a visible manner. For example, in October 2011, National Development and Reform Commission identified China railway bonds as "government-backed bonds", using state credit to boost the acceptability of railway bonds, and then the three short-term railway bonds were subscribed over 10 times. In addition, in November 2011, National Development and Reform Commission, Ministry of Finance, and banking system provided 200 billion yuan as emergency funds to support the financing. MOR stepped out of the fund dilemma temporarily under the support of the country. However, in the peak period of China's railway construction, China railways faced the urgency to reform its seriously backward railway investment and financing system so that trillions of railway investment would be constantly guaranteed.

The Central Economic Work Meeting, held from 12 to 14 December 2011, announced that in 2012 we would maintain appropriate investment, optimize investment structure and focus on projects under construction and those to be constructed, ensuring that fund demands were met for water conservancy, railway construction, major equipment, and other projects approved by the state. On National Railway Work Meeting, held on 23th December 2011, MOR emphasized that "we should provide supporting measures and adequate funds for the commenced major projects".[54] At the same time, "we should actively strive for the support of relevant

[53] Peng Kaizhou. (2001). Transforming the development mode of railways and further promoting diversified operation to improve the quality and efficiency of railway development. *Chinese Railways*, (5).

[54] Sheng Guangzu. (2012). Efforts to create a new situation of scientific development of railway, better services for economic and social development and the people. *Railway Economics*

departments…and we also need to attract the investment from local government and private sectors, expanding the capital sources of railway construction".[55] Local governments further identified their role in railway investment. In 2012, MOR gradually strengthened responsibilities of local governments for railway investment, increased their proportion of investment, and steadily withdrew from investment in regional railways.

In 2013, a new round of institutional reform of the State Council began. In March 2013, China Railway was established. Highway, civil aviation, port and many other sectors once implemented the system featuring "integration of government and business functions", and after years of market reform, all sectors had basically completed the separation of government and business functions. However, the railway sector was the only one with the integration of government and business functions. Establishing China Railway is the significant measure to meet the demands of socialist market economic system, further railway mechanism and institutional reform, solve the problems of railway investment and financing, and promote the modernization of railway construction.

On 16 August, the State Council issued *Opinions on Reforming the Investment and Financing System to Promote Railway Construction*, requiring that the railway should follow the philosophy of coordinated guidelines, diversified investment, market-oriented operation, and supporting policies. "We need to improve railway development plan, open railway construction market in all aspects, classify investment for newly constructed railways, and make local government and social capital accessible to ownership and management rights of intercity railway, suburban railway, railway for exploiting resources, and branch railway, so as to encourage the move of investment into railway construction".[56] With the guidance of central fiscal funds, we also plan to establish railway development fund to attract investment from corporate capital.

In addition, we should establish the dynamic mechanism of railway freight price following high-way freight one, set up railway transportation subsidy system as a public and strategic undertaking, vitalize railway land resources, and encourage banks and other fiscal institutions to further support major railway projects to accelerate the pace of railway investment and financing systematic reform. Since bank loans and issuing bonds are still the main resources of railway investment, we should create more laxer environment for social capital flowing into railway construction faster, which is the key direction of our railway investment and financing reform during the latter half part of the *Twelfth Five-Year Plan* Period. That indicated the strong desire of railway departments to adjust investment structures and create benefits through increasing the proportion of social capital.

Research, (1).

[55] Ibid.

[56] The State Council. (2013). Opinions on reforming the investment and financing system to promote railway construction (GF [2013] No. 33).

At the end of 2013, representatives of National Development and Reform Commission mentioned that we should establish "diversified, risk controlled, and sustainable" railway investment and financing system under the rule of "pursuing progress and making achievements while ensuring stability". That means we need to continually expand railway investment and speed up railway construction, while it is also essential for us to pay attention to advance sound expansion and prevent investment and financing risks. To address the problem of deficient railway funds at the very beginning, we should fundamentally get down to railway investment and financing reform by means of taking sustainable system as guarantee, expanding fund resources, and increasing the proportion of diversified financing in the whole railway investment.

On 2 April 2014, the Railway Development Fund was established by the State Council at its executive meeting.[57] On 25 June, according to the *Approval of Plans for Establishing Railway Development Funds* (GH [2014] No. 52) by the State Council, the National Development and Reform Commission, together with the Ministry of Finance and the Ministry of Transportation issued *Measures for the Management of Railway Development Fund*, clarifying that the investment should be oriented to the approved projects by the country, and social investors should be able to gain stable and reasonable benefits, though they do not participate in railway construction directly.

On the one hand, there was a large amount of idling social capital which can be drawn to the railway sector. On the other, railway projects demand for funds but have no efficient connections to obtain them. Railway Development Fund, supported by the Central Government, is a governmental investment and financing platform for private capital. The establishment of the Fund is regarded as the substantial progress of railway investment and financing system, accelerating the reform of corporate governance, automatic market-oriented coordination and the separation of government and business functions. After that, Railway Development Fund was established separately in all provinces. In 2014, Jiangsu Railway Construction and Development Fund was established. In 2015, Guizhou Railway Development Fund Management Co., Ltd. and Guangzhou Railway Development Fund were set up. In 2016, Shandong Railway Development Fund Co., Ltd. was established. In 2017, the signing ceremony for the establishment Beijing-Tianjin-Hebei Intercity Railway Development Fund was held in Beijing.

In 2015, National Development and Reform Commission issued *Opinions of Encouraging the Private Capital to Invest in Railway Construction*, clarifying that we should comprehensively open the railway investment and operation market, and make railway projects listed on the *Mid- and Long-Term Railway Network Plan*, special and regional plans approved by the country, all accessible to social capital, except for those prohibited by laws and regulations. In September 2015, China Railway initiated new measures concerning the reform of railway investment and

[57] (2014, April 3). Railway Development Fund established by the State Council attracting social capital: 400 billion social capital invested in railway within two years. *Economic Information Daily.*

financing system, issued *Guidelines to Make the Management of Railway Construction Projects by Non-Holding Joint Ventures More Regulated*, and regulated the classification of railway investment.

"For railway construction projects held by non-China Railway holding joint ventures, according to *The Company Law* and the cooperation agreement between China Railway and local governments, investors and their authorized representatives should set up joint venture company for railway construction. As the legal entity of railway construction projects, the company should be responsible for the whole process of railway construction".[58] The three listed companies affiliated with China Railway, i.e., China Railway Construction Corporation Limited, China Railway Group Limited, and CREGC Architectural and Construction Engineering Co., Ltd., have benefited directly from the Guideline.[59]

In conclusion, railway investment and financing systematic reform during the Twelfth Five-Year Plan Period focused more on the coordination between government bodies, industries, and markets, providing more support regarding price reform, enterprise system, classification guidance, incentive mechanism, and resource utility. Such a reform was aimed at cultivating and improving the self-development of enterprises, to better associate and complement with other national economic strategies. Those features reflect the reality of railway construction at that time and also represent the resolution of the railway industry to address the dilemma of railway investment and financing through larger, sounder, and more practical reform initiatives.

The period from the Tenth Five-Year Plan to the Twelfth Five-Year Plan witnessed the formulation of a diversified investment and financing framework, and the pursuit of balanced development of various investment and financing models to attract social capital to railway sectors. During this period, railway investment and financing systematic reform gradually found its philosophy and path, exhibiting a progressive, step-by-step manner in depth and breadth. Although there are still challenges, those features align well with Chinese economic and social development and the historical process of China's Railway from weakness to mightiness. When stepping into the Thirteenth Five-Year Plan Period, the country saw progress from high-speed growth to high-quality development. The Central Committee has made major decisions that we should take supply-side structural reform as our main task and act on a new development philosophy to promote high-quality development. Railway investment and financing systematic reform, together with supply-side structural reform will also cross the threshold into a new stage.

In September 2015, to advance the supply-side structural reform of state-owned enterprises, State Council released *Suggestions on the Development of Mixed-ownership Economic Entities in State-owned Enterprises*, driving the social capital into railway sectors. In 2016, "the National Development and Reform Commission

[58] Qi Hui. (2015, September 7). New measures of China Railway on implementing the reform of investment and financing system. *Economic Daily.*

[59] Ibid.

listed 8 demonstrative projects with investments from the first batch of social capital in railway sectors. Those projects include 3 high-speed railway projects, 4 intercity railway projects, and 1 local railway project".[60] Thus, the "last kilometer" of railway construction invested with social capital was covered, encouraging and expanding the investment of social capital primarily by joint ventures, sole proprietorship, and Public-Private Partnerships in railway construction and operation.

Mid- and Long-Term Railway Network Plan, revised in 2016, emphasized that we should further reform in railway investment and financing system, attract diversified investors, ease market access, and encourage and support local governments and private sectors to invest in railway construction. *Guidelines for Further Reform in Railway Investment and Financing Reform* (CFF [2016] No.27) issued by The People's Government of Sichuan Province on 30 May 2016, clarified that "we should completely open railway construction market, develop land resources along the railway stations, and encourage social investors to participate in railway investment, construction, and operation".[61] After the issuance of the document, the People's Government of Sichuan Province in western China planned to invest 37.7 billion yuan in railway construction in 2016, with 38.2 billion yuan invested, ranking the first in Chinese railway construction investment.

At the Central Economic Work Meeting in 2017, the central government required that we make substantial progress in the reform of mixed-ownership economic entities in 2017; At the 2017 Work Meeting of China Railway held on 3 January 2017, China Railway made it their main task to promote the capitalization of railway assets and the reform of mixed-ownership economic entities. At the railway construction work meetings for some regions held on 23 February 2017, Ma Kai, then Vice Premier of the State Council at that time, also emphasized that the railway sector should reform in depth the railway investment and financing system, take targeted approaches, and raise funds through various channels to establish a sustainable framework of win-win cooperation among China Railway, local governments and social investors in railway construction.[62]

Under the conditions of further reform in the supply-side structure of the Chinese Railway, it was required at the Central Economic Work Meeting that the railway sector should put into place the reform of mixed-ownership economic entities through diversified equity investment, so as to reform Chinese railway investment and financing system and introduce market-oriented financing model.

---

[60] The eight projects are: Jinan-Qingdao high-speed railway, Wuhan-Shiyan railway, Hangzhou-Wenzhou railway, Langfang-Zhuozhou intercity railway, Central district of Chongqing-Hechuan railway, Hefei-Xinqiao International Airport-Liuan railway, Hangzhou-Shaoxing-Taizhou intercity railway, and Sanmenxia-Yuzhou railway.

[61] The People's Government of Sichuan Province. (2016). Guidelines for further reform in railway investment and financing reform (CFF [2016] No. 27).

[62] Ma Kai. (2017, February 23). Implementation of railway construction in a sound and smooth way and completion of annual task with safety and high-quality. www.xinhuanet.com. http://www.xinhuanet.com/2017-02/23/c_1120519889.htm.

During the Thirteenth Five-Year Plan Period, many successful cases showed that local governments and social investors invested in railway construction. For example, investment from the People's Government of Shandong Province accounted for 80% of the total investment in the Jinan-Qingdao high-speed railway, and the People's Government of Sichuan Province and China Railway respectively invested 90% and 10% of the total shares in intercity railway in the southern Sichuan province. The Mianyang-Suining-Neijiang Railway was the first railway project open to social capital in 2017. The government of Taizhou in Zhejiang province issued the first letter of pre-acceptance of the Public-Private-Partnership Project of suburban railways in China with 20% investment from the government and 80% from social investors. Moreover, the innovation of investment and financing models of "intercity railway construction with land development" was actively applied in the Guangzhou-Foshan-Jiangmen-Zhuhai Intercity Railway Project, featuring the philosophy of "equal shares enjoying equal rights" to attract the investment of social capital.

As the government is promoting the reform of the investment and financing system, the Public-Private-Partnership Project model for infrastructure construction has become an important measure to leverage social capital, address railway debt pressure, and push ahead the reform of railway mixed-ownership economic entities. As one with private investors holding majority shares in the history of Chinese railway investment, the Hangzhou-Shaoxing-Taizhou Intercity Railway Project, which commenced at the end of 2016 and received an investment of 40.9 billion yuan,[63] is called the "ice-breaking project" for social capital invested in railways, representing that the reform of investment and financing system has stepped into a new phase. However, the immature return-on-investment mechanism of China's Public-Partnership-Project (PPP for short) model became a significant factor in suppressing the enthusiasm of social investors. Especially in the earlier period, the railway PPP project was characterized by large-scale investment, long payback periods, and low benefits. How to increase the benefits and attract social investors in the next step would become an important breakthrough for the reform of railway mixed-ownership economic entities.

As shown in its financial report, China Railway's debt stood at 5.28 trillion yuan at the end of September 2018 and 1.89 trillion yuan at the end of 2010. Though reform on the railway investment and financing system continued during this period, the question of high debt in railway sector still remained. Investment and financing reform, as a significant part of railway reform, was implemented with "the separation of government and business functions", "the separation of infrastructure management from transport operation", and other comprehensive reforms to adjust investment and financing structure and increase the benefits by means of the reform of mixed-ownership economic entities. The process of such reform is an arduous task. It is stated in the report of the 19th National Congress of the Communist Party

[63] Wu Zhengyi. (2017, September 12). Hang-Shao-Tai project financed by private capital, the reform on railway investment and financing system further promoted. *Shanghai Securities News*.

of China that China should establish a global network of trade, investment and financing, production and service, and gather new strengths in international cooperation and competitiveness. That statement reflects a new orientation for Chinese economy and sets out new requirements for Chinese railway construction to optimize global resource configuration. In the future, the reform of Chinese railway investment and financing system will be continuously promoted in depth.

5.3 Dissolution of MOR

5.3.1 Process of "The Separation of Government and Business Functions" in MOR

Due to historical and internal factors, railway institutional reform has always been regarded as the key part of whether Chinese railway can be open to the market, as well as the crux of the challenges between railway and government, railway and market, railway organizations with other transportation ones. It is vital for railway institutional reform to distinguish the role and function MOR and its subordinate departments play in state politics and economy at different historical stages. Taking Chinese railway institutional reform as an example, its focus was on "the separation of government and business functions" in MOR.

MOR was established on 1 October 1949 according to Article 18 of the *Organic Law of the Central People's Government of PRC* passed by the Chinese People's Political Consultative Conference. In September 1954, as required in *Notice on Some Issues Concerning the Establishment and Adjustment of Central and Local Administrative Institutions by the State Council*, MOR was founded by the state following provisions on *The Organic Law of the State Council of the People's Republic of China,* to meet the demands of China's military and economic development.

During the industrialization in the early period, PRC needed a powerful and centralized governmental department to unify the construction, distribution, and operation of railways in China's territory, providing extensive transport capacity for industrial construction and solid transport support for the flow of passenger and social prosperity. Under such circumstances, MOR continued to develop and expand the scale of its business. Since its establishment in 1949, the Ministry has always played a significant role in national defense, which was strengthened by the large-scale "Third Front Construction" with the guidance of combat readiness during the 1960s. From 1970 to 1975, MOR, Ministry of Transport, and Ministry of Posts and Telecommunications were merged into the Ministry of Transport. However, MOR resumed its status as an independent ministry again and continued to develop after 1975. In 1982, it set up a three-level railway transport procuratorate and railway transport court. In 1984, railway corps were merged into MOR, thus an "independent kingdom" came into being, providing comprehensive functions in military, government, enterprise, public security, procuratorate, and court areas.

Since the reform and opening up, MOR has played a weaker role in national defense but a more important role as the main artery of transportation. During a long period, however, the railway supply-demand situation had been under extreme strain, because the inadequate railway capacity could not meet the needs of economic and social progress, but the deep reason was that the monopoly of railways limited its own development so that railway industry failed to meet the demands of market-oriented economic development, resulting in the gap between supply and demand.

After the reform and opening up, MOR always maintained the system of a "combination of government and business functions". It was normal to see railway departments had their own schools, hospitals, factories, and even the system of public security, procuratorate, and court, which reflected that the Ministry had become the largest state-owned enterprise at that time. Before the reform and opening up, the system of "combination of government and business functions" in railways indeed pushed ahead its development in the special historical period. With the reform of the economic system implemented in all sectors and the socialist system's ongoing self-improvement, however, that system became a large barrier impeding the sound and sustainable development of railways.

Railways indeed implemented self-reform but the process was slower than that in other sectors. The incomplete reform in "the separation of government and business functions" failed to solve the fundamental problems. The consequence was that railways encountered various challenges in the fierce market competition and lacked vitality in market-oriented production and operation. Activities of railway business operations were arranged under the administrative instructions and plans issued by MOR. Therefore, railway enterprises did not have real autonomy in operation and bore no responsibility for profits and losses. Under such circumstances, railway departments had inefficient productivity and failed to make immediate and effective responses to changes in the market. Without dynamism from inside and pressure from outside, its business condition was in a fix.

In terms of the market share of transportation, the turnover of passenger and freight traffic accounted for 50% in 1990, but in 2011 the proportion of the turnover of passenger traffic dropped to less than one-third and that of freight traffic decreased to less than one-fifth. At the same time, highway and civil aviation had a rapid development with an increase in the volume of passenger and freight traffic turnover. Highway alone accounted for more than half in the turnover of passenger traffic and nearly one-third of freight traffic. Railway enterprises were overstaffed in departments concerning internal construction and external relations because of the burdens of achieving various social functions. Although Chinese railways had begun to reform the separation of sideline businesses from core businesses in 1990s, not until 2003 did its sideline institutions like hospitals and schools get separated. In 2012 the reform in public security, procuratorate, and court eventually made a substantial breakthrough. All of those problems seriously limited the progress of railways at different levels, at the same time causing a series of new problems.

In fact, the separation of government and business functions determines whether railway enterprises, as operators of state-owned assets, can exist as independent legal entities in market economy. In short, the separation makes government and enterprises perform their own duties and exercise their rights without intervention and overlap. For example, the government, as the administrator and supervisor, formulates railway development strategies and plans, as well as various supervision laws and regulations, while railway enterprises, as the operators, are responsible for specific operations and transportation and assume corresponding responsibilities. Over the past 40 years of the reform and opening up, China's economic system reform has taken great strides forward, and its corresponding political system reform has also steadily advanced in a direction conducive to the development of a socialist market economy. The railway system, a significant representative of the planned economy, also needed to accomplish the transformation of its functions by means of its own institutional reform, so it was a necessity to separate government functions from business ones. There had always been calls from the state and society for the reform in railway system, but for a long time no consensus had been reached within railway departments on how to do it.

Reform is a gradual process and cannot be achieved in only one step. Deng Xiaoping once said, "Empowering enterprises and separating government and business functions are an economic system reform and also a political system reform".[64] Railway institutional reform and transformation of its functions are closely related to Chinese governmental structural reform, which is characterized by stepping into the market and gradually achieving the separation of government and business functions. The First Session of the Ninth National People's Congress on 10 March 1998 passed the *Decisions on the Proposals for the Institutional Reform of the State Council*, clarifying that "the nation should establish an efficient, coordinated, and normative government administration system, and improve civil servant system to set up professional administration group with high quality and gradually establish government administration system with Chinese characteristics suitable for socialist market economy".[65]

The basic principles of reform are as follows: "As required by socialist market economy, the nation should separate government functions from business ones; adjust government organizations and institutions to streamline administration with fewer but better staff in accordance with the principle of simplification, unification and efficiency; adjust duties and authorities of government departments, clarify the division of duties among departments, and improve administration operation mechanism following the principle of unification of rights and responsibilities; strengthen law enforcement of administration system to meet the requirements of law-based governance and administration".[66]

Under the guidance of the principles, MOR pushed forward institutional reform and transformed the functions of institutions to accelerate the separation of

---

[64] Literature Research Office of the CPC Central Committee. (1993). *Selected works of Deng Xiaoping* (Volume III) (p. 192). Beijing: People's Publishing House.

[65] (1998, March 11). Proposals for the institutional reform of the State Council. *People's Daily.*

[66] Ibid.

government functions from business ones, aiming to change the integration of functions and the government's direct intervention in business operation. Therefore, the party leadership group of MOR set four reform principles. "The reform should be conducive to separating government and business functions to ensure the dominant role of enterprises in the market; streamline administration with fewer but better staff to improve administration efficiency; optimize staff structure and reform in the personnel system; and accomplish two major tasks of railway reform and railway construction in the new era".[67] Then, MOR issued institutional reform proposals and put them into practice by establishing two new institutions to address in particular the difficult problems such as how to manage railway operation and construction in the process of railway reform. One of the institutions was the Transport Command Center for unified management of railway operation in five fields: train depot, locomotive management, railway line maintenance, railway signal engineering, and train maintenance. The other one was the Engineering Management Center to straighten out the relations of railway construction and management system. Some phased achievements were made during this round of reform, such as the transformation of institutional functions, clarification of responsibilities of different departments, laying off redundant staff, and optimization of staff structure.[68] It was also an initial attempt of MOR to separate government and business functions since the practice of function integration dates back to the period of the planned economy.

To some extent, the attempt promoted the separation in railway sectors. However, determined by long-term management inertia, special industry background, complex internal relationships, and tremendous workload, railway system reform could not be accomplished in a short period. Those were not only the challenges for railway reform to overcome but also the significant reason why the reform had always been stagnant, inefficient, and even failed. Railway reform involved a wide range of departments and was influenced by various factors. Therefore, not only a horizontal transformation of government functions, reform of institutions, and adjustment of staff structure but also a vertical hierarchy reduction or a separation of different departments, was tantamount to a "critical surgery" for the railway sector. For example, the failed "Three Separation Reform" (the separation of railway from transport network, the separation of transport network and transportation, and the separation of freight and passenger transport) in 2003 provided a great experience for railway reform. No matter how hard the difficulties were, railway reform about the separation was always on the way. Although it was a long-term and arduous task, the nation's resolute in promoting railway reform never wavered.

On 15 March 2008, the First Session of the Eleventh National Congress passed *Decisions on the Proposals for the Institutional Reform of the State Council*, clarifying the reform task as transforming government functions and optimizing responsibilities of different departments to integrate departments with similar functions into

---

[67] Sun Yongfu. (1999). Promoting the reform on the institutions, the transformation of governmental functions, and the separation of government and business functions. *Railway Economics Research*, (1).

[68] Ibid.

a big one. The functions of macro-control departments should be structured rationally, the management institutions of energy and environment strengthened, the management of industrialization, informatization, and transportation integrated and perfected, and social management and public service departments reinforced and integrated, with great importance attached to improving people's well-being".[69]

At that time, it was obvious to see the trend of establishing a larger Ministry of Transport with comprehensive departments, but MOR was not integrated into it like the civil aviation department, postal department, and others. "In terms of the particular conditions in railway construction and management, MOR was retained, and meanwhile, continued to advance such reform".[70] The reasons why MOR was not abolished temporarily were based on two judgments by the state. On the one hand, the reform in railways was not so mature. With the development of a national integrated transport system, it was urgent to consider the coordination of railways, highways, civil aviation, waterways, and other transportation modes. However, railway reform lagged behind that of other industries, so railways itself had to further the research on "how to reform". On the other hand, Chinese railways witnessed rapid growth in large-scale construction at that time, and a fundamental railway network was being constructed. Besides, passenger and freight double-track railways had shortcomings in the criteria set for them, failing to meet the requirements of the establishment of the integrated transport network. Therefore, it was expedient to retain MOR to complete great missions faster and better, pool professional resources and strengths, ensure the great development of railways in a certain period, and finally meet the demands of our country's economic and social progress as soon as possible. But in terms of the broad landscape of national institutional reform, it was necessary to merge MOR into the Ministry of Transport. So, the revocation of MOR is only a matter of time.

On 14 March 2013, the First Session of the Twelfth National Congress passed *Decisions on the Proposals for the Institutional Reform and Transformation of Functions of the State Council (Draft)* and decided to abolish MOR, achieving the separation of government and business functions. "The administrative duties for making railway development plans and strategies by MOR are redistributed to the Ministry of Transport. The National Railway Administration will be established. Under the management of the Ministry of Transport, this newly established department will undertake other administrative duties of MOR".[71] Besides, an enterprise — China Railway — will be founded to perform business functions of the former MOR. Other than that, considering the facts that railways were still in an important period of opportunities for construction and development and that there were national demands for railways enterprises to undertake public welfare tasks, the state will continue to support the construction and development of railways, the

---

[69] (2008, March 16). Proposals for the institutional reform of the State Council. *People's Daily.*

[70] Ibid.

[71] (2013, March 15). Proposals for the institutional reform and the transformation of functions of the State Council. *People's Daily.*

establishment of modern enterprise system and other goals of the railway sector, on the basis of the balance among infrastructure, operation and railway network. It was widely believed in the industry that the abolition of MOR was not an end, but the beginning of a better future. It represented the revolutionary reform of the railway system.

The abolition of MOR lifted the curtain of the fundamental and systematic reform of Chinese railways from top to bottom, showcasing the Central Party Committee's keen understanding of the complex dimensions of issues and progressive steps to the times as well as its courage and resolution to renovate and further railway development. After that, the Ministry of Transport performed administrative duties to make railway development plans and strategies, rectifying the defects of simplification and monopoly caused by MOR itself. Therefore, the integrated development of various forms of transportation was facilitated, and the great transport landscape was formed under unified plans, coordinated development, and "seamless" connections; the newly established National Railway Administration is subordinate to the Ministry of Transport, which has changed the weird situation of "MOR supervising itself" and truly put into practice the sound interaction among government supervision, business operation, and social overwatch. The establishment of China Railway represents the birth of a new "giant" state-owned enterprise and a true beginning for the railways to enter the market. As for the operation and management of railways, the most distinct changes were reflected in the establishment of two "major roles" of China Railway: one is that China Railway has become the market entity with operation autonomy within the socialist market economic system, further invigorating railway enterprise's vitalities; the other is that China Railway has become the entity responsible for safe operation. All in all, the establishment of China Railway removed barriers to the introduction of competition in marketization and enterprization.

5.3.2 Establishment of Modern Railway Enterprise System in the Post-MOR Era

Modern enterprise system features clear property rights, distinction of right and responsibility, and separation of government and business functions. The dissolution of MOR and the establishment of China Railway mark the beginning of railway institution reform and function transformation in the new era, and the first important move, also the prerequisite foundation for Chinese railways, towards the market in a real sense. However, building a market-oriented and modern railway enterprise with a new mechanism is by no means something that can be accomplished by just renaming MOR as China Railway. How to take the next step in line with the requirements of modern enterprise systems is a new project on reform for the newly-established China Railway. It is crucial to maintain the continuity of reform, improve the effectiveness of the reform and incentivize the industry to reform in an attempt to achieve the ultimate goal. After the dissolution of MOR in 2013, China Railway proceeded to carry out a series of reforms in institution and function transformation.

In the second year after the dissolution, China Railway was committed to sorting out the existing interdepartmental relations under the new management system, by issuing *Provisions of China Railway on the Clarification of the Management between Two-level Enterprises* and advancing the "1 + 23 + N" management system. The *Provisions* ordered the company to establish the basic requirements of the operation mechanism under the new system and expedite the formulation of a system according to the framework of the new management system. To be more specific, "1" refers to the parent document of the management system, *Provisions of China Railway on the Clarification of the Management between Two-level Enterprises.* The document defines the function positioning of China Railway and National Railway Administration, the basic principle of the division of management rights and responsibilities for the legal entities in the two-level enterprises, and the management mode of China Railway to its subsidiaries. "23" refers to the basic management system in each business field, and "N" to several implementation measures matching the basic management system of relevant business fields. Through such a new system, the function positioning and division of rights and responsibilities of both institutions have been defined to enhance China Railway's operation and management of railways.

On May 24, 2015, China Railway issued *Proposals of the China Railway for the Further Reform of Railway Enterprises* which specified the basic principles, ideas and tasks of enterprise reform and proposed the deepening of railway enterprise reforms in the following six aspects: "first, Comprehensively strengthening operational mechanisms to enhance enterprise vitality; second, deepening the reform of railway transportation organizations to improve market competitiveness and transportation service level; third, boosting asset operation and development to improve the efficiency and benefit of railway resource utilization; forth, implementing policy measures for railway investment and financing system reforms to ensure the high-quality and efficient completion of railway construction; fifth, deepening the railway management reforms for joint ventures to promote standard and smooth operations; implementing the strategy of China railways' "going global" to improve the international competitiveness of Chinese railway enterprises".[72] Overall, the proposal specified the basic directions and main methods of railway enterprise reforms, laying an important policy foundation for the establishment of the modern enterprise system.

On December 5, 2018, China State Railway Group Co., Ltd. officially completed the registration of its name change, marking a further step in China's railway enterprise reforms and a market-oriented system upgrade of China Railway and its subsidiaries. Sustained efforts were made in the reform: first, 17 non-transportation

[72] The Archive Center of China Railway. (2016). *China Transportation Yearbook (2015)* (p. 67). Beijing: China Railway Publishing House.

companies affiliated to China Railway were restructured, including China Railway Investment Co., Ltd., China Academy of Railway Sciences Corporation Limited, and People's Railway Newspaper Office; second, in November 2017, a total of 18 railway affiliated administrations officially completed the change of business registration for the corporate system reform, each being renamed Railway Administration Group Co., Ltd. in succession, and all entities of railway transportation were officially marketized; finally, China State Railway Group Co., Ltd. was established after the corporate system reform of China Railway itself. In the long run, implementing a corporate governance structure is a major strategic decision for Chinese railways to build a modern enterprise system and completely "decouple" from the traditional development mode. The corporate system reform is also an inevitable choice for Chinese railway enterprises' upgrading from a traditional transportation mode to a modern one.

After the dissolution, China has made great strides in railway reforms, especially since the 18th National Congress of the Communist Party of China. For example, in August 2018, the semiannual report released by the China Railway showed that in the first half of the year, the company had a total revenue of RMB 498.831 billion and an after-tax loss of RMB 368 million, down 87.6% compared with the RMB 2.968 billion of the previous year. Nevertheless, the remaining losses decreased significantly while the revenue increased. In the future, China State Railway Group Co., Ltd. will continue to reform. Only by accelerating the market-oriented and corporate reform can Chinese railways integrate into the modern transportation system in the new era as soon as possible.

5.4 Building of Railway and Integrated Transportation System

5.4.1 The Beginning and Development of Integrated Transportation in China

The concept of integrated transportation originated from the socialist countries led by the Soviet Union to avoid free competition in the context of the planned economy, and these countries proposed the building of an integrated transportation system. In the late 1950s, China introduced such concept from the Soviet Union into the PRC's first medium and long-term scientific and technological program, *The Long-term Program for Developing Sciences and Technology from 1956 to 1967* which specified the research on integrated transportation, especially the development plan of integrated transportation network. Following the Soviet Union's mode, the Institute of Comprehensive Transportation, China's first national integrated transportation research institution, was established in March 1959, which represented the official beginning of theoretical research on integrated transportation. Around 1960, China introduced specific practices in integrated transportation, such as the coordinated construction of multiple transportation modes, including railway-water intermodal transport and corridor transport. However, under the influence of the ideology and development model of the planned economy, the theoretical research and practice of integrated transportation were constrained by the low-level economic and social foundation in the early days since the founding of PRC in 1949, especially the lagging development of the overall national transportation. As a result, the research and practice in the initial phase were weak and slow.

The management departments of various transportation modes had for a long time not reached a consensus on the development of China's transportation. Meanwhile, the national fiscal investment was mainly channeled into the construction of transportation infrastructures such as railways, highways, and waterways, among which railway was a priority. Railways had served as the "single log bridge" and "main artery" in the transportation industry until the beginning of reform and opening up. In such a specific historical period, it was undoubtedly correct to center on the railway and "principal contradiction" in the development of transportation after the founding of PRC, whether in consideration of China's geographical and economic characteristics or the technical and economic advantages of the railway itself.

In the 1980s, with the development of the commodity economy, the prominent bottlenecks in the transportation industry severely constrained the development of the national economy. At the 12th National Congress of the Committee Party of China, it was determined that transportation would work as the strategic priority to achieve the goal at the end of the century. As suggested by the Institute of Comprehensive Transportation, the blue book of *Technological Transportation Policy* compiled by the National Scientific and Technological Commission in 1984 proposed that a modern integrated transportation system with reasonable economy and coordinated development shall be established as soon as possible.

In 1987, a discussion on the development of an integrated transportation system was held nationwide. In the same year, the concept of "integrated transportation system" was used for the first time in the 13th National Congress of the Communist Party of China, in an attempt to highlight that strengthening the construction of an integrated transportation system was the basic direction of the adjustment and transformation of transportation industrial structure for a long time in the future, but without an explicit policy for implementation. Subsequently, a series of in-depth discussions were held in the academia and management departments on several issues of integrated transportation system, of which whether railway should be the priority in the construction of integrated transportation system became a bone of contention. For example, some articles on the comparison of technology and economy pointed out that "it was debatable to consider railway to be dominant in the transportation industry", "railway was reduced to 'sunset industry' in the world" and "the leading role of Chinese railway was not stable any more".[73] Li Peng, then Premier of the State Council, proposed that "proactive actions should be taken in the

[73] (1989, August 26). It is better to establish integrated transportation system. *Economic Daily.*

development of integrated transportation system with the railway as its priority" at the Second Session of the Seventh National People's Congress. In August 1988, when inspecting the ring road test of the China Academy of Railway Sciences, Li Peng further pointed out "We shall give priority to the railway, the main artery of China, in such a landlocked country by promoting integrated transportation and multiple transportation modes". He also confirmed that China's integrated transportation system should focus on the railway.

In the 1980s, at the beginning of reform and opening up, given the overall lagging of the transportation industry, various transportation modes were of great importance in the development of the national economy under the effect of "bottleneck". Therefore, it was still the top priority to increase the capital investment and to accelerate the construction of infrastructure. Although the research and practice of an integrated transportation system were propelled, efforts were restricted by some aspects like rational division of labor, proportional development, comprehensive balance, and capital channels. Thus, some advanced ideas such as large channels and function combinations in integrated transportation were still in the initial and controversial stage, especially since plans often regarded integrated transportation as a simple sum of various modes of transport.

After the rapid development in the early phase of the reform and opening up, and especially after the South Tour Speeches by Deng Xiaoping in 1992 and the proposal of building a socialist market economy at the 14th National Congress of the Communist Party of China, the transportation modes which used to account for a small proportion of the integrated transportation had been developing at a rapid pace, including highway transportation, water transportation, and civil aviation, by virtue of the relevant policies, investment channels, and open markets. The proportions of multiple transportation modes thus varied significantly. Under such circumstances, the demand for transportation also had changed accordingly, but transportation had been all the more needed in general.

In 1991, the Eighth Five-Year Plan proposed to "construct a satisfactory integrated transportation system by prioritizing the growth of railway transportation capacity, actively giving full play to the advantages of multiple transportation modes to match each other". In 1996, the Ninth Five-Year Plan further clarified that on the basis of the last Five-Year Plan, "the construction of integrated transportation system shall be accelerated to form a number of east-west and north-south corridors with strong transportation capacity. On November 23, 1996, *Constructing a Unified Comprehensive Transportation Network*, an article signed by Premier Li Peng, was published in *China Transport News*, describing "the basic framework of China's integrated transportation system for the first time. In this article, China's transportation development shall be designed to center on railways and based on highways, by fully utilizing the resources of inland, coastal, and ocean transportation and actively developing aviation to form a unified integrated network system for both short and long distance transportation with different functions and easy accesses from all directions throughout the country".[74] After the Asian financial crisis in 1997, driven by the "proactive fiscal policy and expansion of domestic demand", the investment amount of infrastructure development in highways far exceeded RMB 60 billion in railways for several consecutive years, and the investment in civil aviation had also soared in comparison with that during the period of Eighth Five-Year Plan. No matter from the perspective of policy guidance or capital investment in the 1990s, the research and practice of China's integrated transportation were upgraded into the integrated and coordinated development of multiple transportation modes from a simple division of labor according to technical and economic characteristics, and the concept of integrated transportation thus begun to gain popularity.

5.4.2 Medium-and Long-Term Integrated Transportation Network Plan

With the debut of the twenty-first century, a new round of large-scale transportation infrastructure construction had started, along with the accelerating process of industrialization, urbanization, economic globalization and regional integration. Propelled by a stronger impetus for development, the multiple transportation modes represented by railways had entered a phase of improving the layout of road networks and the structural level. A comprehensive development pattern had therefore emerged, exceeding the initial scope of individual transportation mode plans.

The Tenth Five-Year Plan published in 2001 proposed to establish an unimpeded, safe, and convenient modern integrated transportation system. The research and practice of domestic integrated transportation had also evolved, and was further upgraded to a new level to meet the needs of diversified and customized transportation. In this way, the connection, combination, and integration of multiple transportation modes have contributed to a new transportation development mode, and the interaction between transportation and economy has presented the coordinated development pattern featuring the continuous optimization of resource allocation by market and the comparative and combined advantages.

On October 31, 2007, as the first national spatial layout plan in overall transportation since the founding of PRC, the *Medium- and Long-Term Integrated Transportation Network Plan* was deliberated and approved at the State Council's Executive Meeting, covering the comprehensive connection of five transportation modes, such as railway, highway, waterway, civil aviation and pipeline. Guo Xiaobei, then Director of the Institute of Comprehensive Transportation of National Development and Reform Commission, suggested that the publication of this plan was the result of the development of total transportation infrastructures, the new

---

[74] Li Peng. (1996, November 23). Constructing a unified comprehensive transportation network. *China Transport News*.

phase of economic and social development, the condition and opportunity of the era and a new leap in the practice of integrated transportation.

The preparation of the plan included four phases over five years: the preliminary research in 2003, the research on special projects in 2004, the preparation of the plan from 2004 to 2005, and the coordination with other plans, expert review, and planning approval from 2006 to 2007. Eight specific aspects were covered in this plan to solve the problems of total traffic volume, layout, and interconnection. The plan focused on the practical problems of insufficient transportation infrastructure and imperfect basic networks, as well as the external constraints from resource, environmental, and ecologic pressure. Through the system optimization, mutual connection, and optimization of the five transportation modes above, the planning, construction, and management of transportation infrastructure should be coordinated. Besides, efforts should be made to adjust structure, optimize configuration, promote efficiency, and shift the mode of transportation development, to achieve intensive, efficient, and sustainable development.

As for the top-level design of transportation, the levels from high to low can generally contain four main frameworks: national overall development strategy, transportation development strategy, development plans of integrated transportation, and all transportation modes. Then a complete planning system shall be formed from the dimensions of the overall landscape, programmatic literature, guiding planning of the sector, and implementation planning of each transportation mode. The plan of integrated transportation at least includes the integrated transportation network plan and system plan, covering three main aspects: transportation infrastructure, mobile facilities, operation management, and control system. To be more specific, the transportation infrastructure network serves as the problem-solving basis of integrated transportation, while the preparation of the integrated transportation network plan is a leading one. The plan aims to design the spatial layout of the main framework for the national transportation network with a focus on corridors and hubs to achieve multi-modal intermodal transport, by constructing "five vertical and five horizontal"[75] corridors and giving full play to the overall advantages of integrated transportation. In essence, the integrated transportation network is built to improve the system efficiency of transportation and reduce the cost of transportation services, contributing to economic and social progress.

The plan centers on building the main framework of the national land transportation network, namely the main lines of national transportation, to provide solutions to the problems of personnel and material exchange in different geographic regions of China. Despite the planned time from 2006 to 2020, China's integrated transportation corridors had been basically finalized as specified in the plan. The large-scale

[75] The "five vertical" includes the North-South coastal transport corridor, the Beijing-Shanghai transport corridor, the Manzhouli-Hong Kong/Macao/Taiwan transport corridor, the Baotou-Guangzhou transport corridor, and the Linhe-Fangchenggang transport corridor. The "five horizontal" contains the North-West sea transport corridor, the Qingdao-Lhasa transport corridor, the land bridge transport corridor, the transport corridor along the river and the Shanghai-Ruili transport corridor.

transportation construction lasted for 15 years or so, basically solving the transportation problems between regions across the country, and the follow-up construction of the integrated transportation network would focus on its supplementation and improvement. It was based on the construction of the integrated transportation network that the plan put forward new development goals for railway construction. In particular, railways played an essential role in the construction of "five vertical and five horizontal" corridors, which directly accelerated the adjustment of the *Plan in 2004* by relevant competent departments. As an upgraded version of the *Plan in 2004*", the *Plan in 2008* after the adjustment was forward-looking, systematic, and scientific under the guidance of the Scientific Outlook on Development, and it was also a programmatic document guiding China's railway construction and development in the future under the overall layout of the integrated transportation network, featuring "connection", "high speed", "eco-friendliness" and "innovation". It fully demonstrated the latest trend and direction of the construction of the integrated transportation network with Chinese characteristics and railway modernization. Focusing on optimizing the overall structure of transportation, giving full play to the comparative advantages of railways and constructing corridors in transportation, the *Plan in 2008* included the 29,000-kilometer-long new line in accordance with the *Medium-and Long-Term Integrated Transportation Network Plan.* Meanwhile, efforts were made in strengthening the major passenger and freight transport hubs and greater attention was paid to the effective connection of each transportation mode to achieve the integration of transportation.

5.4.3 Development of Railway and Modern Integrated Transportation System under the Strategy of Building China's Strength in Transportation

On December 29, 2016, the Chinese government issued a white paper entitled *Development of China's Transport* which was the first of its kind focusing on the construction and development of China's transportation network. According to the white paper, "Through years of reform and development, China has initially formed a multi-nodal and full-coverage transport network and opened up 'five vertical and five horizontal' corridors in integrated transportation".[76] The white paper concluded the milestone achievements in the development of five transportation modes respectively.

First, by the end of 2015, China's total railway operation mileage ranked second in the world, and high-speed railway mileage ranked first in the world. Second, a full-coverage highway network has been set up, with the expressway mileage

[76]The State Council Information Office of the People's Republic of China. (2016). Development of China's Transport. Retrieved August 13, 2020. http://www.gov.cn/zhengce/2016-12/29/content_5154095.htm#1.

ranking first in the world. Third, a water transport network connecting main and branch lines has been established, and China topped the list of inland waterway navigable mileage worldwide. Fourth, a civil airport system has taken shape. Fifth, China's express delivery volume led the world, and oil and gas pipelines formed a backbone network. China thus has formed an integrated transportation infrastructure network. Since the problems of the integrated transportation infrastructure network have been solved, China's integrated transportation system plan should be compiled to improve the development quality and comprehensive efficiency and accelerate the transformation and upgrading of the integrated transportation network.

On February 28, 2017, the *Plan Outlining Major Targets for Transportation Network Development in the Thirteenth Five-Year Plan period (2016–2020)* (GF [2017] No. 11) was unveiled by the State Council, marking a new stage for the development of China's integrated transportation. This plan was introduced with the distinctive backgrounds and characteristics of the times. According to the report released by the Chinese Academy of Social Science in 2014, China's potential economic growth could range from 6.4% to 7.8% in the next five years. When China entered the new normal of economic development in 2014, transportation could play a prominent role in supporting economic transformation and upgrading. With the continuous improvement of transportation structure and the growing demand for personalized transportation, intelligent transportation, and emerging transportation, how to provide quality transportation services for building a moderately prosperous society in all aspects under the new economic conditions matters in the plan, in response to prominent transportation problems in function, structure and region. Notably, the concept of "the modern integrated transportation system" was proposed in this plan for the first time as a qualitative leap in comparison with the previous "integrated transportation system". It mainly highlights the modernity in management method, market demand, and development mode, comprehensiveness in layout, connection, and service, and systematicness in passenger and freight transportation system and the improvement of safety system".[77]

As socialism with Chinese characteristics has entered a new era, the evolution of the principal contradiction facing Chinese society has placed new and high demands on the development of a transportation system due to "its failure to meet the people's ever-growing needs for a better life in the quality and efficiency of transportation services".[78] The report of the 19th National Congress of the Communist Party of China put forward the goal of building China's strength in transportation, as well as a new mission for transportation in the new era. At the working meeting of China Railway held on January 2, 2018, China Railway proposed for the first time to shoulder the historic mission of "transport empowering China with the railway as

[77] Gao Yan. (2017). On the integrated development of railway and modern comprehensive transportation system. *Theory Studying and Exploration*, (3).

[78] Wang Zhanquan, Liu Tao, & Guo Peihong. (2018). Research on planning thought and concept innovation of comprehensive transportation system based on building a powerful transportation country. *Journal of Transport Management Institute Ministry of Transport*, (4).

the leading part" to achieve "the goals of 'leading in the world from three respects and three further improvements'".[79] At the working meeting in 2019, the company specified that "efforts should be made to promote the seamless connection between railway hubs and other transportation modes for the construction of a modern integrated transportation system, from the perspective of further improvement for the position and role of the railway in this system".[80] Since the railway has served as the main artery of the national economy, the national key infrastructure, and the people's livelihood project in the new era, we should give better play to its comparative advantages and critical role in the further promotion of constructing modern integrated transportation system. The integrated development of railways and the modern integrated transportation system has become the overarching theme of the reform and opening up for railways in the new era.

The transportation industry has been the main source of energy consumption and emission, and the exhaust emission of road freight vehicles, especially medium heavy-duty diesel vehicles, has been a major source of air pollution. The energy consumption and pollutant emission of the turnover per unit of goods in railways accounted for 1/7 and 1/13 of those in highways respectively, and railways were more advantageous in long-distance transportation simply in terms of environmental protection and resource conservation. However, according to the statistics released by the Ministry of Transport in 2019, from 2008 to 2017, among all modes of transportation, the proportion of road freight traffic rose from 74.1% to 78%, while that of railway freight traffic dropped from 13.2% to 7.8%. In 2017, the railway freight traffic increased by 10.7% compared with that in 2008,[81] while the road freight traffic increased by 91%. In consideration of the status quo, the comparative advantages of railways like low cost, low energy consumption, and high coverage were not utilized in the whole integrated transportation system.

This phenomenon resulted from the following two aspects: first, the insufficient capacity and accessibility of railway transportation in some regions; second, the inverted transport prices between highways and railways due to weak transportation management and imperfect mechanisms for transportation departments. Therefore, the Ministry of Transport issued a three-year action plan to restructure transportation and the six nationwide actions, requiring that by 2020 the railway and waterway transportation volume would see an increase of 1.1 billion tons and 500 million

[79] Lu Dongfu. (2018). Railway leading the transportation power to make greater contribution to the sustainable and sound development of economy and society: Report on the working meeting of China Railway (Abstract). *Chinese Railways*, (1).

[80] Lu Dongfu. (2019). Shoulder the historic responsibility of playing the leading role of railway in building China into a strong country in transport and work hard to make progress in China's railway reform and development in the new era (Abstract). *Chinese Railways*, (1).

[81] According to the *2018 China Railway Transport Industry Analysis Report—Market Operation Situation and Investment Prospect Research*, from 2012 to 2015, China's railway freight traffic decreased in succession and even showed an accelerated downward trend in 2015. In February 2016, the railway freight traffic decreased by 12.3%. In 2017, the national railway freight transportation showed an obvious recovery, with a total freight traffic of 3690 million tons, up 10.7% year on year, reversing the decline for three consecutive years.

tons respectively whereas the harbor and road transportation volume a reduction of 440 million tons.

It was noted that the National Development and Reform Commission, Ministry of Transport, and China Railway jointly issued *the Thirteenth Five-Year Plan of Railway Container Multimodal Transport* in April 2017. This plan is the first special plan of railways formulated at the national level and also the programmatic document of multimodal transport led by railways, reflecting the integration with other documents of national logistics development and multimodal transport in terms of directions and measures. It indicates that the Chinese government has determined to support the development of multimodal transport with railway transportation as the backbone by fully utilizing the advantages of railway concentration and heavy-haul railways and promulgating preferential policies at the institutional level. In this way, railway freight transportation will play an increasingly important role in the national integrated transportation system.

As China evolves from the role of falling behind to co-leading or now leading the world in high-speed railways, its rail passenger traffic volume, especially in high-speed electric multiple units, has increased year by year despite a continuous decline in national passenger traffic volume in recent years. According to the statistics from the *Statistical Bulletin on the Development of the Transport Industry* and National Bureau of Statistics, the national total passenger traffic volume in 2016 reached 19.002 billion, with a year-on-year decrease of 2.2%; the volume in 2017 hit 18.486 billion, down 2.7% year on year; the volume in 2018 amounted to 17.9 billion, with a continuous downward trend. However, the railway passenger traffic volume steadily grew during these three years in general. In 2016, 2.814 billion passengers traveled by railways, with a year-on-year increase of 11%; In 2017, the national railway passenger traffic volume reached 3.08 billion, with a year-on-year increase of 9.5%; In 2018, the volume hit 3.37 billion, up 9.4% year on year. Conversely, the road passenger traffic volume saw an obvious downward trend due to the diversion of high-speed railways, the growth of private cars, and the continuous disappearance of demographic dividends. In 2016, the road passenger traffic volume was 15.428 billion, down 4.7% year on year; In 2017, the volume was 14.568 billion, down 5.6% year on year; From January to November 2018, China's road passenger traffic volume reached 12.6 billion, with a year-on-year decrease of 6.4% in total. According to the analysis based on the proportion of passenger traffic volume of various transportation modes in the integrated transportation system, railway transportation accounted for 14.8%, 16.6%, and 18.82% while road transportation for 81.1%, 78.8%, and 71.5% from 2016 to 2018 respectively.

With the continuous improvement of the rapid passenger transport network jointly formed by high-speed railways and other railways, the pulling effect on passenger flow will be expanded by virtue of the operation of high-speed railway network, improvement of high-speed railway transport capacity and optimization of railways services, and the development of high-speed railways will continue at a faster speed. In the construction of the modern integrated transportation system, a sharing mechanism of passenger traffic volume relatively suitable to a specific development stage will be established under the mutual influence and continuous optimization of different passenger transport modes, such as railway, highway, waterway, and civil aviation. Especially, in constructing integrated transportation hubs, each transport mode will give full play to its comparative advantages and synergistic effects to provide convenient, fast, green, safe, intelligent, and diversified mode combinations for people's travel.

"For transportation is a subsystem in the social and economic system, it should be coordinated with the development requirements of other industries and departments in terms of the final size of resources and transportation infrastructure".[82] The relation between the total stock of infrastructures and the passenger and freight traffic volume in various transportation modes is not tantamount to simple aggregate or in proportion. However, in the integrated transportation system, various transportation modes shall be used to couple the advantages and benign competition among different modes under the scientific division of labor and cooperation with the principle of "greatest common divisor" and through the market based on networks and guaranteed by mechanisms. In this way, the resource utilization and economic benefit of multiple transportation modes can be maximized in the existing transportation network.

In the context of the overall development trend of China's transportation industry and the inherent characteristics of railways in land development, resource utilization and environmental protection, Chinese railways will be designed for sustainable development from now on under the modern integrated transportation system in the following two aspects: on the one hand, Chinese railways shall jointly develop and continuously optimize the basic land transportation network and transportation organization system suitable for China's national conditions in combination with the interconnected, coordinated and balanced development of other transportation modes, and support the development of the socialist market economy with Chinese characteristics via the benign interaction between transportation and economy; on the other hand, Chinese railways shall give full play to the comparative advantages of railways in consideration of the quantity and quality of transportation to meet the diversified demands of transportation in a scientific and efficient manner, and railways shall take the lead in the construction of integrated transportation system to propel the high-quality synchronous development of other transportation modes.

[82] Luo Renjian et al. *Strategic planning of comprehensive transport development in Beijing*. The Institute of Comprehensive Transportation of National Development and Reform Commission. http://www.ndrcict.org.cn/node/245.

Chapter 6
"Overnight Trains": Advance of Chinese High-Speed Railway

In the process of China's reform and opening up of railways and even the whole development process of Chinese railways, the emergence and rapid growth of high-speed railways has been undoubtedly notable. In other words, it has been the large-scale construction of high-speed railways that fundamentally changed the negative image of Chinese railways in the eyes of the public. However, there have been considerable controversies and misunderstandings surrounding China's command of high-speed railway technology. According to some mass media, China has accomplished what took the developed countries several decades in less than ten years. While such statements are exaggerated, the greater issue lies in the misconception they created, which separated the Chinese railway from its high-speed railway. No matter how great the achievements of the Chinese high-speed railway are (the fact is such), its history should be regarded as a part of the landscape of the Chinese railways instead of an independent one. Some have voiced opposition to the independent innovation of Chinese high-speed railway technology, considering it as an imitation of the originators.

However, neither is impartial. The development of the Chinese high-speed railway is continuous, which can be traced back to the 1960s, even prior to the period of reform and opening up. In the meantime, thanks to the increasing technical exchange and cooperation with other countries after the reform and opening up, China has commanded high-speed railway technology at the beginning of the twenty-first century. As for the latter view, neither MOR nor the railway enterprises denied the importance of "the innovation based on the introduction, assimilation, and re-innovation of international technology" when summarizing the experience of independent innovation of high-speed railway technology.

Another widely discussed question is why China is the only country, apart from the technical originators like Japan, France, and Germany, to have developed its own high-speed railway technology system, though high-speed railways have been put in operation in more than a dozen countries and regions. Scholars have put forward not a few explanations and new propositions from different perspectives and disciplines.[1] The construction of high-speed railways in China is not only a major

[1] Lu Feng. (2019). Break through the fog: reveal the origin of Chinese high-speed railway technology advances. *Journal of Management World*, (9).

X. Zhang, *China's Railway Expansion in 40 Years of Economic Reform and Opening Up*, Understanding China,
https://doi.org/10.1007/978-981-95-0333-9_6

event for Chinese railways but also for the reform and opening-up process. In addition to technology, it can also be interpreted from other dimensions, such as economy, politics, culture, and society. It can be predicted that the discussion on the way to construct Chinese high-speed railways will continue for many years later. From a historical perspective, this chapter tries to fully describe the process of technical preparation, decision-making and implementation of Chinese high-speed railways in the overall development of Chinese railways without any presupposition.

6.1 Proposal and Preparation for the Concept of High-Speed Railway

6.1.1 Proposal for the Concept of High-Speed Railway

As early as the 1960s, the advent of high-speed railways in Japan attracted the attention of China. The construction of high-speed railways running from Beijing to Tianjin was taken into account in the general layout planning of the Beijing transportation hub approved by MOR.[2] In 1978, Deng Xiaoping, then Vice Premier of China, took the Shinkansen and gave it high praise during his visit to Japan. Soon MOR set up a specialized class on high-speed railways at Beijing Railway Institute (the present Beijing Jiaotong University) to cultivate talents for the construction of high-speed railways. However, at that time, Chinese railways were far from meeting the technological, financial, and socio-economic requirements for high-speed railways, so the construction was not officially put on the agenda. In the late 1980s, with the development both in China and beyond, it was inevitable to start discussions on the construction of high-speed railways.

Domestically, the contradiction between railway transportation capacity and the growing economic and social demands was increasingly tense. According to a report of MOR in 1990, the growing strain on railway transportation capacity was described in detail as follows.

"Chinese railways have acted as the main artery for the development of the national economy and the backbone of the transportation system, carrying 60% and 70% of the national passenger and freight traffic volume respectively. The contradiction between railway transportation volume and capacity has long been very acute, especially for some busy sections of major lines with saturated capacities, such as the Beijing-Guangzhou, Beijing-Shanghai, Beijing-Shenyang, and Harbin-Dalian railway lines. China has almost been on top of the world in the load capacity of railways, and one busy railway line has to support nearly 40 pairs of passenger trains alongside cargo transport with a one-way freight density surpassing 70

[2] Guo Lin, Fu Zongliang, & Peng Jiaxian. (1991). Beijing-Shanghai Railway Line with Beijing-Tianjin Section in particular should be given priority in being upgraded into high-speed railway. *Chinese Railways*, (11).

million tons. Nevertheless, some salient social problems arising from the inaccessibility of ticket booking, and insufficiency in passenger and freight transportation capacity have constrained the growth of the national economy. As might be expected, the competition between passenger and freight transportation on major lines between metropolises will become more severe. By 2000, it was predicted that the number of passenger trains on the busy sections of major lines above would rise to more than 60 pairs, and the freight density would exceed more than 80 million tons. During the Eighth Five-Year Plan and Ninth Five-Year Plan period, even with continued strengthening and renovation of these lines, it is difficult to accomplish such onerous tasks in passenger and freight transportation with the current traditional technology".[3]

According to the report, MOR reckoned that the "tapping the potential and expanding capacity" campaign had lasted for more than 10 years and almost reached its limits. We should then make technological breakthroughs in railways to get out of the predicament, meet the current and future needs of transportation, and maintain its critical role in the transportation system. So, what is the modern railway technology that can replace "traditional technology"? The answer is high-speed railways. At this time, the expansion of high-speed railways from Japan to several European countries showed strong competitiveness, which brought vitality to this industry and also gave China future hope for its railways.

After the Second World War, the railway development in western developed countries generally had been at a low ebb. Contrary to the socialist countries favouring railway development, such as the Soviet Union and China, the Western developed countries vigorously focused on the expansion of aviation and highways in transportation, underestimating railways as a backward and outdated "sunset industry". In particular, the United States, due to the decline in railway transportation volume, even demolished the existing railways on a large scale. After the 1960s, especially following the oil crisis in the early 1970s, the high energy consumption of highways and aviation led these countries to revisit how to upgrade and transform their railways to revitalize this industry. In 1964, as the world's first high-speed railway, Japan's Tokaido Shinkansen was open to traffic, marking a new turnaround for railways.

Of course, the high-speed railway runs faster than the traditional one. The 515-kilometer-long Tokaido Shinkansen links Tokyo and Osaka, two of Japan's major cities, connecting the most important economic corridor of Japan. With a maximum speed of 210 kilometers per hour, it shortens the travel time from Tokyo to Osaka to 3 hours and 10 minutes, which enables travelers to make a round trip in a single day. In addition, compared to highways and aviation, the high-speed railway boasts several advantages, such as large transportation capacity, excellent safety performance, less land occupation, low energy consumption, and negligible

[3] Report on tackling key technical problems of high-speed railway during the *Eighth Five-Year Plan* Period (TKJH [1990] No. 84).

environmental impact. It combines the strengths of traditional railways with the modern trend of resource-saving.[4]

Following Japan, many countries in Europe have developed high-speed railways. In 1981, the first high-speed railway in France, with a total length of 417 kilometers, was put into operation to connect Paris and Lyon, the two most important cities in France. In 1992, Germany launched its first high-speed railway, connecting Hannover and Wurzburg with a total length of 327 kilometers. Since then, high-speed railways have been built successively in other countries and regions such as Spain, Italy, Sweden, Denmark, Belgium, South Korea, and the Taiwan region of China. As more high-speed lines were put into service, countries like Japan, Germany, and France developed integrated high-speed rail networks. By 1990, Japan had developed approximately 1,900 kilometers of high-speed railways, creating a nationwide railway network, namely the Shinkansen. In 1999, Germany's high-speed railway system, combining both new and existing lines, reached a total length of around 2,280 kilometers. France also formed a high-speed railway network stretching across the country.

Therefore, in the above-mentioned report, MOR officially put forward the concept of China's high-speed railways, "It is imperative to build high-speed passenger dedicated lines between major cities in a planned way, and develop separate lines for passenger and freight trains. These moves will be the key to improving the transportation capacity of busy sections in main lines and ultimately addressing the problem of passenger transportation between major cities".[5] The phrase "ultimately addressing" reflects the expectation of MOR to fundamentally solve the imbalance between supply and demand of railway transportation through the construction of high-speed railways.

MOR decided to build China's first high-speed railway on the Beijing-Shanghai corridor, considering that the high-speed railways in Japan, France, and Germany had all been constructed along corridors with the most developed economy, the largest population, and the most limited transportation capacity in these countries. In December 1990, MOR organized a group of railway experts to compile *A Preliminary Study on the Scheme of a High-Speed Dedicated Passenger Railway from Beijing to Shanghai* (hereinafter referred to as *The Study*), which described the significance, main technical standards, design principles, route scheme, project investment estimation and economic benefit evaluation for the construction of the

[4] In fact, the relentless pursuit of high speed has long been the theme since the advent of railways. Throughout the development of railways, a number of high-speed tests have also been recorded with the aid of traditional steam and internal combustion traction technology. It is not only a single technological advancement but also a comprehensive and systematic technological improvement to transform traditional railways into high-speed railways, involving a long technological and industrial chain across various fields, such as civil engineering, mechanical engineering, electrical engineering, information technology, and materials science. The differences between traditional railways and high-speed railways can be distinguished by the common sense of the public, just as we would not confuse the compass with modern global positioning system (GPS) technology.

[5] *Report on tackling key technical problems of high-speed railway during the Eighth Five-Year Plan Period* (TKJH [1990] No. 84).

Beijing-Shanghai high-speed railway. *The Study* highlights that the Beijing-Shanghai corridor is the most developed area in China with strong economic strength and advanced technology, abundant human resources, diverse cultures, excellent education resources, and higher living standards.

With the deepening of reform and opening up and the establishment of the Pudong Development Zone in Shanghai, new and major achievements in the economic development along this corridor would be secured. The growing personnel exchanges and the increasingly improved living standard would create higher demands for railway passenger and freight transportation. Therefore, it was necessary to develop a railway transportation system with large capacity and high speed, separate the passenger and freight transportation lines in busy sections of the Beijing-Shanghai railway, and build dedicated high-speed railways for passenger trains. *The Study* analyzed the overwhelming transportation situation of the Beijing-Shanghai high-speed railway and then concluded that "only by building a high-speed dedicated passenger railway can the contradiction between transportation capacity and traffic volume be fundamentally solved".[6]

The construction of major national railway engineering projects shall conform to a strict decision-making process. Generally speaking, two stages shall be covered: pre-feasibility study and feasibility study. The former shall focus on whether the project should be implemented, while the latter mainly on how to do it. After the completion of the pre-feasibility study, MOR shall submit the project proposal to the National Development Planning Commission and National Development and Reform Commission successively. Upon approval, the project shall proceed to the feasibility study phase, and the construction may commence once approval is granted at this stage. Compared with general engineering projects, MOR has made more adequate preparations before submitting the *Proposal for the Beijing-Shanghai High-Speed Railway Project*. Since 1991, MOR has carried out the investigation of Beijing-Shanghai high-speed railway lines and held a series of domestic and international seminars to study technical standards, route planning, financing, and other issues.

During the preliminary study, MOR also cooperated with other Ministries in an attempt to pool a wide range of ideas and wisdom. On April 24, 1993, the group of "Preliminary Research on Major Technical and Economic Issues of the Beijing-Shanghai High-speed Railway" was jointly established by the National Scientific and Technological Commission, National Development Planning Commission, National Economic and Trade Commission, National Commission for Economic Restructuring, and MOR. In addition to the above-mentioned "four Commissions and one Ministry", more than 120 experts from 47 institutions participated in the research, including the Development Research Center of the State Council, the Chinese Academy of Sciences, China Communications and Transportation Association, as well as administrative departments, research institutes, and colleges

[6] Writing group. (2014). *Summary on the construction of Beijing-Shanghai High-Speed Railway (Decision-Making Volume)* (p. 23). Beijing: China Railway Publishing House.

and universities from the provinces and municipalities along the Beijing-Shanghai railway line. The research focused on major technical and economic issues related to decision-making, such as project construction schemes, fund-raising and operation mechanisms, international cooperation, and economic evaluation. In 1994, the 500,000-word *Report on Preliminary Research of Major Technical and Economic Issues of the Beijing-Shanghai High-speed Railway,* with a large amount of data analysis, demonstration, and validation, was officially published by the "four Commissions and one Ministry", highlighting that "It is imperative and reasonable to build the Beijing-Shanghai high-speed railway in consideration of existing technologies, economic status, national strength, and available construction funds".[7]

During the research process, the Premier's Office Meeting of the State Council and the Leading Group on Financial and Economic Affairs debriefed MOR and discussed relevant issues about the construction of the railway. The Leading Group agreed in principle to the Ministry's proposal to conduct the pre-feasibility study for the construction of the Beijing-Shanghai high-speed railway and pointed out that "the implementation of such a major national project should strictly conform to procedures of infrastructure development". In October 1994, the Ministry submitted the *Report on Organizing the Pre-feasibility Study of Beijing-Shanghai High-speed Railway* to the leading officers of the State Council in charge of railway construction and received approval in the same month.[8] Subsequently, the Pre-feasibility Study Office of the Beijing-Shanghai High-speed Railway was established by the Ministry and the *Pre-feasibility Study Report on the Beijing-Shanghai High-Speed Railway* was completed in May 1996. In March 1997, the project proposal was submitted after the conclusion of the *Supplementary Report of Pre-feasibility Study Report on the Beijing-Shanghai High-Speed Railway,*[9] manifesting that the decision-making process began for the project.

6.1.2 Technical Preparation for High-Speed Railways

In the decision-making process of more than ten years, at the centre of the controversy was technical competency. In the 1980s, the average travel speed of Chinese railway was less than 50 kilometers per hour, far from the required minimum speed of 250 kilometers per hour for the new high-speed railway, indicating that it was necessary to make a breakthrough in technology. The feasibility of the technologies has always been a major concern throughout the discussions.

---

[7] Shi Dinghuan, Jin Lvzhong, & Zhang Bingfu. (2015). Outline of preliminary research on major technical and economic problems of Beijing-Shanghai High-Speed Railway. *China Soft Science,* (9).

[8] Writing group. (2014). *Summary on the construction of Beijing-Shanghai High-Speed Railway (Decision-Making Volume)* (p. 34). Beijing: China Railway Publishing House.

[9] Letter on submission of proposal for the new project of Beijing-Shanghai High-Speed Railway (TJH [1997] No. 72).

MOR recognized that it was hard to have a good command of high-speed railway technology, as proved in its report to the State Council, "High-speed traffic, like heavy-haul transportation, is the leading technical field of railways in today's global railway industry, covering a wide range of departments, sophisticated technologies as well as numerous key technologies, the technical challenges of which cannot be overcome in the short term. Therefore, we should prepare to launch research projects over a long period to tackle technical constraints with the concentrated input of human and financial resources. This has been our wish for many years. Railway scientists and technicians are determined to contribute all their wisdom to the construction of China's high-speed railway in collaboration with other departments. To commence the work from now on, we hereby apply for the inclusion of high-speed railway technology into the Project of National Key Technical Equipment During the Eighth Five-Year Plan".[10]

According to the report, it would take 10 years for China to master high-speed railway technologies. "China will spend 10 years developing operational high-speed railways due to its weak industrial foundations despite the available international experience". The specific schedule was as follows: Firstly, during the Eighth Five-Year Plan period, we shall tackle key technical challenges in the development of high-speed trains with a maximum speed of 200 kilometers per hour, spare no effort to develop the prototypes of rolling stock, and build test sections. On this basis, the goal would be to achieve an operating speed of 160 kilometers per hour for passenger dedicated trains on the Guangzhou–Shenzhen railway in 1993, to accumulate operation experience and set the stage for the development of other lines with the maximum speed of 200 kilometers per hour.

Secondly, from the end of the Eighth Five-Year Plan period to the beginning of the Ninth Five-Year Plan period, busy sections on the three main lines—the Beijing-Guangzhou Railway, Beijing-Shanghai Railway, and Beijing-Shenyang Railway—shall be selected for the construction of test lines of high-speed passenger dedicated lines. During the Ninth Five-Year Plan period, it was expected that all technology research and development programmes would be completed, achieving the goal of building railways at a maximum speed of over 200 kilometers per hour around 2000. All such experiences gained would then be introduced to other areas at the beginning of the twenty-first century.

The "ten years" timeline mentioned above reflects the prudence of MOR. Moreover, according to the plan, the target speed of the high-speed railway to be built was set at 200 kilometers per hour, which was equal to that of the world's high-speed railway when it first came into being. Although the speed of the international high-speed railway had been upgraded to 300 kilometers per hour at that time, it was still an arduous task given the current situation of Chinese railway technology. As for implementation steps, MOR, following the general mode of new railway technology research and development, divided the overall goal into two stages: first,

[10] Report on tackling key technical problems of high-speed railway during the Eighth Five-Year Plan Period (TKJH [1990] No. 84).

mastering railway technology capable of reaching a speed of 160 kilometers per hour, and then advancing to a speed of 200 kilometers per hour. Regarding the technical approach, efforts in basic research and breakthroughs in key technical hurdles remained the top priority, and engineering tests should be completed and projects should commence in succession. In the 1990s, however, the steps followed by the Ministry were not completely consistent with the initial ones due to some internal and external factors, especially disputes in the decision-making process. With the rapid development of the economy and rising public demand in the twenty-first century, the goal for the development of high-speed railways had shifted significantly to achieving world-leading standards. Consequently, the methods for technical preparation should be adjusted accordingly. But in general, the technical preparation for high-speed railways can be divided into two distinct stages in order.

Basic research and academic exchanges were first carried out. In fact, as soon as the high-speed railway emerged, Chinese communities in railway science and technology paid great attention to it and began to monitor its technological progress. Since the 1980s, Chinese railway technicians have collected and compiled a large number of relevant information and materials on international high-speed railways. Those engaged in the field had also published hundreds of papers in professional journals such as *Chinese Railways* and *Railway Knowledge*, covering the progress of high-speed railways in various countries and regions. In addition, non-professional media outlets, like *Outlook Weekly,* also published articles such as "The Rise of High-speed Railways in the World" (No. 39, 1993) and "Countries Around the World are Racing to Develop High-speed Railways" (No. 26, 1994), introduce the development status and emerging trend in the field, describe the characteristics of high-speed railways and analyze motivations behind its rapid application in countries like Japan, the U.S., and throughout Europe.

International academic exchanges on high-speed railway technology could be traced back even earlier. In the 1960s, China started exchanges with Japan in the field after the completion of Japan's Tokyo-Osaka Shinkansen Line. Japan Railway Engineers' Association (JREA) and other institutions dispatched multiple delegations for the Japan-China Railway Technology Exchange to China, and Chinese institutions like China Railway Society also sent delegations to Japan to learn about the procedures, main features, and trends of Japanese railways. In June 1994, Japan Railway Technical Service (JARTS) convened a high-level and comprehensive "seminar on China's high-speed railway development" in Beijing. In addition to exchanges with Japan, China also maintained frequent academic exchanges with France, Germany, and other pioneering countries in high-speed railway technology. For example, MOR held the "Sino-French Technical Seminar on High-speed Railway" in June 1991, and the Department of Science and Technology of the Ministry hosted the "Sino-German Technical Seminar on High-speed Railway" in November 1993. From 1990 to September 1993, academic organizations such as China Railway Society and China Communications and Transportation Association held nearly 10 academic seminars to discuss various issues related to the construction of high-speed railways in China.

The phased tasks to tackle key technical hurdles in the development of high-speed railways have been specified in two consecutive Five-Year Plans for Railway Science and Technology.

In 1990, MOR formulated the *Plan in Advance—Outline of the Ten-Year Programme of Railway Scientific and Technical Development as well as the Eighth Five-Year Plan* (hereinafter referred to as the *Advance Plan*), which explicitly specified the guiding policies and goals for railway scientific and technical development, and the *Key Points of Railway Technology Progress Plan during the Eighth Five-Year Plan*. These documents served as the general outlines for railway development in the 1990s.

The policy outlined in the *Advance Plan* focused on "leveraging reform as a driving force, with safety in transportation and modern management as its foundational principles. The priorities for railway scientific and technical development were to advance the heavy-haul technologies, upgrade the passenger transport equipment, and develop high-speed technologies for passenger transport with locomotives, vehicles, and information technology as key fields to improve the transportation capacity. Different levels of technologies and equipment shall be used to adapt to varying transportation needs, aiming to transform new technologies into strong productivity as soon as possible based on available funds, feasible technologies, and favourable economic conditions".

The goals for the 1990s can be concluded as the following three: "Firstly, we shall promote the progress of railway technologies and develop railways with modern technology, equipment, and management methods. This included the electrification and internal combustion of traction power, the development of information technology, and the application of computer networks in the transportation systems of all lines, thereby modernizing both transportation control and management. Secondly, we shall strive to develop the quality of track structures, focusing on heavy-duty and high-quality designs, and promote mechanization in railway maintenance, loading and unloading operations of cargo, as well as construction processes. Thirdly, we shall improve the technical performance, quality, and service life of major industrial products such as locomotives, trains, equipment for public works section and electricity services, making them approach or reach the advanced level of similar international products".[11]

Covering four major technical projects, 20 technical systems and 100 programs, the *Advance Plan* should be implemented accordingly. The four major technical projects include the "capacity expansion technology project" to improve the railway transportation capacity, the "safety technology project" to ensure transportation safety, the "pilot technology project" to accumulate technical reserves, and the "technology integration project" to promote the localization and adaptation of imported technologies.

[11] Department of Science and Technology of MOR. (1992). Plan in advance——Outline of the "Ten-Year Programme of railway scientific and technical development" as well as the Eighth Five-Year Plan. *Chinese Railways*, (1).

The Key Points of Railway Technology Progress Plan during the Eighth Five-Year Plan covers twelve full-scale plant projects (dubbed as "Twelve Dragons" in Chinese), including (1) the application of 10,000-ton heavy haul transportation technology of the Datong-Qinhuangdao Railway, (2) the operation of 5000-ton heavy-haul trains running between the Beijing and Shanghai Railway and the Beijing and Wuchang Railway, (3) the improvement of the transportation capacity of busy single lines such as the Baoji-Chengdu Railway, the Xiangyang-Chongqing Railway, (4) the operation of passenger trains with a speed of 160 kilometers per hour on the Guangzhou-Shenzhen Railway, (5) the comprehensive automation of major marshalling yards of railway network and the simple automation of small and medium-sized marshalling yards, (6) the all-round promotion of the "Three-protection Technology" to ensure operation safety, (7) the supporting application of new railway construction technologies in the construction of new lines with the Nanning-Kunming Railway as priority, (8) the application of computer technology to accelerate the technological transformation of traditional railway industries and gradually modernize railway management, (9) the accelerated modernization of the railway communication network; (10) the development of new passenger trains and new facilities to improve the quality of passenger service, (11) the development of container intermodal transport, (12) the utilization of large-scale maintenance machinery in conjunction with small and medium-sized machinery clusters on busy main lines.[12]

It shows that although the "research and development of high-speed technologies for passenger transportation" was the "major goal", numerous R & D projects could not be implemented at this stage due to the low level of Chinese railway technologies and the urgent needs of transportation. Major tasks at this stage were the technical transformation and upgrading of ordinary railways.

The proportion of research and development of high-speed railway technology increased in the *Outline of the Ninth Five-Year Plan for Railway Science and Technology Development and the Long-term Plan for 2010* which proposed to implement four major technical projects, namely, (1) the "cross-century key projects" covering researches for the technical foundation for the long-term development of railways, (2) the "key technology development projects" comprised of urgent projects to solve the current hotspot and difficult problems related to railways, (3) the "projects to strengthen the basic technology" made up of major projects that could upgrade the railway basic technology, and (4) "projects of the transformation of scientific and technological achievements", including promotion projects that could accelerate the transformation of mature technologies into real productivity.

In his explanation on the *Outline*, Hua Maokun pointed out that "The five technical fields, including high-speed technology, heavy haul transportation technology, safety, information technology, and soft science, are research priorities for the cross-century key projects". "The railway shall be built in sections following

[12] Ibid.

resolution of key technical problems at different levels and investment of substantial funds. Sustained efforts shall be made to research and develop key technologies that facilitate the construction of test sections of high-speed railways in this century to lay a solid technical foundation for the construction of high-speed railways for passenger transportation at the beginning of the next century". The "key technology development projects" cover ten programs, focusing on the improvement of passenger trains speed, freight train weight, service quality of passenger and freight transportation, modern operation and management.[13]

According to MOR, greater attention should be paid to the "Top 10 Landmark Projects" and the development of four major technological innovation projects during the Tenth Five-Year Plan period. Among the top 10 landmark projects specified in the *Outline of the Tenth Five-Year Plan for Railway Science and Technology Development and the Long-term Plan for 2015*, half of them were related to high-speed railway, including (1) the construction of the Beijing-Shenyang High-Speed Railway with a speed of more than 160 kilometers per hour, (2) solving the technical problems related to high-speed railways and commencing the construction of the Beijing-Shanghai High-Speed Railway, (3) the completion of an express passenger transport network of 14,000 to 16,000 kilometers covering major cities in China, (4) the successful development of EMUs, tilting trains and advanced urban rail EMUs at a speed of 270 kilometers per hour, and the mass production of AC drive electric locomotives and diesel locomotives, (5) the successful development of an integrated automation system in trains at a speed of 200 kilometers per hour consisting of train dispatching command system (TDCS) and train operation control system.

Major technological innovation projects include (1) speeding up the development and application of information technology and making every effort to promote railway informatization, (2) strengthening independent innovation of high-speed technology and continuously implementing the speed increase strategy, (3) building a traffic safety monitoring network and a technical system to safeguard operation; (4) overcoming technical difficulties to further improve the level of construction.[14]

In order to implement the above plans, MOR has invested a huge amount of money in basic research and technological research and development of high-speed railways since 1990. In the past decades since the 1990s, the Ministry had successively approved more than 460 research project proposals, of which about 50 were listed as National Programs for Science and Technology Development during the Eighth Five-Year Plan, Ninth Five-Year Plan, and Tenth Five-Year Plan Periods, covering transportation economy, public works, locomotives and trains, communication signals, traction power supply, material technology, and operation management. Thousands of people from about 40 institutions participated in these projects

[13] Hua Maokun. (2001). Outline of the Ninth Five-Year Plan for railway science and technology development and the long-term plan for 2010. *China Railway*, (6).

[14] (2002). Outline of the *Tenth Five-Year Plan* for railway science and technology development and the long-term plan for 2015. *Beijing Railway Science and Technology newsletter: Taiyuan Journal*, (3).

to solve the technical problems of high-speed railway and formulate the technical system of China's high-speed railways.

The construction of test lines and the tackling of key technical problems were proceeding in a staggered manner. The rennovation and expansion of the Guangzhou-Shenzhen Railway to reach a speed of 160 kilometers per hour was completed on schedule, while the construction of the test line, with a speed of 200 kilometers per hour, was not implemented as planned until 2003.

In March 1990, MOR included "Research on the Technical Scheme for the Guangzhou-Shenzhen Railway to Reach the Maximum Speed of 160 Kilometers per Hour of Passenger Trains" in the programs for science and technology development that year, and held a large-scale expert demonstration and validation meeting to approve the feasibility study report in July. In October, the project proposal was approved.

The construction of the Guangzhou-Shenzhen Quasi-high-speed Railway was not only a new attempt but also a great challenge as railways are a complex system where all sectors are closely intertwined. Therefore, there were high requirements for railway technical equipment, engineering technology level, test capacity, management and organization, etc. in the attempt to raise the maximum speed of trains from less than 120 kilometers per hour to 160 kilometers per hour. At that time, most of the traditional technologies of Chinese railways, such as locomotive and vehicle performance and maximum running speed, track structures, turnout switching equipment, signal system, and parameters of railways in horizontal and longitudinal sections, could not meet the technical requirements of 160 kilometers per hour. To overcome the technical difficulties in the construction of the quasi-high-speed railway, MOR, on the basis of demonstration and validation, determined fourteen scientific and technological research projects such as "DF11 high-power passenger diesel locomotive", three "quasi-high-speed passenger trains" and "speed hierarchical control system of quasi high-speed passenger train", and organized dozens of institutions such as railway administrations, factories, scientific research and design institutes and engineering Departments to jointly tackle the key problems.

From 1992 to 1994, many comprehensive tests were carried out on the basis of technology research, covering the main technical parameters for a variety of types of technical equipment. These tests revealed some new findings after the increase of train running speed, and furthered the understanding of the driving technology of 160 kilometers per hour, thus providing the theoretical basis to optimize product performance and design parameters, which was essential to promote the construction of the quasi-high-speed railway. In the summer and autumn of 1994, the main line test, lasting for more than 3 months, was performed to inspect the configuration of the reconstructed infrastructure and to determine the safety and comfort index when the speed was raised to 160 kilometers per hour. A large number of data obtained in the test proved that the Guangzhou-Shenzhen Railway was technically safe and reliable. On December 15, 1994, MOR held the "meeting of safety acceptance and evaluation for the Guangzhou-Shenzhen quasi-high-speed railway" to comprehensively evaluate the safety of this line, and then it was open to traffic on December 22.

The construction of the Guangzhou-Shenzhen quasi-high-speed railway, the first step in the development of high-speed railways, had provided valuable experience for building a quasi-high-speed railway with a speed of 160 kilometers per hour in China and offered crucial research and testing conditions for the subsequent development of high-speed railways with the speed of 250 kilometers per hour in the second step.

From the speed of 160 kilometers per hour to that of 250 kilometers per hour, the transition took another five years of technical preparation. In the late 1990s, MOR decided to build a new passenger-dedicated railway in compliance with the high-speed railway standard.[15] The 404.64-kilometer-long main line connects the Qinhuangdao Station in the south to the Shenyang North Station in the north.[16] In February 1999, the National Development Planning Commission submitted the project proposal to the State Council and obtained official approval.[17] This line, finally designed as double-track electrified railways with a design speed of 200 to 250 kilometers per hour,[18] commenced its construction on August 16, 1999, and was put into operation in January 2003.

The Qinhuangdao-Shenyang Passenger-Dedicated Railway was built in full compliance with the technical requirements of the high-speed railway on design theories and methods, construction methods and technologies, as well as the operation and management modes.[19] According to the requirements set during the decision-making process, the characteristics of the project shall embody the "Three Highs and Three News", namely, high running speed and new rules and regulations, high technical level and new design standards, as well as high-quality requirements

[15] The fundamental standard for defining a high-speed railway is its operational speed, yet it is characterized by other attributes as well. For example, the high-speed railway is generally designed for passenger transportation (only Germany maintains a mixed operation), which is different from the traditional railways that carry both passengers and freight. Therefore, the name of passenger-dedicated railway was adopted during the initial stage of China's high-speed rail development, while specific railways such as the Beijing-Tianjin High-Speed Railway were referred to as "inter-city railways". There were also instances where names derived from high-speed railways, such as "EMU (Electric Multiple Unit)" were adopted. In fact, high-speed EMUs refer to power-distributed high-speed trains, which correspond to traditional power-centralized high-speed trains. The reasons for the emergence of multiple names are very complex, stemming not only from different perspectives of high-speed railway technology but also the need to avoid disputes and gain recognition and support of all parties in the decision-making process of high-speed railway.

[16] Yang Bin, & Zhang Fuchun. (1999, August 18). Another important battle for railway modernization at turn-of-the-century, construction of China's first express passenger-dedicated railway: Qinhuangdao-Shenyang Passenger-Dedicated Railway. *People's Railway Daily.*

[17] Lei Fengxing. (2013). *Speeding ahead: China's rapid transit railways* (p. 27). Beijing: China Intercontinental Press.

[18] Construction headquarters of Qinhuangdao-Shenyang Passenger-Dedicated Railway, Project Management Center of MOR. (2003, January 14). Towards high speed: documentary on the construction of Qinhuangdao-Shenyang Passenger-Dedicated Railway. *People's Railway Daily.*

[19] Ibid.

and new construction technologies.[20] To meet these standards, MOR specially established a Science and Technology Leading Group of the Qinhuangdao-Shenyang Passenger-Dedicated Railway, which brought together the prime forces of scientific research, design, construction, supervision and construction management in this field. Based on the introduced international advanced technologies and experience, the leading group aimed to make breakthroughs in key scientific research projects and tests, and promoted all-round technological innovation in many aspects, such as subgrade, bridge, track, "four electrical projects" and the development of locomotives and vehicles.[21]

As for subgrade engineering, the Qinhuangdao-Shenyang Passenger-Dedicated Railway traverses the Liaoxi Corridor, an area characterized by complex engineering geological conditions, with 40 kilometers of marshland and soft soil in the east. Therefore, subgrade structures were designed to cross soft soil layers, reinforced by measures such as composite foundation and drainage consolidation methods, and bridges were erected in some sections. The subgrade of the Railway should be firm and stable with the required value of post-construction settlement lowering than 15 cm. The construction company adopted a number of subgrade reinforcement measures with standards higher than the expressway ones to cope with the soft soil foundation along the line. The graded crushed stones were used as the bed course in subgrade filling for the first time to improve the bearing capacity of the foundation bed; the transition sections of the road and bridge were filled with graded crushed stones in cascade to ensure the uniform transition of stiffness; the soft soil subgrade and mollisol subgrade were strengthened by the application of composite foundation deep treatment method. These technologies provided a uniform and stable foundation with high strength and stiffness for the track.

In terms of bridge engineering, the length of all bridges and culverts accounts for 23.1% of the total length of the whole line, and there are 211 bridges with a total length of 59.9 km and 1,278 culverts and drains with a total length of 33.4 km. The difficulties of the construction of the Railway lie in the bridges. Six types of beams such as new box girder, prestressed continuous concrete beam, and steel-concrete composite beam were adopted for the bridges along the whole line, and advanced bridge construction methods like cast in situ as well as section erection and combination were employed.

For track engineering, the construction technology of laying the inter-district continuous welded rails was adopted for the first time, and the rails and turnouts were welded and fastened into 371-kilometer-long rails. The world's most advanced single-sleeper continuous track-laying machine was first introduced to lay the 250 to 300-meter-long track strings. The standard-gauge tracks were welded in the

[20] Luo Qian, & Wang Jian. (2003, January 14). Supporting high technology of railway: documentary on creating excellence of China Railway Electrification Engineering Group Co., Ltd. for the construction of Qinhuangdao-Shenyang Passenger-Dedicated Railway. *People's Railway Daily.*

[21] Construction headquarters of Qinhuangdao-Shenyang Passenger-Dedicated Railway, Project Management Center of MOR. (2003, January 14). Towards high speed: documentary on the construction of Qinhuangdao-Shenyang Passenger-Dedicated Railway. *People's Railway Daily.*

factory. The following works such as the welding of rails link on site, the fastening-down and stress dissipation of rails, and careful clearing and grinding of rails were completed during the construction process. In addition, internationally advanced No. 38 turnouts were installed for the first time to ensure the precise alignment of straight and curved rails, as well as the accurate fitting of industrial and electrical equipment.

The "four electrical projects", including communication engineering, signal engineering, electrical power engineering, and electrification engineering, played a pivotal role in the application of high-speed technologies in the construction of the Qinhuangdao-Shenyang Passenger-Dedicated Railway. The functions such as remote control, self-inspection, self-diagnosis, safety monitoring, and an integrated automation system developed by China were adopted in the traction substation. The overhead catenary was built in the form of China-designed simple catenary suspension and stitched catenary suspension which were suitable for the high-speed operation of railways. In addition, the simulation calculation independently developed by China facilitated the safe adjustment of catenary wires at one time. Furthermore, new technologies were utilized for the first time to assist the construction of the whole railway, representing the highest level of railway communication and signal technology in China at that time. For example, the radio frequency over fibre optics repeater was first used in the communication system with wireless technology, and the advanced data-voice compatible wireless train dispatching system was first adopted. The display of railway speed and block signals on the ground were no longer used and substituted with the on-board display.[22]

China not only made new technical achievements in bridge engineering, subgrade engineering, track engineering and so on, but also accumulated new experience in transportation organization. The construction of the Qinhuangdao-Shenyang Passenger Dedicated Railway was an important engineering practice, which provided a major platform for the construction of high-speed railways with a speed of 300 to 350 kilometers per hour,[23] paving the way for the large-scale construction of high-speed railways in China later. It was the first railway passenger-dedicated railway independently researched, designed, and constructed in China, signifying that China initially possessed the ability to design and construct railways at a speed of more than 200 kilometers per hour with independent intellectual property rights, complete equipment manufacturing, and comprehensive system integration. In general, this project offered rich technical experiences and laid a solid foundation for the growth of China's high-speed railways.

When it comes to importance, locomotive manufacturing is second only to the construction technology in the railway system. Since 1992, MOR has carried out

[22] Luo Qian, & Wang Jian. (2003, January 14), Supporting high technology of railway: documentary on creating excellence of China Railway Electrification Engineering Group Co., Ltd. for the construction of Qinhuangdao-Shenyang Passenger-Dedicated Railway. *People's Railway Daily.*

[23] Sun Chunfang. (2011, June 11). Chronological history of China's high-speed railway: how "the first high-speed railway" Qinhuangdao-Shenyang Passenger-Dedicated Railway was obliterated". *21st Century Business Herald.*

research on key technologies and main components of different types of high-speed trains in Japan, Germany, and France. The main research content includes overall technical conditions of the high-speed train, the bogie and body structure of the high-speed train with concentrated independent power, the bogie and body connection structure of the high-speed train with concentrated articulated power, the high-speed train with distributed power, AC-DC-AC traction drive system, braking system and key parts of the high-speed train, train control, diagnosis and monitoring system, sealing technology and interior environmental control technology of trains, base for repairs and maintenance of trains, aerodynamic performance of the train and lightweight design of the train body.

Based on the research of key technologies, the EMUs were designed and developed, represented by DJF2 "Xianfeng" and DJJ2 "China Star" which had been tested many times on the Qinhuangdao-Shenyang Passenger Dedicated Railway. The DJF2 "Xianfeng" EMU, developed under a key program for science and technology development of China during the Ninth Five-Year Plan Period, was the power-decentralized type with a maximum speed of 250 kilometers per hour and an operating speed of 200 kilometers per hour. The DJJ2 "China Star" EMU, under a national high-tech industrialization program during the 10th Five Year Plan Period, was a power-centralized type with an aluminum alloy body, reaching a maximum speed of 300 kilometers per hour and an operating speed of 270 kilometers per hour.

The operation speed of traditional railways was increased while the construction of test lines of high-speed railways was underway. From 1997 to 2004, the railway authorities implemented five rounds of campaigns to raise the speed of railway travel in China.[24]

At 00:00 on April 1, 1997, the initiative of the first speed increase and adjustment of the railway operating plan on a large scale was fully implemented in China to significantly raise the operating speed of trains, marking a historic breakthrough. The maximum operating speed of trains reached 140 kilometers per hour on the three main lines, the Beijing-Guangzhou Railway, the Beijing-Shanghai Railway, and the Beijing-Harbin Railway, and the operating speed of passenger trains nationwide rose from 48.1 kilometers per hour in 1993 to 54.9 kilometers per hour, an increase of 6.8 kilometers. In addition, fast trains and overnight trains began operation for the first time in big cities including Shenyang, Beijing, Shanghai, and Wuhan. Dubbed as "mobile hotels" by passengers, 40 pairs of fast trains with a maximum speed of 140 kilometers per hour and an operating speed of more than 90 kilometers per hour, and 78 overnight trains were dispatched from these cities. The numbers of nonstop express trains, nonstop fast trains, local express trains, and local fast trains had increased while those of nonstop slow trains, local slow trains, and suburban trains had been reduced, and the number of berths had been increased while that of seats appropriately reduced.

[24] For related studies, refer to: Ju Jiaxing. (2002). Speed increase, a grand achievement at the turn of the century: a review of the course of railway speed increase. *China Railway,* (11)

At 0:00 on October 1, 1998, the second speed increase and adjustment of the railway operating plan on a large scale was implemented in China. Focusing on the three main lines, the Beijing-Guangzhou Railway, the Beijing-Shanghai Railway, and the Beijing-Harbin Railway, the initiative further extended the length of lines with increased speed and optimized the transportation organization as well as the allocation of transportation capacity resources. The maximum operating speed of the fast train reached 160 kilometers per hour, and that in other normal speed sections reached 120 kilometers per hour. The operating speed of passenger trains increased to a certain extent compared with those in 1997. The average speed of nonstop fast trains and express passenger trains reached 71.6 kilometers per hour, an increase of 4.5 kilometers. Furthermore, a total of 80 pairs of fast trains and 228 overnight trains were dispatched. The intercity passenger trains running between Beijing and Tianjin, West Beijing and Shijiazhuang, short-distance passenger trains, holiday trains, and trains tailored for migrant workers were operated to meet the demand of intercity passenger traffic. Nonstop trains running on popular tour lines from Beijing to Xiamen and Harbin to Wuchang departed for the first time, attracting a large number of tourists.

At 0:00 on October 21, 2000, the third speed increase and adjustment of the railway operating plan on a large scale were implemented on the Lanzhou-Lianyungang Railway, the Lanzhou-Xinjiang Railway, the Beijing-Kowloon Railway, and the Zhejiang-Jiangxi Railway to form China's speed increase railway network on the basis of the previous two initiatives. The four south-north "major arteries" including the Beijing-Guangzhou Railway, the Beijing-Shanghai Railway, the Beijing-Harbin Railway and the Beijing-Kowloon Railway, and the two east-west main lines consisting of the Lanzhou-Lianyungang Railway, the Lanzhou-Xinjiang Railway and the Zhejiang-Jiangxi Railway were sped up. By then, the railway mileage of the speed increase lines nationwide had nearly reached 10,000 kilometers, and a speed increase railway network was formulated, featuring a "four vertical and two horizontal" layout and covering major regions of China.

New gains were made in the improvement of train speed while the range of speed increase lines was expanded. The average speed of passenger trains nationwide increased by 5.1 kilometers per hour to 60.3 kilometers. Up to then, the structure of transportation had been further optimized and a full range of railway passenger transport types had been formed, which brought the total number of overnight trains to 266 and that of dedicated trains for inter-administration tourism lines to 28 pairs to match the development of holiday economy and tourism. Besides, MOR revised the types and numbers of trains and rolled out a more accurate version of operation plan to bring convenience to passengers. In the past, the seven types of trains in the previous plan included fast trains, express trains, nonstop fast trains, ordinary passenger trains, mixed trains, suburban trains, and military personnel trains, which were readjusted to three types, namely, express passenger trains, fast passenger trains, and normal passenger trains including general fast train and general slow train. Two train marshalling plans and two ticket prices were set for inter-administration passenger trains in peak season and low season respectively, which was conducive to resource utilization.

At 0:00 on October 21, 2001, China started to implement its fourth large-scale speed increase and adjustment of the railway operating plan. The key sections of this adjustment were along the Beijing-Kowloon Railway, the Wuchang-Chengdu Railway (encompassing the Hankou-Danjiang, Xiangyang-Chongqing, and Dazhou-Chengdu Railways), the Zhejiang-Jiangxi Railway, the Harbin-Dalian Railway, and the southern section of the Beijing-Guangzhou Railway. After this adjustment, the Railway Network with Accelerated Speed covered most provinces nationwide and the extended mileage reached 13,000 kilometers. Moreover, the number of express passenger trains further increased, and the running schedule was optimized. For example, the Beijing-Shanghai Railway operated four pairs of overnight express trains consecutively between Beijing and Shanghai within the two hours from 18:00 to 20:00. Among them, Trains T13 and T21, as well as Trains T22 and T14, had a mere eight-minute interval between their departures. Some transportation types such as "Overnight Train" demonstrated the advantages due to the optimized running schedule. The departure time was set from 17:00 to 23:00 while the arrival time was from 5:00 to 10:00. To meet the demand for tourism passenger flow and prop up the railway tourism, 28 pairs of inter-administration tourist-dedicated lines were planned and mapped out, providing better conditions for the operation of special tourist trains.

The fifth large-scale speed increase and adjustment took place on April 18, 2004, manifesting the dramatic progress of the railway transportation capacity. The speed of some sections of major trunk railways reached up to 200 km/h. The total mileage of the involved railway networks topped 16,500 kilometers, of which the speed of over 7,700 kilometers of railways could be accelerated to 160 km/h and above. The average speed of the national passenger trains reached 65.7 km/h with an increase of 5.4 km/h compared with the average speed in 2001, and the speed of direct express trains reached 119.2 km/h and that of express trains 92.8 km/h. The running speed of passenger trains in the major cities further increased, which shortened the journey time dramatically. Taking the opportunity of this adjustment, MOR launched a few new transportation types, such as 19 pairs of direct express passenger trains. All trains were equipped with the most advanced Bombardier and 25 T rolling stock, with a maximum running speed of 160 km/h. The direct express passenger trains shuttled point-to-point without stopping in operation. The departure and arrival stations were located in 13 cities with large passenger flow volumes, including Beijing, Shanghai, Hangzhou, Yangzhou, Nanjing, Suzhou, Hefei, Wuhan, Changsha, Harbin, Changchun, Xi'an, and Tianjin to ensure rapid passenger transportation among big cities. The method of tracking and continuous departure was adopted to help more passengers take the train in prime time.

For example, there are 11 direct express passenger trains on the Beijing-Shanghai Railway, with an interval of only 7 minutes. MOR vigorously improved the quality of the overnight train through this speed adjustment and further optimized the departure and arrival time. The overnight trains were operating in the prime times. The departure time was set between 17:00 to 23:00, and the arrival time between 5:00 to 10:00 on the next day, and passengers could choose the travel time according to their needs. To meet the increasing demand for passenger transportation, the number of direct overnight trains further increased to 169, and that of overnight trains within the jurisdiction of one railway administration reached 136. Moreover, the inter-administration tourist-dedicated lines increased from 28 to 39 pairs due to the strong growth of the tourism economy and holiday economy. Overall, the increasing train density of trunk lines contributed to the growth of transportation capacity.

Although the maximum design speed of 160 km/h had not reached the standard one of a high-speed railway, the five times speed increases and adjustments in railways played a key role in promoting the development of the Chinese high-speed railway. The speed adjustment was eventually manifested in the modification of the train diagram, but it involved all aspects of the railway system including the line transformation, improvement of equipment, optimization of management, etc. When the technical transformation of the Guangzhou-Shenzhen Railway was completed in 1994, only this short line of the entire network had a running speed of 160 km/h, just like a small boat in the ocean. By 2004, there were 7700-kilometer railways whose design speed had reached this level, and these railways were mostly on the trunk lines. Though the mileage of these railways only accounted for a little over 10% of the total of Chinese railways at that time, the proportion of traffic volume was much higher. It was the overall improvement of infrastructure, equipment, and management that had its clout in the rapid development of the high-speed railway network. Otherwise, the completion of one single world-class high-speed railway could only bring about little practical significance.

The impact and changes brought by the speed adjustment not only increased the speed but also fueled the revival of railway development and transformed its social image. For Chinese passengers, taking the train was often a necessary but frustrating choice due to its poor experience. The five speed increases in railways demonstrated that the management of railway transportation in China stepped into a new stage. Specifically, passengers received better services and enjoyed a wonderful train travel experience. This in turn promoted the improvement of railway economic benefits.

The first adjustment in 1997 helped the railway passenger volume edge up and end the three-year consecutive decline. In the whole year, 919.193 million passengers were dispatched, a rise of 3.7% over the previous year, of which the through passenger volume climbed about 9.7%. In 1998, the number of passengers was 929.909 million with an increase of 1.2% over the previous year, of which the number of through passengers increased by 5.4%. In 1999, 977.251 million passengers were sent with a growth of 5.1% over the previous year, of which the number of through passengers rose 10.1%. In 2000, the number of passengers was 1.03 billion with an increase of 5.4% over the previous year. In 2001, the passenger volume was the same as that in 2000, but the passenger volume grew significantly, with an increase of 5.2%. The passenger volume had soared for five consecutive years, with an average annual growth rate of 6.9%. The transportation revenue of the whole railway had a huge jump for five consecutive years with an average annual increase of 9.83 billion yuan, and the cumulative increase amounted to 49.16 billion yuan. After the significant losses for two consecutive years in 1997 and 1998, the railway revenue stopped to suffer from losses one year ahead of schedule in 1999 and had maintained its profitability for many consecutive years thereafter.

The speed adjustment of existing lines was also directly related to the development of high-speed railways, providing support in multiple aspects. The five-speed adjustments within just seven years brought the potential of Chinese existing railways almost to the extreme. With the sustained and rapid growth of China's economy, the requirements of the whole society for improving the quality of railway transportation were increasing year by year, but the speed adjustment of the railway was not unlimited, which was not only constrained by the existing railway conditions but also affected by the competition between different modes of passenger and freight transportation. The successful experience of developed countries proved that the mode of passenger and freight separation should be adopted in railway transportation on inter-city transport lines with large passenger and freight volumes. Namely, high-speed passenger transportation and heavy haul freight transportation are the fundamental ways to solve the problems of capacity and quality.

In fact, the sixth large-scale speed increase and adjustment was also implemented on April 18, 2007. This speed adjustment involved 18 lines including the Beijing-Harbin Railway, the Beijing-Shanghai Railway, the Beijing-Guangzhou Railway, and the Lanzhou-Lianyungang Railway. Train speed on a total of 22,000 kilometers of railways was accelerated to 120 km/h and above, representing an increase of 6,000 kilometers over the fifth speed adjustment. The new train diagram showed that the number of direct express trains increased by 7 pairs, reaching 26 pairs in total, and the number of overnight trains also increased from 305 to 337.

The sixth speed adjustment marked the emergence of high-speed EMUs. The operation of EMUs indicated that China had its high-speed railways. Except for the Qinhuangdao-Shenyang Passenger Dedicated Railway, other railways were upgraded through technological transformation. Because the Beijing-Shanghai High-Speed Railway had not been approved for construction, the diagram adjustment was still included in the speed adjustment for the existing line at that time. A total of 257 pairs of EMUs was scheduled to be put into operation according to the new train diagram, which would be phased in four stages. 140 pairs of EMUs were first dispatched on April 18, 2007. In terms of mileage, the high-speed railways with a speed of 250 km/h extended to 900 kilometers. In contrast, in 1964, the mileage of the world's first high-speed railway was only 515 kilometers.

The technical route of the high-speed railway underwent a fundamental transformation during the completion of the Qinhuangdao-Shenyang Passenger Dedicated Railway. The expression of independent innovation did not change, but the connotation was completely different. Independent innovation was the same as our intuitive understanding previously, that is, the technology of high-speed railway was finally mastered through continuous accumulation of new technology on the basis of existing technology. International advanced technology should be used for reference, but our own strength "played a key role in researching and developing the high-speed railway technology system with independent intellectual property rights and forming a high-speed transportation development model with Chinese characteristics".[25] The new interpretation divided independent innovation into three types including original innovation, integrated innovation, and introduction-absorption-re-innovation in which re-innovation is the key point.

A popular saying was that the minister and deputy ministers of MOR were about to witness a test run of China Star (DJJ2 EMU) on the morning of November 18, 2002. However, the train malfunctioned and the reasons could not be ascertained immediately, which resulted in the cancellation of the event. Although this accident was proved to be caused by an imported bearing, this technology was still considered by the leading officers of MOR as unreliable. "Compared with the internationally advanced technology, there was still an obvious gap in technical level, product maturity, and reliability". Although this train had been dispatched along the Qinhuangdao-Shenyang High-speed Railway since then, the plan for independent research and development of high-speed trains had been temporarily canceled.[26]

It sounds inconceivable that a malfunction had led to the suspension of a decennial R&D plan for large-scale equipment. Attributing blame solely to accidental events is evidently unconvincing. In-depth research needed to be performed to reveal what had really happened. However, the research route of high-speed railway technology, including high-speed trains had indeed undergone a fundamental transformation.

In 2004, the State Council put forward the general requirements of "introducing advanced technology, jointly designing and producing, building Chinese brand" for EMU R&D. The target of R&D for EMU was to command the manufacturing technology through stages including technology introduction, joint design, digestion and absorption, integrated innovation, system improvement and comprehensive innovation.

The first stage was technology introduction and joint design. Based on the introduction of EMU technology with a speed from 200 km/h to 250 km/h, the localization rate reached 70%. The optimized design for the imported EMU successfully addressed the issue of acclimatization between the imported technology and the Chinese railway environment. Through comprehensive simulation analysis, ground test, and line test of the imported EMU, Chinese railways had mastered the design and manufacturing technology of the EMU with a speed of 200 km/h, and constructed the manufacturing platform of the EMU.

The second stage was digestion, absorption and integrated innovation. Based on the technical platform of EMU with a speed of 200 km/h to 250 km/h, we had independently developed the long-formation seated EMU with the design speed of 250 km/h and the world's first long-formation sleeping EMU, through optimization of the design for marshalling type, dynamic configuration, network control system, vibration and noise reduction and passenger interface. The EMU with a design

[25](2000, March 31). China will independently develop high-speed railway technology system. *Guangming Daily*

[26]Wang Qiang. (2006, March 5). Why did China Star become a transience? *Business Watch Magazine*

speed of 300 km/h to 350 km/h had been developed based on the optimization and improvement of traction performance, vehicle strength, mode, bogie, and so on.

The third stage focuses on system improvement and comprehensive innovation. Based on the technology of high-speed EMUs with a design speed of 350 km/h, through multiple analyses, comparative demonstrations, and test verification of simulation calculation, bench test, and line test, Chinese railways commanded the key technology of high-speed EMUs and improved the R&D, design, manufacture, test platform, and standard system of high-speed EMUs.

On February 26, 2008, the Ministry of Science and Technology and MOR jointly signed the *Joint Action Plan for Indigenous Innovation of China's High-Speed Train* to support the independent innovation of China's high-speed trains. In April of the same year, the Ministry of Science and Technology rolled out the project of Key Technology Research and Equipment Development of China's High-speed Train in line with the National Science and Technology Support Program under the Eleventh Five-Year Plan. The total investment of the project was 3 billion yuan, including 1 billion yuan allocated by the Ministry of Science and Technology and 2 billion yuan jointly invested by MOR and enterprises. All sectors collaborated to support the R&D for the new generation of high-speed trains in technology and funds. To actively promote the implementation of the Joint Action Plan, the National Science and Technology Support Program under the Eleventh Five-Year Plan was open to professionals of the entire country. Nationwide scientific and industrial resources were utilized to establish the highest-level national research and development teams, and the "production-academia-research-application" model was employed with over ten thousand scientists, engineers, and technicians participating in the endeavour.[27]

The CRH380A (Hexie) train high-speed train, developed for the operation of the Beijing-Shanghai High-speed Railway, shall meet three requirements. Firstly, the Beijing-Shanghai High-speed Railway has a long distance of 1318 km. The top-level technical indicators of the new generation of high-speed EMUs are set to operate at a continuous speed of 350 km/h and a maximum speed of 380 km/h, with a non-stop travel time of 288 minutes. The EMUs shall be able to withstand the test of long-distance continuous high-speed operation.

Secondly, the operating conditions are complex. The Beijing-Shanghai High-speed Railway passes through 244 bridges and 22 tunnels, and the length of bridges accounts for 80.6% of the total length. Ballastless tracks are laid along the whole line. The complex operating conditions resulted in huge technical challenges for high-speed EMUs passing through tunnels, high-speed EMUs encounters, and riding comfort.

Thirdly, the operating environment varies greatly. The Beijing-Shanghai High-speed Railway crosses four provinces and three municipalities in complex terrain and geology. The temperature and humidity change substantially due to geological

[27] Writing group. (2014). Summary on the construction of Beijing-Shanghai High-Speed Railway (pp. 168–170). Beijing: China Railway Publishing House.

locations both in temperate and subtropical zones. Therefore, the operating environment undermines the dynamic performance, drag reduction, noise reduction, and operation reliability of high-speed trains.

To satisfy the operation demands of the Beijing-Shanghai High-speed Railway, the CRH380A (Hexie) train has demonstrated technological innovations in nine aspects, such as aerodynamic locomotive, gas intensity, and tightness, vibration mode of vehicles, high-speed bogie, noise control, traction drive system, braking system, current collection with double pantographs at high speed and passenger interface. On December 3, 2010, CRH380A high-speed EMU broke the record of the testing train speed, achieving a maximum speed of 486.1 km/h in the pilot section from Zaozhuang to Bengbu.[28]

In September 2007, the *Decision of Party Group of MOR of PRC on Strengthening Independent Innovation of Railway and Promoting Harmonious Railway Construction* was issued, which comprehensively put forward the objectives of technological innovation of high-speed railways.[29] The *Decision* pointed out that in order to continuously enhance the independent innovation of railways and modernize the construction of the railway, the overall requirements of railway technological innovation include promoting the original innovation, integrating innovation, re-innovating through introduction, digestion, and absorption, making breakthroughs in a number of major key technologies, constantly developing the railway scientific and technological talent team, completing the railway innovation system, and advancing railway technology and providing strong technical support for accelerating the Chinese railway modernization.

To achieve the overall goal, during the *Eleventh Five-Year Plan* period, 10 specific goals were listed for the development of railway science and technology.

Firstly, a complete set of technical systems should be established for the construction and operation management of the Chinese passenger dedicated railways. It was necessary to systematically master the technologies of passenger-dedicated railways, such as engineering construction technology, traction power supply technology, communication signal technology, operation dispatching technology, passenger service technology, system integration technology, and comprehensive maintenance technology.

Secondly, the technological system of Chinese EMUs and high-power AC locomotives is the core of 200 km/h to 250 km/h EMUs and high-power locomotives that can provide the platform for re-innovation. The independent design and manufacturing capacity for building EMUs and high-power locomotive series products demands for research on 300 km/h to 350 km/h EMUs, sleeping cars, dining cars, luggage cars, and other products.

---

[28] Wang Jun. *CRH380A high-speed train technology innovation.* Report on International Forum of High-speed Train Technology Innovation. Retrieved August 12. 2020, from http://finance.china.com.cn/roll/20120530/760188.shtml

[29] Party Group of MOR of PRC. (2007). Decision of Party Group of MOR of PRC on strengthening independent innovation of railway and promoting harmonious railway construction. *Management of Railway Engineering Enterprises*, (5).

Thirdly, the complete set of speed-raising technology systems of Chinese existing railway lines should be systematically improved. The critical points are improving the safety guarantee system for speed adjustment of existing lines and ensuring a continuous and stable speed increase. Collinear operation of EMUs with a speed of 200 km/h to 250 km/h, freight trains with a speed of 120 km/h, and double-layer container trains with the minimum tracking interval of 5 minutes shall be available on the main trunk lines.

Fourthly, the complete set of heavy haul transportation technology should be mastered. China would formulate technical standards for heavy haul trains, and develop new heavy haul trains with axle load of more than 25 tons and special trains with load of 100 tons. Moreover, the existing freight train technology needs upgrading. 10,000-ton heavy haul and 20,000-ton heavy haul combined trains would be dispatched in the main coal transportation channels, and the technology of dispatching a 30,000-ton heavy haul combined train should be mastered.

Fifthly, a set of technical systems for the Chinese plateau railway should be completed. This goal is to modernize the engineering technology of plateau frozen soil, systematically master the safety monitoring and guarantee technology of plateau railway operation, and formulate and improve the technical standards of plateau railway.

Sixthly, a fully functional railway information system should be in place. The intellectualization of dispatching and command, marketization of passengers and freight sales, and modernization of operation and management play important roles in promoting the implementation of the overall plan of railway informatization.

Seventhly, a railway safety technology system should be established. The relevant railway safety technology system contains comprehensive detection technology, safety monitoring technology, safety assessment technology, disaster early warning technology, and emergency rescue technology.

Eighthly, energy-saving and environmental protection technologies should be promoted. It was necessary to comprehensively carry out the application research and promotion of resource conservation, comprehensive utilization technology, and environmental protection technology, and improve the technology for railway energy conservation and environmental protection.

Ninthly, the scientific and technological resources of the railway industry should be integrated to strengthen the construction of innovation platforms and bases, and a relatively perfect railway technological innovation system should be formulated.

Tenthly, a group of world-class experts and innovation teams should be cultivated and formed to help the railway basic theory and railway application technology reach the world's advanced level. Among the total ten goals, apart from heavy haul railway and plateau railway technology, the others are all targeted at high-speed railways and their relevant technologies.

6.2 Long and Arduous Decision-Making Process

The decision-making process of constructing the Beijing-Shanghai High-speed Railway was beset with twists and turns because this Project influenced the future development of Chinese Railways and involved high technologies and a large amount of funds. All sectors including the government, luminaries, and ordinary citizens participated in the decision-making process. Generally speaking, the debates focused on the following two issues. Whether the high-speed railway should be built as soon as possible and what kind of high-speed railway should be built. Around the two issues, the first round of decision-making mainly probed into whether to build a high-speed railway as soon as possible, and the second inquired into what technology should be adopted to build the high-speed railway.

6.2.1 The First Round of Decision-Making: "Reconstruction & Expansion" and New Construction of High-Speed Railway

MOR had always been the most important and active promoter of high-speed railway construction since its conception was put forward. However, some experts in MOR expressed disagreement with the construction of the Beijing-Shanghai High-speed Railway at the initial demonstration and validation stage of the decision-making, because the relevant issues of the high-speed railway had not attracted much attention from the public and a large number of technical details were difficult for laymen to understand.

Two factors against the immediate construction of the high-speed railway could be concluded. Firstly, MOR underestimated the potential of the existing Beijing-Shanghai Railway, whose capacity expansion could be achieved by the electrification transformation and the operation of tilting trains, instead of building a new high-speed railway. Since the tilting train had been running at a speed of 200 km/h in the United States, it could reach a higher speed technically. Secondly, as a developing country, China's per capita GDP was far below US $1,000. The construction of a high-speed railway would bring a heavy financial burden to the underdeveloped national economy and could not meet the consumption level of passengers. These opponents accused scholars who supported the construction of the Beijing-Shanghai High-speed Railway of overestimating traffic volume and benefits but underestimating transportation capacity and investments to make the project feasible.

The problems mentioned above were emphatically discussed in the early preparation stage. In the *Report on Preliminary Research of Major Technical and Economic Issues of the Beijing-Shanghai High-Speed Railway*, the main capacity expansion schemes of the Beijing-Shanghai Railway were compared and studied.

On the one hand, tapping the potential and expanding the capacity of the existing Beijing-Shanghai Railway was considered. To expand the transportation capacity of the Railway, a series of technological transformations were carried out before 1994, and remarkable achievements were made. By 1994, the technology of internal-combustion traction had been used on the whole line. Except for the 10-minute train tracking interval between Beijing and Dezhou, intervals in other sections had been compressed to 8 minutes, and the number of passenger trains had been increased from 18 to 20. The weight of freight had reached 4,000 tons in most sections, and some 5000-ton trains had been operated.

The further expansion of capacity would be explored mainly by increasing the load for all heavy cargo trains to 5,000 tons and the number of cars to 20 for passenger trains, and by cutting the tracking interval to 6 minutes. In accordance with *Policies on Key Railway Technologies*, if these measures were to be put into practice under the circumstances of normal transportation orders and a reasonable load of the railway, the maximum transportation capacity of the existing Beijing-Shanghai Railway would climb to 70.4 million tons of freight capacity in the southbound direction, 44.8 million person-times of two-way passenger transportation, and a maximum capacity of 52 million person-time for 50% of double-deck passenger trains. It should be noted that train weight and traffic density constrain each other under certain conditions. In spite of the realization of the most possible expected goal, the limited capacity expansion could not meet the needs of transportation development at the end of the twentieth century and the beginning of the twenty-first century without new railways.

On the other hand, using tilting trains was also considered. Through the efforts of nearly 20 years, tilting trains had been successfully applied in Sweden, Italy, and other countries. This kind of train could automatically tilt inward when passing through the curve line to partially compensate for the unbalanced centrifugal acceleration, which generally increases the speed of the train passing through the curve by 20% ~ 40% and the average speed by 15% ~ 20%. However, the titling trains were still generally used on existing lines with surplus transportation capacity to increase the speed of passenger trains and reduce the investment in the technological transformation of curves. However, the application of this technology on such busy lines as the Beijing-Shanghai Railway could only increase the speed of passenger trains, but could not expand the capacity. Moreover, the higher the speed of these passenger trains, the greater the deduction coefficient for other trains, which was bound to occupy more of the tight transportation capacity. In addition, the costs of technological transformation would be considerable due to the higher costs for manufacturing and maintenance and seats/unit on such trains than ordinary and high-speed trains. As there are many level crossings along the Beijing-Shanghai Railway, the transformation cost would also be substantial. Therefore, tilting trains were not considered feasible for capacity expansion and speed adjustment on the Beijing-Shanghai Railway.

The scheme of tilting trains was rejected in the *Report on Preliminary Research of Major Technical and Economic Issues of the Beijing-Shanghai High-Speed Railway*.

In this Report, four schemes related to building the second double-track of the Beijing-Shanghai Railway were proposed, including building four lines in parallel to the existing Beijing-Shanghai Railway, building a new dual-track for both passenger and freight transportation, building a new normal speed passenger dedicated line, and building a new high-speed railway passenger dedicated line. The investment budget of the four schemes was 53.7 billion yuan, 40.1 billion yuan, 43.4 billion yuan, and 52.3 billion yuan respectively according to the cost estimate at the end of 1992.

The comparison and selection of the four schemes lead to the discussion of the following two issues.

The first was related to whether to follow or break the traditional model of the railway with mixed passenger and freight transportation. The railway with such a traditional mode occupied a major position in the world. However, this transportation mode had its disadvantages. Because the passenger and freight trains with different speeds run together at the same time on busy railways, their operations would constrain each other, which was not only inconvenient for traffic organization but also difficult to effectively utilize technical equipment, maximize transportation capacity, improve transportation quality and modernization level on the basis of the different features of passenger and freight transportation. Adherence to the capacity expansion scheme of the mixed passenger and freight transportation mode could not meet the huge increasing passenger and freight transportation demands on the Beijing-Shanghai Railway in the twenty-first century. Therefore, the new second double track of the Beijing-Shanghai Railway should be constructed to break through the traditional mode and separate passenger and freight transportation. The passenger dedicated line could be developed with high-speed technology, while the heavy haulage technology could be adopted in the construction of freight dominated line to greatly expand the transportation capacity, and make the railway leap to a new level in terms of transportation capacity, speed, safety, quality, and efficiency, which was the fundamental way to build a high-capacity passenger and freight transportation corridor and solve the "bottleneck" problem of busy trunk transportation.

Through the above analysis, from the perspective of operation mode, it was inappropriate for the new second double track of the Beijing-Shanghai Railway to follow the first and second schemes with the traditional mixed passenger and freight transportation modes. In addition, the disadvantage of the two schemes, especially the first one, was reflected in the large workload of building station hubs, making it difficult to relieve the work pressure, remove old facilities, and transit into the new station. Moreover, the disturbance to construction and operation was also severe, so that building parallel lines along the existing line was unfavorable for the existing Beijing-Shanghai Railway or almost impossible, particularly in some busy sections.

The second issue was whether to choose normal speed or high speed for the new passenger dedicated line. In consideration of initial project investment, the new passenger dedicated line with the normal speed would save nearly 9 billion yuan compared with the high-speed passenger dedicated one, but in terms of transportation speed, efficiency, service quality, and economic benefits, the ordinary speed passenger dedicated railway would lag far behind the high-speed one. With the reform and opening up and the establishment of the socialist market economic system, the acceleration of urbanization, and the improvement of people's living standards, and time values, people were bound to make higher requirements for travel speed, travel time, comfort, safety, and other service quality. Judging from the demand for passenger transportation in this region in the twenty-first century, it was not suitable for the construction of a normal-speed passenger dedicated line.

Through the comparison of various schemes, the "four commissions and one ministry" reached a consensus that the Beijing-Shanghai High-speed Railway was the best choice for the expansion and modernization of the existing one because the technology was feasible and the state could afford the reasonable cost.[30]

In 1994, another important research project, the "Main Analysis on Technological Conditions and Cost-benefit of Beijing-Shanghai High-speed Railway", was launched. Jointly sponsored by the China Academy of Railway Sciences and the World Bank, this Project was listed in the seventh batch of loan projects of the World Bank, and it covered four main research areas. Firstly, in accordance with the recommendation of the expert group from the World Bank, the traffic volume measurement method commonly used in the world at that time should be applied to the traffic volume measurement of various transportation modes along the Beijing-Shanghai Corridor including the high-speed railway. Secondly, the transportation organization issues were discussed, such as making rational use of the capacity of high-speed railway and existing lines, choosing the appropriate transportation organization mode, determining the right target speed, and the specific passenger train operation scheme. Thirdly, through analysis, research, and computer simulation, the main technical conditions and parameters of public works, rolling stock, signal, and control of the Beijing-Shanghai High-speed Railway were put forward on the basis of the comprehensive collection and understanding of the technical parameters of high-speed railways in other countries to provide a technical foundation for the construction of Beijing-Shanghai High-speed Railway. Fourthly, the comparison and selection for the schemes and the cost-benefit analysis of Beijing-Shanghai High-speed Railway were carried out.

After three years of efforts, a *Report on the Research of Main Technical Conditions and Cost-benefit Analysis of the Beijing-Shanghai High-Speed Railway* was submitted. The report concluded that the transportation capacity of the Beijing-Shanghai Railway could not meet the current needs of passenger and freight transportation, and railway transportation had become a constraining factor for the economic and social progress of the regions along the line, so it was imperative to expand the capacity of the Beijing-Shanghai Railway. As the best capacity expansion scheme, the construction of the Beijing-Shanghai High-speed Railway satisfied

[30] For related studies, refer to: The joint preliminary research group on major technical and economic issues of Beijing-Shanghai High-Speed Railway by the state Science and Technology Commission, the state Development Planning Commission, the state Economic and Trade Commission, the state Commission for Restructuring the Economic System and MOR. (1994). *Report on Preliminary Research of Major Technical and Economic Issues of Beijing-Shanghai High-Speed Railway.*

not only the need to expand railway capacity, but also the demand in economic and social progress of the regions along the line.[31]

Respective research conclusions were not convincing, and the problem was solved along with the passage of time. In the twenty-first century, with China's economic development, sufficient financial resources and consumption upgrading, as well as the expansion of the international high-speed railway network, the debate on whether to build the railway gradually disappeared. Although the issue of capacity expansion through technological renovation was still presented in the report on the feasibility study of the Beijing-Shanghai High-speed Railway in 2006, it was no longer the focus of discussion.

6.2.2 The Second Round of Decision-Making: "High-speed Wheel-rail Railway" or "High-speed Maglev Train"

Maglev technology was not considered a technical option for the construction of the Beijing-Shanghai High-speed Railway until 1998. However, the maglev technology had been taken into account in the demonstration and validation process.

In 1922, Hermann Kemper, a German engineer put forward the conception of maglev train technology aimed to make the train float without contacting the ground. By the 1960s and 1970s, more than a dozen countries had invested a large amount of human and material resources to develop such technology, and in 1969, the world's first small model of maglev train appeared in Germany, and Japan also successfully developed the maglev trains three years later. In 1979, the maglev train technology set a speed record of 517 km/h, but no country really built an operation line with it.

In June 1994, focusing on the development strategy of high-speed railway technology and the prospect of high-speed maglev trains and related superconducting technology, the 18th Academic Conference of Xiangshan Science Conference, with the theme of the Development Strategy of High-speed Railway Technology, was successfully held in Beijing.[32] This Conference gathered 39 top experts in the field of railway, maglev technology, applied superconductivity, and science and technology management to discuss five topics including the Demand and Programme of High-speed Railway Development, the Condition of High-speed Railway Development, the Prospect and Suggestions of High-speed Maglev Train Technology, Superconductive Magnet & Application and the Development Strategy of Chinese High-speed Railway. Following fierce discussions, the participating experts had affirmed the technical achievements of the two high-speed

[31] For related studies, refer to: CETE-2 project team of MOR, 95-F53 research group of China Academy of Railway Sciences. (1997). *Report on the research of main technical conditions and cost-benefit analysis of Beijing-Shanghai High-Speed Railway.*

[32] Yan Luguang. (2007). *Maglev traffic anthology* (p. 1). Beijing: China Electric Power Press.

transportation modes—high-speed railway and maglev train but had not made a decision on to which priority should be given. The *Report on Preliminary Research of Major Technical and Economic Issues of Beijing-Shanghai High-Speed Railway* completed in 1994 mentioned a type of technology under development, a new maglev train technology without wheel-rail contacting, but the experts preferred high-speed wheel-rail technology.

On August 11, 2003, the National Development and Reform Commission entrusted China International Engineering Consulting Corporation to carry out a special demonstration on the issues related to the construction of the Beijing-Shanghai High-speed Railway and presented the *Report on Issues Concerning the Construction of Beijing-Shanghai High-Speed Railway*. The main content of the demonstration included "the necessity and urgency of the construction of Beijing-Shanghai High-speed Railway" and "the comparison and selection for High-speed Wheel-rail Railway Scheme and High-speed Maglev Train Scheme".

From September 1 to September 5, 2003, entrusted by the State Council, the National Development and Reform Commission, together with China International Engineering Consulting Corporation, held a conference on the comparison and selection for technological schemes of the Beijing-Shanghai High-speed Railway, aiming to widely consult the experts for suggestions. On September 3, at the Conference, detailed introductions and analyses were presented on both the High-speed Wheel-rail Railway Scheme and the High-Speed Maglev Train Scheme.

The participating experts made a comparative study on the two reports from the following ten aspects.

The first was the comprehensive comparison of technology. Both high-speed wheel-rail technology and maglev technology were new and high technologies in development. To be specific, high-speed wheel-rail technology was an innovation of traditional railway technology. The high-speed maglev transportation system shows its advantage in handling the wheel-rail interaction and pantograph-catenary relation, the two major obstacles of the wheel-rail system to increasing the speed. The electromagnetic principle provided the theoretical basis for high-speed operation without mechanical friction or contact power collection, and some modern technologies were continuously developed and maturing.

The second was the comparison of the compatibility of the railway network. At that time, the length of the wheel-rail transportation network reached 72,000 kilometers. The Beijing-Shanghai Railway was a trunk line in the network, with 25 trunk and branch lines connected to it, of which more than 40% of the passenger flow was cross-line passenger flow. The high-speed wheel-rail railway and the existing railway belong to the same compatible wheel-rail transportation system, so the cross-line passengers can go directly to their destinations without changing trains. At that time, the maglev transportation system was still in a phase of small-scale testing, and the scope of transportation services was relatively limited because of the uncompleted support of the railway network.

The third is the comprehensive engineering comparison. At that time, China accumulated rich experience in constructing and operating long trunk lines thanks to the construction of high-speed wheel-rail railway with 250 km/h and the test operation of railways of 320 km/h, but lacked engineering practice in the construction and operation of high-speed wheel-rail railway at 300 to 350 km/h. It had tested the maglev system with an operation speed of 430 km/h, and such a great achievement had a big clout around the world. However, China didn't apply the maglev system on long trunk lines.

The fourth is the comparison of comprehensive benefits. The maglev transportation system had not been applied in the construction and operation of medium and long-distance trunk lines in the world, nor had the high-speed wheel-rail railway been operated in practice in China. Therefore, the benefits of the two schemes cannot be compared, and further comparative research needs to be conducted.

The fifth is the comparison of investment. The unit cost of a high-speed maglev system may be higher at that time. However, the capital investment could not be accurately estimated and compared due to the lack of experience in engineering design for the application of maglev technology on long trunk lines and practice in the construction of high-speed wheel-rail with the speed of 300–350 km/h in China at that time.

The sixth is the comparison of basic work. MOR had completed more than 350 special studies on the construction of the Beijing-Shanghai High-speed Railway, 15 Provisional Regulations on survey, design, construction, and management were compiled, and the preliminary survey and feasibility study were basically completed. The type selection of high-speed trains had been studied, and the development of high-speed trains with speeds of 200 km/h and 270 km/h was carried out. The comprehensive speed adjustment test of a domestic D-Series High-Speed Train with a speed of more than 200 km/h was carried out on the Qinhuangdao-Shenyang Passenger Dedicated Railway, and a 66.8 km long railway was selected as a 300 km/h high-speed comprehensive test section. The high-speed maglev system had been put into practice in China. On the basis of the German normal conducting high-speed maglev technology system, achievements had been made through the demonstration project of the Shanghai maglev line and the "863" maglev special research program. Based on transferred technology, China had mastered all the technologies for the design and manufacturing of the track system. Moreover, China had completed the installation of system equipment, with the competency to decompose and integrate the system. Through the decomposition of vehicles, traction power systems, operation control, and other subsystems, some verification prototypes and practical sample machines were developed. The application of these technologies to medium and long-distance trunk lines needed more exploration and practice.

The seventh is the comparison of international experiences. By the end of 2002, nine countries had built high-speed railways and put them into operation, with a total length of 5,435 kilometers. Another 9 countries and Taiwan province were building 16 high-speed railways, totaling 3,267 kilometers. By 2020, Europe planned to build 10,000 kilometers of new high-speed railways and transform 15,000 kilometers of existing lines to formulate a high-speed railway network. The compatibility with the existing railway network was taken into account in the construction of high-speed railways in these countries and regions, and the cross-line operation of high-speed trains and network efficiency were important factors in the planning. Abundant experiences in the construction and operation of high-speed wheel-rail railways were accumulated and could be widely used for reference in our country. A large number of suppliers were engaged in the development and manufacture of high-speed trains and their control systems in the international market, which is conducive to China's technology introduction. However, there was no commercial operation line for maglev transportation systems around the world. At that time, the international community had not reached a common understanding on the technical route and technical specifications of the maglev transportation system.

The eighth is the comparison of environmental impact. On September 1, 2003, China began to roll out the *Law of the People's Republic of China on Environmental Impact Assessment*. The law stipulates that environmental assessment is not only required for construction projects, but also for regional planning. The environmental protection department needed to carefully evaluate the quality of the environment, water and soil conservation, waste gas emission, electromagnetic radiation, and other aspects along the line. According to the technical data at that time, the two technical schemes were almost the same.

The ninth is the comparison of operation experience. The high-speed wheel-rail traffic system had been put into operation for 40 years around the world. At that time, the trial operation mileage of domestic 200–250 km/h EMUs, with stable and safe performance, reached 350,000 km, which demonstrated that the design goal had been achieved. Nevertheless, raising the speed from 200–250 km/h to 350 km/h still needed further performance tests. The high-speed maglev train had been put into trial operation of nearly 90,000 km and could also offer valuable experience.

The tenth is the comparison of safety and reliability. An important symbol of mature and reliable high-speed railway technology is to ensure the safety of its operation. Both high-speed wheel-rail and maglev technologies proved to be safe and reliable.

The Report held that new high technologies, high-speed wheel-rail and high-speed maglev had a certain foundation and development prospect in China. The development of these two technologies was not either-or and incompatible with each other. Instead, they needed mutual promotion and complementary interaction to jointly promote the modernization of China's rail transit.[33]

The Report suggested that the high-speed wheel-rail technology had been proven mature, safe, and reliable in the world, and it should be the first choice for the Beijing-Shanghai High-speed Railway so long as the policies and measures, based on the existing abundant reserve of technologies, talents and over 10-year tracking research, were appropriate and suitable for the recent construction of the railway. Magnetic levitation is a promising rail transit technology because of its high speed and fast starting-up. Thus, the Report also proposed that the state set up a special group to conduct research and select a test line with a length of 200 to 300 km for

[33] China International Engineering Consulting Corporation. (2003). *Report on issues concerning the construction of Beijing-Shanghai High-Speed Railway.*

magnetic levitation to complete commercial demonstration, making technical reserves for the construction of medium and long-distance lines in the future.

Although it seemed that the Report did not completely repudiate the maglev scheme, in fact, the proposal about the "200 to 300 km line" was not finalized in the end. On January 7, 2004, the State Council's Executive Meeting discussed and approved in principle the *Mid- and Long-Term Railway Network Plan*, which proposed to establish a "four vertical and four horizontal" high-speed railway network, with the Beijing-Shanghai High-Speed Railway ranking first in the four vertical lines. On February 22, 2006, the project of Beijing-Shanghai High-speed Railway was officially approved after the State Council's Executive Meeting discussed and approved in principle the *Proposal for Beijing-Shanghai High-Speed Railway Project* which made clear that "high-speed wheel-rail technology should be adopted for Beijing-Shanghai High-speed Railway".

6.3 Beijing-Shanghai High-Speed Railway

According to the standards of the International Union of Railways, new lines with a design speed of 250 km/h can be regarded as high-speed railway. In this regard, the Qinhuangdao-Shenyang Passenger Dedicated Railway is the first high-speed railway in China. After the Qinhuangdao-Shenyang Passenger Dedicated Railway was completed, the Beijing-Tianjin Intercity Railway, the Zhengzhou-Xi'an and Wuhan-Guangzhou High-speed Railways were also built before the construction of the Beijing-Shanghai High-speed Railway. However, the Beijing-Shanghai high-speed railway is recognized as the first national high-speed railway, since it both marks the formation of the technology system and highlights the image of the Chinese high-speed railway. The preceding railways mentioned above provide technical preparation and verification for the construction of the Beijing-Shanghai High-speed Railway. However, from the perspective of building China's high-speed railway network, the Zhengzhou-Xi'an and Wuhan-Guangzhou high-speed rail lines also have their unique significance.

6.3.1 Starting Laying out Rapid Railway Network

Before 2004, the so-called preparation for a high-speed railway was de facto the same as the preparation for the Beijing-Shanghai High-speed Railway. In February 2004, the *Mid- and Long-Term Railway Network Plan* decided to build a 12,000 km passenger dedicated line, namely, a high-speed railway, while the Beijing-Shanghai High-speed Railway only covers 1,300 km. Although the approval of the Plan means that the Beijing-Shanghai High-speed Railway had made crucial progress in the prolonged decision-making process, there was still a lot of work ahead before the project was officially approved. To ensure that the goal of 2020 can be achieved on schedule, the construction of other lines was also in progress alongside the preparations for the Beijing-Shanghai High-speed Railway.

The Beijing-Tianjin High-speed Railway was built as a section of the Beijing-Shanghai High-Speed Railway. Perhaps inspired by Japan's case of promoting the Shinkansen in support of the Tokyo Olympic Games, MOR also proposed to support the Beijing Olympic Games with the construction of the Beijing-Tianjin High-Speed Railway. Since the Beijing-Shanghai High-speed Railway had not been approved, MOR proposed to use the name of the intercity railway and planned to construct it on the route prepared for the Beijing-Tianjin section of the Beijing-Shanghai High-Speed Railway. But in the final decision, the Beijing-Shanghai High-speed Railway chose another route instead. Therefore, there are two high-speed railways between Beijing and Tianjin. Later, it was proved that this seemingly extravagant decision was correct, and a third railway may be needed between the two cities in the future. The construction of the Beijing-Tianjin High-Speed Railway started in 2005 and it was put into operation on the eve of the Beijing Olympic opening ceremony in August 2008.

By examining the framework of the "four vertical and four horizontal" high-speed railway network proposed in the *Mid- and Long-Term Railway Network Plan*, we can feel its historical influence on the present. Among China's existing railway network, the Lanzhou-Lianyungang Railway crosses China's hinterland from east to west, and the Beijing-Guangzhou and Beijing-Shanghai Railways run from north to south. They are the three most important trunk lines of the "four vertical and four horizontal" high-speed railway network. In history, neither the Beijing-Guangzhou Railway nor the Lanzhou-Lianyungang Railway was built at one time, both of which started from the most important section and then extended forward. The same method was also applied in the implementation of the *Mid- and Long-Term Railway Network Plan*. The difference is that the earliest section of the Lanzhou-Lianyungang Railway started from Kaifeng to Luoyang (the Bianlaing-Luoyang Railway), and now it begins from Zhengzhou to Xi'an. The northern section of the Beijing-Guangzhou Railway was built at first (from Beijing to Wuhan), and now the southern section starts to be constructed (from Wuhan to Guangzhou).[34]

In November 2005, the National Development and Reform Commission approved the *Report on the Feasibility Study of New Passenger Dedicated Line from Zhengzhou to Xi'an*. On September 25 of the same year, the construction of Zhengzhou-Xi'an High-speed Railway commenced and the railway was officially put into operation on February 6, 2010. MOR launched the construction of the Wuhan-Guangzhou High-speed Railway on June 23, 2005, and it was open to traffic on December 26, 2009.

One of the reasons several high-speed railways were approved so quickly was because the technology prepared for the construction of the Beijing-Shanghai

[34] These two changes also reflect the transition of China's regional politics and economy in the past century. Kaifeng is no longer the political center of Henan Province. Its position was replaced by Zhengzhou which rises due to Beijing-Guangzhou Railway.

High-speed Railway could be used directly, which in a sense, also provides technical validation for the Beijing-Shanghai High-speed Railway.

6.3.2 *Decision-Making and Construction of the Beijing-Shanghai High-Speed Railway*[35]

With a total length of about 1,318 km, the Beijing-Shanghai High-speed Railway completed in one single stage, is the longest high-speed railway line with the highest standard in the world at present. The project of Beijing-Shanghai High-speed Railway was approved with a total investment of 220.94 billion yuan, making it the largest construction project ever invested since the foundation of PRC. Its construction was finished in November 2010. On December 3, 2010, the test train on the Beijing-Shanghai High-speed Railway operated at the speed of 486.1 km/h, creating the fastest speed record in Chinese high-speed railway history. The Nanjing Dashengguan Yangtze River Bridge is currently known as the world's first six-track railway bridge with the world's highest design load, and its 336 m main span is the longest among the same-class high-speed railway bridges in the world.

The Beijing-Shanghai High-speed Railway line traversing North China Plain and the Middle-Lower Yangtze Plain crosses four major river systems including Haihe River, Yellow River, Huaihe River, and Yangtze River, and the length of its bridges and tunnels accounts for 81.6% of its total length. It connects Beijing, Tianjin, Hebei, Shandong, Anhui, Jiangsu, and Shanghai, linking the Bohai Economic Rim and the Yangtze River Economic Zone. The railway starts from Beijing South Railway Station. It ends at Shanghai Hongqiao Railway Station, connecting with the existing railway lines under the jurisdiction of Beijing, Jinan, and Shanghai Railway Administrations. The designed one-way transportation capacity reaches 80 million passengers per year.

The decision-making process about the Beijing-Shanghai High-Speed Railway lasted 18 years from MOR's proposal for its construction in 1990 to the approval of its commencement report at the 205th State Council's Executive Meeting on January 16, 2008.

With the announcement of then Premier Wen Jiabao at the opening ceremony on April 18, 2008, the construction of the Beijing-Shanghai High-Speed Railway project officially started. On June 12, concreting for the first 900 t and 32 m ballastless box girder was successfully completed in the Dadian girder fabrication and storage yard. On December 2, the construction of the Huaihe River Super Large Bridge was finished. The track-laying work began on July 19, 2010, and was completed on November 15. The subgrade was ready on November 9. On December 3, the new generation high-speed EMU, the CRH380A (Hexie) created a maximum speed

[35] For related studies, refer to: Writing group. (2014). *Summary on the construction of Beijing-Shanghai High-Speed Railway*. Beijing: China Railway Publishing House.

record of 486.1 km/h in the pilot section from Zaozhuang to Bengbu. The test data showed that the subgrades, bridges, and tracks in the pilot section all met the operational requirements of high-speed trains, and the catenary quality reached the world level. The joint debugging and commissioning tests were carried out for the section from Dezhou East Station to Zaozhuang Station (313 km) and Bengbu South Station to Shanghai Hongqiao Station (471 km) on February 20, 2011, and for the section from Beijing South Railway Station to Dezhou East Station (314 km) on March 20. From May 11 to June 10, the trial operation of the whole line was conducted.

On May 31, the Beijing-Shanghai High-speed Railway passed a preliminary acceptance inspection after MOR invited 30 academicians and experts to inspect and evaluate this high-speed railway project from May 25 to 26. Experts concurred that the railway track had reached a state of high smoothness and stability; the communication signals and traction power system were stable and reliable, and the passenger service system met the demands for operation. It was also agreed that CRH380A had achieved the goal of high speed and comfort, and the support facilities for operation were complete, safe, and reliable. Owing to world-class engineering technology and quality management with technological innovation and standard management, the high-speed railway was fully qualified for operation. On June 12, the Beijing-Shanghai High-speed Railway passed a safety assessment and met the conditions for operation. At 15:00 on June 30, 2011, the railway was officially put into operation with a speed of 300 km/h in the initial phase.

The Beijing-Shanghai High-speed Railway is not only an important national strategic transportation project, but also a major national innovation project with its long lines, high standards, complicated technology, large quantities, and wide coverage. During the construction of the Beijing-Shanghai High-speed Railway, China innovated its design concept, construction technology, and management system, integrated domestic scientific and technological resources, expanded international exchanges and cooperation, learned from the innovative achievements and experience of the Beijing-Tianjin Intercity Railway and other high-speed railways, and made great efforts to promote the achievements of technological innovation, forming a novel mode for Chinese high-speed railway construction and a technical standard system of Chinese high-speed railway.

The completed Beijing-Shanghai High-Speed Railway has become the main framework of China's high-speed railway network which runs through North and East China. On September 2, 2012, the joint debugging and commissioning tests for the Hefei-Bengbu High-speed Railway connected with the Beijing-Shanghai High-speed Railway succeeded, and the railway was officially put into operation on October 16. Thanks to the connection of the two lines at Bengbu South Station, the shortest running time only takes 3 hours and 50 minutes from Hefei Station of the Hefei-Bengbu High-speed Railway to Beijing South Station of the Beijing-Shanghai High-speed Railway, which is 5 hours and 42 minutes shorter than that of passenger trains on the existing lines. In March 2013, an average of 80 pairs of trains were dispatched on the Beijing-Shanghai High-speed Railway, including 68 pairs of high-speed EMUs with a speed of 300 km/h and 12 pairs of EMUs with a speed of 250 km/h, greatly shortening the travel time of trains on this line.

The Beijing-Shanghai High-speed Railway has set a new record for railway construction. On December 3, 2010, the speed of test trains reached 486.1 km/h, creating the fastest record for Chinese high-speed railway. The Nanjing Dashengguan Yangtze River Bridge claimed four firsts in the world. The largest volume of works: It has the world's first six-track railway bridge, and the total amount of steel equates to four times that of the Wuhan Yangtze River Bridge. The largest span: The 336-m main span makes it rank first among the same-class high-speed railway bridges over the world. The highest load capacity: The bridge bears the largest design load in the world, with the dead load plus the live load of the main bridge being about 120 tons/m, and the load of the bridge reaching 35 tons/m, equivalent to a 35-lane highway bridge. The fastest design speed: The design speed of the railway on the bridge reaches 300 km/h, reaching the top level of the world. The bridge has also set the precedent for accommodating three double-track railways the Beijing-Shanghai High-speed Railway, Shanghai-Wuhan-Chengdu High-speed Railway, and Nanjing Metro, making the Beijing-Shanghai High-speed Railway the first in the world. On February 25, 2013, the Beijing-Shanghai High-speed Railway passed the acceptance inspection by the National Acceptance Committee.

6.3.3 Forming Chinese High-Speed Railway System

The process of preparing for the Beijing-Shanghai High-speed Railway also witnessed the formation of the whole high-speed railway system. In this process, the basic document of high-speed railway construction—the *Code for Design of High-Speed Railway*—was gradually formed and improved.

On the basis of scientific research and experience of construction on test lines in the 1990s, MOR completed the compilation and review of codes for the design of civil works. After many rounds of discussion based on the key achievements made in public works of the Eighth Five-Year Plan and Ninth Five-Year Plan, MOR compiled the Interim Provisions on the Design of Bridge and Tunnel Stations of Beijing-Shanghai High-Speed Railway on January 15, 1999, which provided a preliminary basis for the design of civil works. Then, MOR began to study the design of electrical and mechanical (E&M) works. On November 14, 2002, MOR held a review meeting on the draft of Chap. 18—disaster prevention, safety, and monitoring of the *Interim Provisions on the Design of E&M Works of Beijing-Shanghai High-Speed Railway*. In 2003, MOR completed compiling the *Interim Provisions on the Design of Shanghai High-Speed Railway* and revised the *Interim Provisions on the Design of Beijing-Shanghai High-Speed Railway* in 2005 after consulting with international experts.

During this period, China had gained a broader understanding of high-speed railway technology, after the construction of the Qinhuangdao-Shenyang Passenger Dedicated Railway, especially the construction and comprehensive test of the Qinhuangdao-Shenyang Passenger Dedicated Railway. In May 2003, MOR organized the compilation of the *Interim Provisions on the Design of the*

Beijing-Shanghai High-Speed Railway, which was issued as Document TJS [2003] 13. The *Interim Provisions* consists of 14 chapters, namely General Provisions, Terms, Train Operation, Electrification, Communication, Signal, Informatization, Communication Signals and Other Provisions, Power, Operation and Equipment Maintenance of EMUs, Water Supply and Drainage, Comprehensive Detection and Maintenance, Disaster Prevention and Safety Monitoring, Environmental Protection, which set the standards and specifications for the design of Beijing-Shanghai High-speed Railway. However, to address numerous technical challenges in high-speed railway construction, we needed to consult international experts due to a lack of readily available solutions.

From July 2003 to May 2005, French SYSTR Group, Japan Railway Technical Service (JARTS), and Germany DEC Deutsche Engineering Consultants GmbH were invited successively by MOR for consultations on the preliminary design of the *Interim Provisions on the Design of Beijing-Shanghai High-Speed Railway* (hereinafter referred to as the Interim Provisions), the Shanghai-Nanjing section and the Beijing-Nanjing section closely related to the Interim Provisions. The consultation results verified the merits and rationality of the Interim Provisions and the preliminary design. In addition to the affirmation of the Interim Provisions and preliminary design (with a total of 12 subjects including civil works and E&M works), the consultation report also gave corresponding improvement schemes and modification suggestions after pointing out the deficiencies in the Interim Provisions, and preliminary design. On this basis, the Interim Provisions, scientific research report and preliminary design were revised and improved.

Through the above three international consultations, Chinese railway science and technology workers have systematically and widely understood the development direction and key points of world high-speed railway technology. They absorbed advanced design concepts and methods, as well as adopted design standards and cutting-edge technologies from international high-speed railway systems. In particular, they learned human-centered design concepts and ideas including giving priority to the safety and comfort of passengers, paying attention to the configuration and systematic requirements of transportation and service, and at the same time, adhering to high standards, making provisions for future development, putting emphasis on the construction of maintenance system, etc.

The international consultancies affirmed the main technical standards and preliminary design documents adopted for the civil works design of the Beijing-Shanghai High-speed Railway and held that following minor modifications the design documents can meet the requirements for the speed of 350 km/h, safety, and comfort. Besides, they also basically agreed to the design parameters and technical standards determined in the Interim Provisions. Their review of the plane and vertical section design, main structure design, electrification and selection of communication signal system, and main design parameters of the whole Beijing-Shanghai High-speed Railway had testified to our design competence, and also further improved the Interim Provisions and preliminary design. Based on international experiences and opinions, the technical principles for the construction of the Beijing-Shanghai High-speed Railway were formed, that is, to build a world-class high-speed railway with high standards, high quality, and high efficiency by adhering to people-centered philosophy, combining independent R & D and innovation with extensive absorption and introduction of international advanced and mature technologies, drawing upon the strengths of others, and pursuing the road of leapfrog development.

MOR organized design institutes to analyze in detail, digest, and absorb the opinions given by international consulting agencies on the preliminary design, including post-construction settlement standard of roads and bridges, turnout girder design, ballastless track and its influence on track circuit transmission and countermeasures, subgrade stability coefficient, catenary suspension type, redundancy mode of wireless communication, calculation of fluctuating pressure on the sound barrier, etc. The 29 substantive revisions of the relevant content of Interim Provisions have optimized the preliminary design. As for some key technical problems involved in the design consultation, the Chinese and international experts carried out extensive and in-depth discussions and exchanges, reached a consensus, and defined the solutions after careful analysis and comparison, thus the relevant technical standards or technical schemes were optimized. In brief, the revised Interim Provisions and preliminary design were more advanced, reasonable, and systematic.

Through consultation, experts from MOR had gained new insights into the technology and respective characteristics of international high-speed railways. Representing the highest level of high-speed railway in the world, the high-speed railway technologies of Germany, France and Japan have their advantages and systems. In addition to general characteristics, the Beijing-Shanghai High-speed Railway also has its distinct characteristics and requirements, which are mainly reflected in the following aspects.

The Beijing-Shanghai High-speed Railway should be connected with the existing lines with a speed of 200 km/h, and an operation program mixing high and medium-speed trains should be implemented in the initial stage of operation, which requires that the communication signals and electrification facilities should be compatible with each other. The construction of the whole line of more than 1,300 km at one time is an arduous task under highly complex geological conditions. To directly copy the technology of any country is infeasible. We should draw on the strengths of others and widely absorb the advanced and mature technologies of various countries. Germany, France, Japan, and other countries have their own high-speed railway technology systems which are incompatible with each other. To leverage the best of each, we should not only have an in-depth understanding of their professional technology but also carefully analyze the technical characteristics of each technical domain, scientifically address the connection and interface issues between different domains, to study and determine the Beijing-Shanghai High-speed Railway technology system suitable for China's national conditions. All these show that systematic integration is the emphasis and key of the Beijing-Shanghai High-speed Railway technology system.

While carrying out international consultations, MOR organized the study on the Beijing-Shanghai High-speed Railway technology system, focusing on world-class standards, upholding the principle for the adoption of "advanced, mature, economical, applicable and reliable" technology, and integrating independent innovation and the introduction of advanced technology based on China's national conditions. Finally, the technology system was preliminarily formed through systematic integration.

From the approval of the Beijing-Shanghai High-speed Railway project to the commencement of construction, MOR also organized experts to prepare a number of supporting technical specifications, which contributed to the high-quality construction of the Beijing-Shanghai High-speed Railway.[36]

6.4 World's Largest High-Speed Railway Network

6.4.1 Short Trough Time

In early 2011, the corruption case of Liu Zhijun and Zhang Shuguang quickly affected a group of middle-level officials, and the development of the Chinese high-speed railway was almost stopped. "In April 2011, the news began to spread that the speed of Chinese high-speed railway would be reduced. On June 30, the news was confirmed that the Beijing-Shanghai High-speed Railway was put into operation at a speed of 300 km/h. On July 1, MOR announced the speed deceleration of the Chinese high-speed railway. The speed of trains on the Wuhan-Guangzhou High-speed Railway, the Zhengzhou-Xi'an High-speed Railway, the Shanghai-Nanjing High-speed Railway, and the Shanghai-Hangzhou High-speed Railway was reduced

[36] The main technical specifications are as follows. In terms of public works, the codes prepared include *Guide for Design of Ballastless Track Railway of Passenger Dedicated Line* (TJS [2005] No. 754), *Interim Provisions on Design of Durability of Railway Concrete Structures* (TJS [2005] No. 157), *Code for Design of Ground Base and Foundation of Railway Bridges and Culverts* (TB 10002.5-2005), and *Code for Seismic Design of Railway Engineering* (GB 50111-2006), etc. In terms of mobile equipment, *Interim Provisions on Design of Electric Multiple Unit Facility* (TJS [2007] No. 89) has been issued. In terms of communication signals, the codes formulated include *Guidance on Implementation of Comprehensive Protection against Lightning and Electromagnetic Compatibility of Railway Signal Equipment* (TY [2006] No. 26), *Method of Implementation of Comprehensive Grounding Technology of Passenger Dedicated Line (Interim)* (TJ [2006] No. 220), *Interim Provisions on Design of Railway GSM-R Digital Mobile Communication Engineering* (TJS [2007] No. 92), and *Technical Specification for Train Control Center of CTCS-2 Train Control System of Passenger Dedicated Line (Interim)* (KJY [2007] No. 158), etc. In terms of environmental protection and safety, the documents complied include *Guidance on Implementation of Railway Green Corridor Construction* (TJSH [2007] No. 472), *Notice on Issues Concerning Adjustment of Design for Green Corridor of New Railway and Railway under Construction* (TJH [2007] No. 544), *Interim Provisions on Lightning Protection, Electromagnetic Compatibility and Earthing Technology of Railway* (TJH [2007] No. 139), *Notice on Printing and Distributing the Minutes of Seminar on Division of Labor and Document Preparation of Design of Railway Lightning Protection and Earthing Engineering* (JX [2007] No. 96), *Notice on Publishing Configuration Standard of Wheel-set Flaw Detection Equipment for CRH series EMUs (Trial)* (YZGY [2009] No. 37), and *Notice on Configuration of Maintenance and Servicing Equipment and Instruments Used in EMU Operation* (YZGY [2007] No. 3235), etc.

to 300 km/h, and that of 250 km/h was reduced to 200 km/h on a large number of high-speed railways".[37]

To exacerbate the situation, the Wenzhou 7/23 rear-end collision accident caused six carriages to derail, resulting in 40 deaths and 172 injuries and breaking off the traffic for 32 hours and 35 minutes, with a direct economic loss of 193.7165 million yuan.

MOR's corruption case and the 7/23 accident brought fatal consequences to the Chinese high-speed railway. "On the one hand, the speed of high-speed railway is further reduced, not only for operating lines but also for the standard of new high-speed railway lines. For example, the civil works of the Datong-Xi'an High-speed Railway and the Lanzhou-Xinjiang High-speed Railway have been almost completed according to the requirements for a speed of 350 km/h, but the E&M works were executed according to the standard of 250 km/h. On the other hand, the further tightening of credit by banks has led to a near depletion of funds for China's high-speed rail lines, resulting in numerous suspensions of construction projects and a mass return of migrant workers to their hometowns, nearly strangling the development of China's high-speed rail system".[38]

In a sense, the setback of the Chinese high-speed railway is the inevitable result of the problems accumulated in the rapid development of the previous years. This claim is hindsight. But it also shows that the Chinese high-speed railway needs to experience such a setback, if we associate with the twists in the previous decision-making process of the Chinese high-speed railway. For railway development, technology is basic at least, if not decisive. The speed and comfort of trains can be demonstrated immediately, while the safety and reliability in performance need time to test. Due to the solid accumulation of technology, it did not take much time for the Chinese high-speed railway to get out of the trough.

6.4.2 High-Speed Railway Construction Boom

Chinese high-speed railway started to come out of the trough in 2013 and a new construction boom was fueled. By the end of 2015, the operating mileage of high-speed railways in China had reached 19,000 kilometers, ranking first in the world. The *Mid- and Long-Term Railway Network Plan* released in 2016 described the innovation ability of Chinese high-speed railway as follows, "Relying on the

---

[37] Gaotie Jianwen. (2015). *High-speed railway in colour* (p. 331). Changsha: Hunan Literature and Art Publishing House.

[38] Gaotie Jianwen. (2015). *High-speed railway in colour* (p. 331). Changsha: Hunan Literature and Art Publishing House. However, some insiders also put forward different opinions on the "speed reduction of the operating line". The design speed of the Beijing-Shanghai High-speed Railway is 350 km/h, but it has been stipulated in the feasibility study that the initial operation speed should be 300 km/h. Later, when trains run at a speed of 350 km/h, it should be called "reach speed", not "speed up".

development of high-speed, plateau, and cold-region railways, significant innovations have been achieved in construction, equipment manufacturing, and system integration. The capabilities for independent development and core competitiveness are continuously strengthened. The overall technical level of Chinese railways has ranked among the top, and high-speed rail has become a shining international calling card for China". The *Mid- and Long-Term Railway Network Plan* highlighted, "On the whole, at present, the shortage of Chinese railway transport capacity has almost been alleviated, and the bottleneck constraints have largely been eliminated, which basically meets the needs of economic and social progress".

However, there was a huge demand for railways, especially high-speed railways in China. The *Mid- and Long-Term Railway Network Plan,* taking five years as a stage, puts forward periodical targets at three stages in turn: By 2020, a number of major landmark projects would be completed and put into operation, and the mileage of high-speed railway network would cover 30,000 kilometers. By 2025, the high-speed railway network would reach about 38,000 kilometers in length. By 2030, a modern high-speed railway network would be built, aiming to connect major urban agglomerations and link provincial capital cities and other large and medium-sized cities with a population of over 500,000, thereby forming a high-speed railway network with megacities as the center covering the whole country and provincial capital cities as the fulcrum covering the surrounding areas. The traffic time would take 1 to 4 hours between adjacent large and medium-sized cities and 0.5 to 2 hours within urban agglomeration, which could provide safe, reliable, high-quality, efficient, comfortable, and convenient passenger transportation services.

In terms of spatial layout, based on the "four vertical and four horizontal" high-speed railway network, high-speed railways would be increased to support passenger flow and meet the demand for development through some railways at a speed of 200 km/h. Then a high-speed railway network would be built with the main channels of "eight vertical and eight horizontal" high-speed railways as the framework, connected by regional lines and supplemented by intercity railways, which makes provincial capital cities accessible through high-speed railways as well as offers efficient and convenient connection between regions.

As described in the previous chapter, the 2016 *Mid- and Long-Term Railway Network Plan* is the third version since 2004. Compared with the previous two versions, the status of high-speed railway in railways, transportation network, and national economic development has changed substantially.

Firstly, instead of being a supplement to the traditional railway, high-speed railways become a substitute for traditional railways in terms of passenger transportation.

In the new plan, the high-speed railway network consists of three parts. The first part is the main channels called "eight verticals and eight horizontals", functioning as the framework and main body of the high-speed railway network to connect the main cities of various provinces. The second part is the regional railway links. Based on the "eight vertical and eight horizontal" main channels, the regional railway links of the high-speed railway are planned for "further improving the network and expanding the coverage". The third part is the intercity passenger railway. While priority is given to the use of high-speed and conventional speed railways for intercity train services, intercity passenger railways for urban agglomeration are planned to support and lead the development of new urbanization and to effectively connect large and medium-sized cities with central towns, serving the commuting function.

Compared with the first two versions, the framework and main body of the high-speed railway network are greatly expanded from "four verticals and four horizontals" to "eight verticals and eight horizontals", covering almost all major cities in each region. In the new plan, medium and high-speed railways play an important role in intercity transportation which was usually supported by roads and traditional railways before. The long-distance railway is like the "artery" of the human body, and the intercity railway is like a "capillary". Now, since the railway is responsible for all the "blood circulation" of the human body, the "regional railway link" would serve to connect "arteries" and "capillaries".

Secondly, high-speed railway needs to support the development of "urban agglomeration". Urban agglomeration, generally regarded as an advanced phenomenon of regional spatial form in the process of industrialization and urbanization, is one of the symbols of the rapid development of the national economy and continuous improvement of modernization level. It is the highest spatial organization form in the mature stage of urban development, which can produce huge agglomeration economies.

After 40 years of reform and opening up, China's urbanization has entered a stage where urban agglomeration becomes the main body of regional development. In the functional description of the main channels of "eight vertical and eight horizontal" high-speed railway network, railways not only connect "cities", but "urban agglomerations".[39] Within the urban agglomeration, the railway is tasked to connect

---

[39] The "Eight Verticals" are listed as follows. Coastal Passageway pass through eastern coast of China and connects urban agglomerations in Beijing-Tianjin-Hebei, Central and Southern Liaoning, Shandong Peninsula, East Longhai, Yangtze River Delta, West Coast of the Strait, Pearl River Delta and Beibu Gulf. The Beijing-Shanghai Passageway connects North and East China as well as urban agglomerations in Beijing-Tianjin-Hebei and Yangtze River Delta. The Beijing-Hong Kong (Taipei) Passageway connects North, Central, East and South China as well as urban agglomerations in Beijing-Tianjin-Hebei, Middle Reaches of the Yangtze River, West Coast of the Strait, Pearl River Delta. The Harbin-Hong Kong (Macao) Passageway connects Northeast, North, Central and South China, Hong Kong and Macao, and connects urban agglomerations in Harbin-Changchun, Central and Southern Liaoning, Beijing-Tianjin-Hebei, Central Plains, Middle Reaches of the Yangtze River and Pearl River Delta. Hohhot-Nanning Passageway connects North, Central and South China, Central Plains, and urban agglomerations in Hohhot-Baotou-Erdos-Yulin, Central Shanxi, Central Plains, Middle Reaches of the Yangtze River and Beibu Gulf. Beijing-Kunming Passageway connects North, Northwest and Southwest China, and urban agglomerations in Beijing-Tianjin-Hebei, Taiyuan, Guanzhong Plain, Chengdu-Chongqing and Yunnan. Baotou (Yinchuan)-Hainan Passageway connects Northwest, Southwest and South China, and urban agglomerations in Hohhot-Baotou-Erdos, Ningxia along the Yellow River, Guanzhong Plain, Chengdu-Chongqing, Central Guizhou and Beibu Gulf. Lanzhou (Xining)-Guangzhou Passageway connects Northwest, Southwest and South China, and urban agglomerations in Lanzhou (Xining), Chengdu-Chongqing, Central Guizhou, Pearl River Delta. The "Eight Horizontals" are as follows. Suifenhe-Manzhouli Passageway connects Heilongjiang and eastern Inner Mongolia. Beijing-Lanzhou Passageway connects North and Northwest China and urban

cities. Due to the different development levels and population density of urban agglomerations, the development objectives of intercity railways in urban agglomerations are also various. An intercity railway network needs to be built in relatively mature urban agglomerations such as Beijing-Tianjin-Hebei, Yangtze River Delta, Pearl River Delta, Middle Reaches of the Yangtze River, Chengdu-Chongqing, Central Plains, Shandong Peninsula and other urban agglomerations. An intercity railway framework needs to be formed in less developed urban agglomerations including the West Coast of the Strait, Harbin-Changchun, Central and Southern Liaoning, Guanzhong Plain, Beibu Gulf, and other urban agglomerations. Main channels of intercity railways need to be constructed in urban agglomerations with small population density, such as Central Yunnan, Central Guizhou, Northern Slope of Tianshan Mountains, Ningxia along the Yellow River, Hohhot-Baotou-Ordos-Yulin and other urban agglomerations.

Such a huge high-speed railway network is planned not only because of the advocacy of the central government but also due to local governments' changed understanding of the role of high-speed railways in regional development. Before and even in the early stage of reform and opening up, local governments did not pay attention to railway construction. With the development of the economy and the increase in land prices, local governments even challenged railway construction because the land through which the railway passes would be expropriated. Local governments were keener to build expressways for toll charging since they were not able to share the railway revenue. This attitude began to change in the decision-making of building the Beijing-Shanghai High-speed Railway, which is highlighted in the change of station selection in various cities through which the Beijing-Shanghai High-speed Railway passed. In the 1990s, the stations of the then proposed Beijing-Shanghai High-speed Railway were basically the existing ones because the passengers had developed their own travel habits in the choice of stations. Building a new station carries the risk of affecting passenger flow and increase the construction cost. But the actual Beijing-Shanghai High-speed Railway has almost completely new stations, and most of them are far from traditional urban areas, causing criticism from passengers once. The main reason for this was that

agglomerations in Beijing-Tianjin-Hebei, Hohhot-Baotou-Erdos, Ningxia along the Yellow River, Lanzhou-Xining. Qingdao-Yinchuan Passageway connects East, North and Northwest China, and urban agglomerations in Shandong Peninsula, Beijing-Tianjin-Hebei, Taiyuan, Ningxia along the Yellow River. Eurasia Continental Bridge Passageway connects East, Central and Northwest China, and urban agglomerations in East Longhai, Central Plains, Guanzhong Plain, Lanzhou-Xining and Northern Slope of Tianshan Mountains. Yangtze River Passageway connects East, Central and Southwest China, and urban agglomerations in Yangtze River Delta, Middle Reaches of the Yangtze River, Chengdu-Chongqing. Shanghai-Kunming Passageway connects East, Central and Southwest China, and urban agglomerations in Yangtze River Delta, Middle Reaches of the Yangtze River, Central Guizhou and Central Yunnan. Xiamen-Chongqing Passageway connects the West Coast of the Strait, Central South and Southwest China, and urban agglomerations in the West Coast of the Strait, Middle Reaches of the Yangtze River, Chengdu-Chongqing. Guangzhou-Kunming Passageway connects South and Southwest China and urban agglomeration in Pearl River Delta, Beibu Gulf and Central Yunnan.

local governments hoped high-speed rail would stimulate the development of planned new urban areas. The plan has indeed taken effect in many cities. After 2013, this trend has further strengthened. In the planning process of a line, all cities along the line made various efforts to make the railway pass through their own cities and set up new stations. Whether a city owns a high-speed railway has even become a symbol of urban development.

7 "Interconnectivity": Chinese Railways' "Going Global"

The People's Republic of China has long witnessed the "going global" of its railways, and the earliest records of such undertakings documented China-aided railway projects in Asian and African countries. During the course of reform and opening up, China radically remodeled its railway construction competency and implemented a multitude of overseas railway projects, rivaling railway giants in the world. With new opportunities brought by the Belt and Road Initiative, Chinese enterprises are commissioned to collaborate with their counterparts to connect infrastructures of partner countries and will contribute to the common development and prosperity of the entire world.

7.1 Adjustment to the "Going Global" Model of Chinese Railways

7.1.1 The First Footprint of "Going Global"

China railways' first "going global" move was the entry of the PLA Railway Corps' spearhead force into North Korea for the emergency repair of war-damaged railways in November 1950, the second year after the founding of PRC. After transitioning to socialism from 1953 to 1957, China had enhanced its railway construction competency and embarked on providing assistance to several developing countries in railway construction.

Prior to reform and opening up, Chinese railway departments provided aid to neighboring countries such as North Korea, Vietnam, Cambodia, and Pakistan, and African countries such as Tanzania, Zambia, Congo, Guinea, and Mali. Chinese aid to North Korea and Vietnam during the Korean War and the Vietnam War is of particular historical significance.

During this period, Chinese soldiers and workers helped North Korea repair about 500 km of railways and rebuild or build another 860 km. In addition, they helped Vietnam repair 971 km of railways, rebuild 324 km, and build 57 km.

Among the Chinese railway aids to African countries, the Tanzania-Zambia Railway exerts the most international influence. Spanning 1,860 km long and

X. Zhang, *China's Railway Expansion in 40 Years of Economic Reform and Opening Up*, Understanding China,
https://doi.org/10.1007/978-981-95-0333-9_7

connecting 11 cities in the two countries, it is not only China's biggest overseas railway construction project before reform and opening up but also the country's biggest completed foreign aid project at that time. To accomplish such an arduous task, China dispatched in total of more than 50,000 engineers and technicians to Africa, with the number of Chinese staff on site reaching 16,000 during the peak construction period. During the process, 64 Chinese workers died at their posts in Africa. China also provided RMB 988 million (equivalent to USD 154 million) in interest-free loans and RMB 100 million (equivalent to USD 16 million) in subsidies for the Railway.

It was during the Cold War when Chinese railways first "went global" to neighboring countries. Provided basically under a state of war, China's railway aid to North Korea, Vietnam, and some other countries during this period had economic as well as political and military significance, and most of the projects involved tight schedules, high risks, political complications, and large quantities of work. In retrospect, attainments and deficiencies coexist in the early course of Chinese railways' "going global".

First, China did not possess world-leading technologies in railway construction despite its comparatively strong competency.

China received many negative comments from Western countries when its overseas railway projects just commenced. Before the construction of the Tanzania-Zambia Railway, for instance, American experts predicted that China would spend ten years to complete the project, and experts in Britain and Canada estimated at least eight years. Around five years later, the Railway opened to traffic and withstood many natural disasters, earning praise worldwide. However, Chinese railway construction technologies could not match the world's leading ones. For example, when the Tanzania-Zambia Railway was being built, China's shield tunneling technology was in its infancy and had been used in the construction of only two short tunnels in Shanghai. The New Austrian Tunneling Method (NATM) was just introduced to China and not widely applied until 1988 when MOR published *the Guidelines on the New Austrian Tunneling Method for Railway Tunnels*. In contrast, Western countries had built a large number of tunnels using many different new technologies during the same period of time. The progress made in the early days is attributable to the strong organizational capability of the Chinese Government, the selfless dedication of workers and technicians, and the huge investment of human resources.

Second, China did not have sufficient reserves for the supply of machinery and equipment despite its competency in the execution of complete projects.

Since the 1960s, China's competency in railway construction has been built up and substantially enhanced, leading to the successful completion of the earliest foreign aid project. For the Tanzania-Zambia Railway, for instance, Wuhan Iron and Steel Works (now known as Wuhan Iron and Steel (Group) Corp.) supplied all 200,000 tons of steel rails, Dalian Locomotive and Rolling Stock Works (now known as CRRC Dalian Co., Ltd.) and Qingdao Sifang Locomotive and Rolling Stock Factory (now known as CRRC Qingdao Sifang Co., Ltd.) provided 102 locomotives, and Shanghai Communication Factory and Beijing Erqi Communication

Factory produced all the communication equipment. The railroad ties were also designed by China and manufactured in Africa. However, China was still weak in manufacturing machinery and equipment. For example, the capacity of homemade "Hongqi" wheel tractor-scrapers was only half that of Japanese and European products, and domestic dump trucks were not fit for long-distance transportation. To complete foreign aid projects, China spent huge sums of foreign currency to purchase construction equipment from Western countries. In 1971, China even shipped almost all its best equipment abroad and had to suspend some domestic railway construction projects. Besides, recipient countries often asked Western countries to upgrade Chinese-made locomotives.

Third, China did not treat the future profitability of railways as an important factor despite its competency in partnering with foreign governments and enterprises.

In the early days, the Chinese government was able to break through the diplomatic siege imposed by Western countries, and its construction teams could also respond effectively to various emergencies at work. However, some railways completed during this period did not generate the expected economic returns. For example, one of the original intentions for China and Africa to build the Tanzania-Zambia Railway was to break the blockade of Western countries against Central Africa and export Zambian mineral products by sea through East Africa. However, changes in international politics gradually eased the tension between newly independent African countries and Western countries, and large amounts of cargo from Central Africa were exported through South Africa. As a result, the traffic volume of the Tanzania-Zambia Railway declined continuously. Statistics showed that the Tanzania-Zambia Railway had suffered losses since it opened to traffic, adding to the economic burdens of China, Tanzania, and Zambia.

Right behind the above attainments and deficiencies, one cannot fail to notice that a strong planned economy predominated Chinese railways' "going global" efforts before reform and opening up.

As for the management system, Chinese railways' foreign aids were directly supervised by the government during this period. When PRC was just founded, the central government established various foreign aid groups or foreign aid committees to deal with different countries. After 1958, the former State Planning Commission was charged with designating a specific department as the administrator to coordinate with local governments. Since 1971, the relevant construction department had acted as the administrator, responsible for selecting domestic partners for the preparation, design, and construction of a foreign aid project in the railway sector. According to this system, foreign aid projects were first categorized by industry, and then the competent construction department was determined. The former Ministry of Economic Relations with Foreign Countries (MOERFC) was only accountable for coordinating with local Chinese governments or foreign governments.

At the very beginning, the International Through-Transportation Bureau[1] attached to MOR administered and coordinated various foreign aid tasks. In June 1965, MOR and PLA Railway Corps Command jointly set up the "Office to Aid Vietnam Railway Construction". In October 1958, to prepare for the construction of the Tanzania-Zambia Railway, MOR officially established the Office of Foreign Aid with the approval of the State Council to manage foreign aid issues in railway construction as a whole.[2]

As for financial returns, China's foreign aid projects in the railway sector were basically provided pro bono, and the railway departments were only responsible for the accomplishment of foreign aid tasks instead of acting as independent operators. Before 1970, the various costs were directly reimbursed by the Ministry of Finance. After 1970, the budget and final accounts were prepared by foreign aid project teams and reviewed by the Office of Foreign Aid of MOR, and then submitted to the Ministry of Economic Relations with Foreign Countries for approval. With the two methods, the central government allocated equipment and materials and made payments, and the railway departments were not accountable for the profit and loss of each project. During the implementation process, MOR, to prioritize foreign aid projects over domestic ones, repeatedly reduced the amounts of domestic supplies in order to cater to the demands of other countries. Under such a financial system, the practices of Chinese railways' "going global" failed to fulfil the need for win-win cooperation between China and foreign countries.

Although Chinese railways' "going global" in the early days left much to be desired, it exerted a profound and far-reaching influence on the whole.

To begin with, many Chinese enterprises have in this period accumulated experience in international engineering contracts, improved their overseas construction competency, and grown into large multinationals during reform and opening up. For example, China Civil Engineering Construction Corporation (CCECC), China Railway No. 2 Engineering Co., Ltd. (CREGC), and China Railway No. 3 Engineering Group Co., Ltd. (CR3)—which all participated in the construction of Tanzania-Zambia Railway—are currently very active in overseas railway construction.

Secondly, foreign aid in the railway sector has also boosted technological innovations of domestic enterprises to remove tough technical obstacles encountered in overseas projects. For instance, Erqi Locomotive and Rolling Stock Factory (now known as CRRC Co., Ltd.) developed a train braking system specially for the Tanzania-Zambia Railway, making vacuum brakes compatible with air brakes in a train. While Qingdao Sifang Locomotive and Rolling Stock Factory improved the engine crankshaft of the locomotive, Wuhan Iron and Steel Works and Luoyang Heavy Machinery Factory (now known as CIC Luoyang Heavy Machinery Co.,

[1] In 1965, the International Through-Transport Bureau changed its name into the International Liaison Bureau.

[2] From July 1970 to December 1974, the Office of Foreign Aid of MOR was once renamed "the Office of Foreign Aid of the Ministry of Transport".

Ltd.) launched specific research projects to tackle technical constraints for overseas railway projects.

Thirdly, China established long-term cooperation with some developing countries in the process and helped the latter cultivate a large number of railway professionals. For instance, among 8,000 employees of the Tanzania-Zambia Railway, more than 1,200 underwent training in China and became the backbone of railway operations after returning to their countries. The Railway helped many African countries better understand China's sincerity in its foreign aid endeavors and recognize China's railway construction competency. These early partnerships laid a solid foundation for Chinese enterprises to expand overseas business.

7.1.2 *Transformation of Business Entities of "Going Global" after Reform and Opening-Up*

Before the Third Plenary Session of the Eleventh CPC Central Committee, the railway departments begun to explore foreign aid approaches from both economic and administrative perspectives. At the end of 1977, in collaboration with the Ministry of Economic Relations with Foreign Countries and the Ministry of Foreign Affairs, MOR submitted a *Proposal for the Implementation of Nigeria Railway Projects* to the State Council, which recommended that China organized construction and provided equipment and materials for Nigeria while Nigeria paid China in spot exchange. The three ministries concluded that "this kind of self-financed project can be a new form of foreign aid and shall be successfully executed to ensure profitability on top of cost recovery".[3] Although the Nigeria railway project was not officially launched until many years later, the idea of "profitability on top of cost recovery" proposed indicated that the target of Chinese railways' "going global" has undergone a substantial change.

On November 13, 1978, at a working meeting of the CPC Central Committee, the Office of Foreign Aid of MOR submitted a report to the Party Group of MOR on the reform of foreign aid approaches, suggesting that a more stable and professional foreign aid technical team in railway construction be assembled and a base established in Beijing. The Office of Foreign Aid might have two identities to serve different purposes. Internally, it was a functional department of MOR. Externally, it was an independent company named China Civil Engineering Construction Corporation (CCECC) with an independent accounting system.

The report was approved by the heads of MOR. About six months later, the Office of Foreign Aid of the Ministry embarked on the restructuring, planning to build a team of 1,000 professionals and another of 14,000 semi-professionals to execute international projects worth USD 500 million. Meanwhile, the Office set up

[3] (2006). *Memorabilia of China Civil Engineering Construction Corporation—Early stage of office of foreign aid of MOR* (p. 201). Internal Resources.

an External Liaison Office to investigate the international construction market and collect relevant economic and technological information.

The business scope and operation mode of China Civil Engineering Construction Corporation (CCECC) differed considerably from those of the Office of Foreign Aid. Firstly, although CCECC had retained the name of the Office of Foreign Aid, it was a functional department of MOR back then. In the meantime, it was an independent state-owned enterprise assuming sole responsibility for profits or losses. Secondly, CCECC continued to execute overseas projects in line with the principles of "operation in accordance with economic laws, management driven by market concerns, and collection of foreign exchange to provide funds for the country".[4] Thirdly, CCECC's initial capital was financed by bank loans, while the Office of Foreign Aid relied on capital of overseas projects obtained with the assistance of government appropriations.

On March 12, 1979, MOR and the Ministry of Economic Relations with Foreign Countries jointly submitted a *Proposal for the Formation of China Civil Engineering Construction Corporation* to the State Council. The proposal was endorsed on April 7 by Chen Muhua, the then Vice-Premier, and was reviewed and signed by other leaders of the State Council, including Deng Xiaoping.[5] On June 1, CCECC was inaugurated. Various foreign aid projects originally administered by MOR were transferred to CCECC, and all the projects newly awarded since then were to be executed in accordance with the contract responsibility system.[6] The founding of CCECC signified the transformation of business entities of Chinese railways' "going global" from government administration departments to corporate entities. Thus, the original aid-oriented strategy was shifted into a market-oriented one, and a political consideration-oriented strategy to a profit-driven one. At the beginning of reform and opening up, among about 150 institutions directly attached to MOR, six were entitled "corporation" or "company". CCECC, the leading corporation of China's global endeavors at that time, had the largest business volume related to overseas markets.[7]

Shortly after the set-up of CCECC, MOR proposed to start a company providing international railway freight services. On November 21, 1981, the China Export-Import Regulatory Commission approved the establishment of the China Railway Foreign Service Corporation (CRFSC). Founded in early 1982, CRFSC aimed to provide foreign customers with different kinds of services within China's national

[4] Ibid.: 230.

[5] Ibid.: 230.

[6] Back then, there were four earliest companies that were approved by the State Council and qualified to execute overseas engineering projects, namely China National Complete Plant Export Corporation, China Road and Bridge Engineering Company, China State Construction Engineering Corporation, and China Civil Engineering Construction Corporation.

[7] By 1986, ministerial companies of MOR included China Railway Engineering Corporation, China National Railway Locomotive & Rolling Stock Industry Corporation, China Railway Materials Commercial Corporate, China Civil Engineering Construction Corporation, China Railway Foreign Service Corporation and China Railway Signal and Communication Company.

railway system, and its major businesses included external consulting and agency services for the subordinate departments of MOR.[8] In October 1982, China National Railway Locomotive & Rolling Stock Industry Corporation was officially inaugurated, and exported a small amount of railway components and parts at the beginning of reform and opening up.

For the subsidiary companies and corporations of MOR, the transformation of operation modes was in general successful. By the end of 1985, CCECC signed 315 international contracts in total, collected a turnover of USD 360 million and net foreign exchange of USD 130 million, and paid RMB 147.1 million (equivalent to USD 23 million) in tax to the State.[9] However, the export of railway equipment lagged. In 1990, China sold 65 wagons to Myanmar and completed its first export order of complete railway equipment. Ever since then, manufacturing enterprises have become a crucial propeller of Chinese railways' "going global" undertakings. Meanwhile, to handle cargo for the New Eurasian Continental Bridge, CRFSC set up branch offices and customs broking agencies at rail ports in collaboration with different local railway administrations, and forwarded international freight at the volume of several million tons every year, making a great contribution to the development of China's international freight transport in the early stage of reform and opening up.

Under the progress of reform and opening up, the bonds between subordinate departments of MOR and the outside world grew much stronger, and international business gradually became routine. Certain international businesses involving MOR were no longer required to be managed by ministerial companies or corporations. Moreover, the ministry wished to introduce a competition mechanism to improve operations and integrate collaborative efforts from other departments. In such circumstances, adjustments were made to the centralized management systems and the internal structures of the companies and corporations.

International freight transport services originally managed by CRFSC were then largely transferred to its branch offices, and its head office was incorporated into China State Construction Railway Investment & Engineering Group Co. Ltd. At the beginning of the twenty-first century, China National Railway Locomotive & Rolling Stock Industry Corporation was divided into two companies, China South Locomotive & Rolling Stock Corporation Limited (CSR) and China North Locomotive & Rolling Stock Corporation Limited (CNR). In 2015, the two companies merged into CRRC Corporation Limited (CRRC). CCECC was restructured into China Civil Engineering Construction Corporation in December 1996, spun off from MOR in September 2000, and put under the administration of the Central Work Committee for Large Enterprises. After reorganization in 2003, CCECC became a subsidiary of China Railway Construction Corporation.

[8] *Articles of Association of China Railway Foreign Service Corporation.*

[9] Li Jixiang et al. (1990). *Contemporary China: railway undertakings* (p. 436). Beijing: China Social Sciences Press.

In the 1990s, CCECC outshone all other Chinese enterprises in overseas railway construction. Statistics released by the Ministry of Foreign Trade and Economic Cooperation showed that among China's largest enterprises for foreign trade and economic cooperation in 1997, CCECC ranked third, only after China State Construction Engineering Corporation and China Harbour Engineering Company Limited, both of which were not engaged in international railway construction then.

The beginning of the twenty-first century saw a remarkable growth in the number of Chinese enterprises in the international railway market. Soon after the founding of China Railway Group Limited and China Railway Construction Corporation Limited, many subsidiaries of the two railway giants sped up their paces in competition for overseas railway projects. When the Belt and Road Initiative was proposed, the Power Construction Corporation of China (POWERCHINA) and China Communications Construction Company Limited followed suit and started to pursue those projects. According to statistics available, China now has more than ten state-owned enterprises directly engaged in overseas railway construction (see Table 1).

To conclude, the business entities of Chinese railways' "going global" have changed substantially over time with the deepening of reform and opening up. Government departments have been replaced by enterprises. Companies and

Table 1 State-owned enterprises engaged in overseas railway construction

Names of Enterprises	Parent Companies
China Railway No. 2 Engineering Group Co., Ltd., China Railway No.3 Engineering Group Co., Ltd., China Railway No. 8 Engineering Group Co., Ltd., China Railway International Group Co., Ltd., China Railway Eryuan Engineering Group Co., Ltd., China Railway Shanghai Design Institute Group Co., Ltd., China Railway Tunnel Stock Co., Ltd., and China Railway Electrification Engineering Group Co., Ltd.	China Railway Group Limited
China Railway 12th Administration Group Co., Ltd., China Railway 18th Administration Group Co., Ltd., China Railway 20th Administration Group Co., Ltd., China Railway 12th Administration Group Co., Ltd., China Civil Engineering Construction Corporation, and China Railway Fifth Survey and Design Institute Group Co., Ltd.	China Railway Construction Corporation Limited
CCCC First Harbor Engineering. Co., Ltd., CCCC Second Harbor Engineering. Co., Ltd., and China Road & Bridge Corporation	China Communications Construction Company Limited (CCCC)
PowerChina Sinohydro Administration 7 Co., Ltd.	Power Construction Corporation of China (POWERCHINA)
China Sinomach Heavy Industry Corporation	China National Machinery Industry Corporation Ltd (Sinomach)
CRSC Xi'an Industry Group Co., Ltd. and China Railway Signal & Communication Shanghai Engineering Administration Group Co., Ltd.	China Railway Signal and Communication(group) Corporation Limited (CRSC)

corporations from different industries have joined the competition for international railway projects, while in the early stage, the Office of Foreign Aid at MOR (CCECC)—a combination of a government functional department and a state-owned enterprise—was the only one entrusted with such undertakings. In addition to state-owned enterprises, private companies also provide services for overseas projects. Thus emerges the new norm: Chinese enterprises from different industries, sectors, and systems participate in overseas railway construction, and they collaborate and integrate their strengths to pursue Chinese railways' "going global" strategy.

7.1.3 Restoration and Expansion of Channels for Chinese Railways' "Going Global" After Reform and Opening Up

Channel for Chinese railways' "going global" can be narrowly defined as the ways in which Chinese enterprises enter the international market, such as by executing overseas railway construction projects or exporting railway equipment. Channel in its broad sense consists of railway connectivity between China and its neighboring countries, intermodal transportation of passenger and freight, and international exchanges and cooperation in railway-related fields. Although construction projects and equipment export serve as the major channels and orientations of Chinese railways' "going global", other channels for international cooperation and exchanges should not be neglected.

Prior to reform and opening up, the channels for Chinese railways' "going global" were relatively clogged and sometimes even obstructed. The destinations were mostly developing countries and projects concentrated in the construction sector. After reform and opening up, the channels were diversified and expanded into the following sectors.

Firstly, international railway intermodal transportation was restored and redeveloped. PRC at the beginning of its founding signed railway intermodal transport agreements with the Soviet Union, North Korea, and Mongolia, and later another eight countries joined in. Against the backdrop of underdeveloped sea shipping, China's intermodal transportation played a significant role in the promotion of its foreign trade and cooperation. Nevertheless, railway intermodal transportation in the early stage was confined within socialist countries and often disrupted after the 1960s. After reform and opening up, international intermodal transportation of passengers and freight was soon restored, the most noteworthy being freight transport. Take the freight volume through the China-Soviet Union Manzhouli Port as an example. The volume of freight amounted to 3.979 million tons in 1959, but plummeted to 0.12 million in 1970, then rebounded to 0.265 million in 1980, and multiplied to 2.56 million in 1985. In addition to the boost in freight volume, China started the trial of providing international container freight service; the volume of freight, as a result, soared from over 200 containers in 1980 to more than 10,000 in 1985.[10]

After reform and opening up, China, along with countries including the Soviet Union, North Korea, and Mongolia, reactivated regular meetings for transport plans to ensure the smooth flow of international intermodal transportation. With the constant increase in its economy after entering the twenty-first century, China has made great strides forward. For instance, the annual volume of China's railway freight rose to tens of millions of tons, hundreds of times more than that at the beginning of reform and opening up. On the one hand, more countries joined in related cooperation. In addition to Russia, Central Asian countries and Eastern European countries, countries including Germany, France, Poland, and the Netherlands have also become vital destinations for China's railway freight. On the other hand, besides border railway ports, inland cities such as Chongqing, Chengdu, Wuhan, Zhengzhou, Yiwu, and Suzhou can dispatch regular direct freight trains of containers to foreign countries and regions as well.

Secondly, China got actively involved in the construction of international railway corridors. Prior to reform and opening up, such passageways were distributed unevenly. Even though there were eight railway ports, only Manzhouli Port and Suifenhe Port provided access to Europe through the Soviet Union, and the trains had to take the Siberian Railway in sparsely populated regions of high altitude and extreme cold. Since 1958, China had engaged in the construction of the Urumqi-Alashankou Section of the Lanzhou-Xinjiang Railway, but the project was suspended twice and not resumed until 1985.

On September 12, 1990, a ceremony was held in Alataw Pass, a city in Xinjiang Uygur Autonomous Region of China, to celebrate the connection of the Lanzhou-Xinjiang Railway in China and the Turkmenistan-Siberia Railway in the Soviet Union, marking the completion of the second Eurasian Land Bridge. Compared with the older one, the new land bridge not only shortens transport distances by 2000 km but is also separated into three lines from Alataw Pass to Europe, covering densely populated areas with far greater intensity of ramification. In 2009, the tracks of the Jinghe-Yining-Horgos Railway in Xinjiang Uygur Autonomous Region reached Horgos, making Horgos the second railway port after Alataw Pass and strengthening the traffic capacity of the New Eurasian Land Bridge. Besides, China also signed the "Intergovernmental Agreement on the Trans-Asian Railway Network" in 2006 and vigorously pushed ahead with the construction of railway corridors between China-Vietnam, China-Laos, China-Thailand, and China-Burma.

Thirdly, China participated earnestly in the activities held by international organizations in the railway sector. In the twentieth century, the Organization for Co-operation between Railways (OSJD) and the International Union of Railways (UIC) were the major international railway organizations. The former was established by socialist countries led by the Soviet Union, while the latter was mainly by European capitalist countries. China, a founding member state of OSJD, attended

[10] Ibid.: 406.

its first Ministerial Conference in 1956 and had always dispatched its standing delegates and resident staff to OSJD. However, its Ministerial Conference had been suspended since 1966 and was not resumed until 1984. OSJD survived the collapse of the Soviet Union and Revolutions of 1989 in East Europe and has developed into an organization with 27 member states. As an important member, China has attended all the following Ministerial Conferences, joined in many research projects and technical seminars, and hosted on OSJD's behalf some meetings and forums. China Railway Publishing House is also authorized to publish and distribute the Chinese version of *OSJD Bulletin.*[11]

China was among the founding member states of the International Union of Railways when it was inaugurated in 1922. However, China's lawful seat in UIC had been occupied by Taiwan since 1949. On July 6, 1979, the 37th International Union of Railways General Assembly restored China's lawful seat in UIC. Back then, China was not a principal member state because it had no intermodal passenger and freight transport link with Western and Middle European countries. However, with the advance of its railways, China has elevated its status in UIC and now holds two seats among the 74 "active members".[12]

Fourthly, China enhances and expands scientific and technical exchanges in railways with foreign countries. After the founding of PRC, such exchanges took place mostly between China and the Soviet Union and other socialist countries but were suspended during the Cultural Revolution. After reform and opening up, Chinese railway departments had striven to resume the exchanges. The Ministers of Railways, from 1979 to 1985, led delegations to visit the United Kingdom, France, Japan, the Federal Republic of Germany, Romania, North Korea, Iran, Libya, and Myanmar. In addition, the Chinese Railway Society cooperated with academic groups from Japan, the United States, Canada, Chinese Hong Kong, and other countries and regions through informal channels. After reform and opening up, China prioritized technical exchanges with Western developed countries. From 1978 to 1985, the Chinese government sent many groups and teams of engineers and managers to Japan, the United Kingdom, Germany, the United States, and Canada to learn from their counterparts. In addition, China fostered technical exchange and cooperation with other countries like Austria, Sweden, Belgium, Italy, Denmark, Swedish, and Australia. In the early period of reform and opening up, those efforts enabled Chinese railway departments to further their understanding on the trend of the world's railway technology and create favorable conditions for China to develop its own.

In conclusion, China has expanded various "going global" channels after reform and opening up. As China has ameliorated the relationship with socialist countries like the Soviet Union and eastern European countries and patched up ties with

[11] OSJD Bulletin was released only in Chinese, German, and Russian.

[12] The member states in the UIC are divided into "active members" and "associate members" at present. The Ministry of Railway of China on behalf of PRC was the only "active member" before 2013, but nowadays both the National Railway Administration of PRC and China Railway are "active members".

Western developed countries, the relatively secluded Chinese railway system since the 1960s was thus opened to the outside world.

While Chinese railway departments continued to carry out overseas railway projects, they developed intermodal railway transportation, constructed international railway corridors, engaged in activities held by international organizations, and conducted technical exchanges with other countries. To further implement the "going global" strategy, Chinese railways changed the destinations from developing countries to both developed and developing ones, paying particular attention to learning railway technology from developed countries.

7.2 The Course of Chinese Railway's Exploration of Overseas Market Since Reform and Opening Up

7.2.1 Early Experiences in the Exploration of Overseas Market

During the early stage of reform and opening up, overseas business was supervised and administered by the Ministry of Foreign Economics and Trade. In 1985, the Ministry delegated part of its authority to the China National Complete Plant Export Corporation,[13] which was empowered to select domestic corporations for the implementation of overseas projects through the negotiation or tendering process. Generally, China Civil Engineering Construction Corporation (CCECC), subordinate to MOR, was the one selected for railway projects. It would sign agreements with China National Complete Plant Export Corporation, and then subcontract the projects to different institutions and companies attached to the railway departments. With dramatic changes to governmental functions after the 1990s, the major tasks of the Ministry of Foreign Economics and Trade (Ministry of Commerce later) were converted into policy-making and improvement of rules and regulations. Gradually, CCECC assumed the responsibility for the exploration of overseas markets.

From the 1980s to 1990s, Chinese railway departments and its subsidiary enterprises represented by CCECC had completed the following overseas projects.

7.2.1.1 Rangoon-Thanlyin Bridge Project in Myanmar

Located at the edge of Rangoon city, the Rangoon-Thanlyin Rail-road Bridge spans 2,939 meters long across the Bago River. In 1980, Myanmar appealed to China for assistance in the construction of the bridge and finally signed the "Contract on the Implementation of Rangoon-Thanlyin Bridge Project" with China in January 1987 after years of consultation and investigation.

[13] China National Complete Plant Export Corporation was founded in November 1959 and changed its name to China National Complete Plant Import & Export Corporation (Group) in 1993.

China National Complete Plant Export Corporation and CCECC initiated the tendering process and awarded the contract to China Railway Major Bridge Engineering Group Co., Ltd. The Chinese government provided RMB 200 million (equivalent to USD 31 million) in interest-free loans for Myanmar, shipped tens of thousands of tons of materials and equipment, and dispatched about 200 engineers and technicians to Myanmar. Chinese leaders like Li Xiannian and Qiao Shi visited the site at different times during the investigation and construction stages. In 1988, the project had to be suspended due to unrest in Myanmar. Construction work resumed in 1990 and the bridge was completed in July 1993.

7.2.1.2 Botswana Railway Renovation Project

Initially, Botswana had only one narrow gauge railway built in 1890, and 70 percent of its population lived along the 714-km-long railway. The rapid growth of freight traffic in the 1980s increased demand for renovation of the railway with serious wear and tear. At the request of the Botswana government, China and Botswana signed the "Protocol on the Renovation of Railway Projects (the Southern Section)" in July 1982, then both decided to further renovate the northern and middle sections respectively in May 1987 and December 1989. The Chinese government has offered four interest-free loans totaling USD 26 million to Botswana since July 1985. CCECC was responsible for the project and some of the engineering divisions subordinate to MOR also sent their staff for the project. The renovation mainly includes providing and changing steel rails of 50 kilograms per meter, welding the steel rails into 150-meter-long rails, changing the turnouts, and replacing the railroad ties with reinforced concrete ones, etc.

7.2.1.3 Nigerian Railway Renovation Project

Nigeria had a narrow-gauge railway built in the colonial period. Since the country's independence in 1960, the traffic volume had declined dramatically because the old railway was out of repair for years. Nigeria started negotiations with China on the renovation of its railway in 1977. However, the commencement of the project was delayed. In December 1995, CCECC finally signed a contract for USD 528 million with the Ministry of Transport of Nigeria to repair some sections in Nigeria.

It was the largest international engineering project awarded to a Chinese railway enterprise in the twentieth century, which included the investigation and design of a 4288-km-long railway, repair and renovation of a 3288-km-long railway, and signal renovation of 180 stations. China would also provide 50 locomotives, 150 passenger cars, 400 freight wagons, and 20 trolleys. By 1999, CCECC had substantially completed the project. After that, the Nigerian government repeatedly entrusted Chinese enterprises to build new railways or renovate old ones, thus opening the railway market in Nigeria and even the whole of West Africa to Chinese firms.

Besides construction projects, China's strength in the export of locomotives and vehicles has also improved since reform and opening up.

In 1980, the world's top 10 locomotive exporters were all developed countries such as the United States (with exports worth USD 150 million), Japan (USD 90 million), and European nations like France (USD 230 million), Germany (USD 210 million) and the United Kingdom (USD 120 million). The United States ranked first in the export of components and parts (USD 380 million).[14]

China's competitiveness in such markets is quite weak, though it had manufactured locomotives and vehicles for other countries in the early stages of foreign aid. At the beginning of reform and opening up, components and parts of locomotives and vehicles accounted for a major portion of China's exports. In 1990, China exported 65 wagons to Myanmar, marking the start of the export of complete vehicles. In 1995, China exported 15 diesel locomotives to Malaysia and Myanmar, and for the first time, Chinese locomotives entered the Southeast Asian market. In 1997, China sold 12 electric locomotives to Iran, thus opening access to the Middle East market. After that, Iraq and Saudi Arabia began to purchase locomotives from China.

By 1998, China had exported in total 73 steam locomotives, 281 diesel locomotives, 125 electric locomotives, 1,206 passenger cars, 10,957 wagons, 580 subway vehicles, 10 railway diesel cranes, 289 container trailers and semitrailers, as well as large amounts of parts, castings and forgings and other products like diesel generator sets, bogies, bolsters, side frames, couplers, axles, rubber components, plasma cutting machines to more than 30 countries and regions, including Thailand, Malaysia, Myanmar, Bangladesh, Iran, North Korea, Botswana, Nigeria, Chinese Hong Kong, Tanzania, and Zambia. The value of exports in locomotives and vehicles and their parts had surged from USD 267,000 dollars to USD 105 million in 1998.[15]

Generally speaking, in the early stage of reform and opening up, the greatest change of Chinese railway enterprises in their "going global" endeavors lay in voluntary exploration of the international market instead of compliance with the government's arrangements for economic-aid projects under the planned economy. Undaunted by intense competition with powerful international contractors and manufacturers, China had taken a significant bite out of the market share. However, before the rise of the Chinese high-speed railway, Chinese railway enterprises did not reach the world's top level in construction competency or locomotive and vehicle quality. Thus, Chinese railways' "going global" was not fast enough at this stage, and the destinations were mostly underdeveloped countries and regions.

---

[14] Wang Xiaohua. (1984). The exports of the locomotives from the world's major countries in 1980. *Foreign Rolling Stock*, (1).

[15] Compilation Committee. (1999). *Fifty years (1949–1999) of Chinese rolling stock industry* (p. 461). China Railway Publishing House.

7.2.2 *Chinese Enterprises Entering International High-Speed Railway Market*

At the beginning of reform and opening up, China's railway construction competency was at a medium level in the world. For example, when China Civil Engineering Construction Corporation (CCECC) was just established, the enterprise aimed to "command the technology of design, construction, and installation for international medium-speed railways (160km/h), general industrial and civil buildings, highway and railway bridges".[16] With the gradual upgrade of their railway construction competency, Chinese railway enterprises started to implement overseas projects of relative difficulty and complexity. In 2005, some packages of the Ankara-Istanbul High-Speed Train Project in Turkey were awarded to China Railway Construction Corporation Limited (CRCC) and other Chinese companies, charting a new path in the construction of overseas high-speed railway projects, especially in medium-developed countries.

The Ankara-Istanbul high-speed railway, Turkey's first one of its kind, was designed to connect the capital Ankara with the largest city Istanbul. The 20 million people living in the two cities relied completely on a normal 576-km-long railway for transport between them before 2006. A one-way trip took six and a half hours. The country had planned to renovate the existing railway since 1994 but failed to put it into practice. Eventually, the government decided to build the Ankara-Istanbul high-speed railway with a total length of 533 km and a design speed of 250 km per hour. It was planned to dispatch 12 trains every day and the one-way-trip time was shortened to three and a half hours.

Immediately after the Turkish Government launched the Ankara-Istanbul High-Speed Train Project, competitors participating in the tendering included 24 companies from 10 different countries: China, Germany, Japan, Spain, Italy, the United Kingdom, Switzerland, India, and Ukraine. Foreign contractors were required to collaborate with local Turkish companies in the bid. Ultimately, the project was awarded to consortia formed between Turkish and Spanish, Chinese and Italian contractors and would be carried out in three phases. The Turkish-Chinese consortium was entrusted with the construction work in the second phase.

The second phase covered a total length of 158 km divided into two sections of 104 km and 54 km respectively. In 2005, the contract for the two sections was awarded to CRCC-CMC-CENGIZ INS.-IC ICTAS INS consortium established by China Railway Construction Corporation Limited (CRCC) and China National Machinery Import & Export Corporation (CMC) in collaboration with Turkish companies Cengiz Construction and Ibrahim Cecen Ictas Construction. The consortium signed the contract with the Turkish government in 2006.[17]

---

[16] (2006). *Memorabilia of China Civil Engineering Construction Corporation—Early stages of office of foreign aid of MOR* (p. 231). Internal Resources.

[17] In 2005, China Civil Engineering Construction Corporation (CCECC), a subsidiary of China Railway Construction Corporation Limited (CRCC), was mainly responsible for the construction

The Ankara-Istanbul high-speed railway is China's first international high-speed railway contract as well as the largest engineering cooperative project between China and Turkey since their establishment of diplomatic relations. Admittedly, the sections assigned to the Chinese enterprises were the most challenging along the entire line and also a test of China's railway construction competency, as it extended through mountainous areas, mostly composed of granite, involving the construction of 31 bridges and 34 tunnels with the bridge and tunnel ratio reaching 42%.

The strengths and weaknesses of Chinese enterprises were conspicuous during the construction of the Ankara-Istanbul high-speed railway.

First of all, Chinese enterprises have strong financing ability, abundant experience in railway construction, and high technical level in design, survey and construction. When bidding for the Ankara-Istanbul High-Speed Train Project, the consortium established by Chinese and Turkish enterprises did not offer the lowest price, but won the bid for its strong comprehensive competency in financing, technology, and management. The total contract price amounted to USD 1.27 billion, of which USD 720 million was from a loan issued by the Export-Import Bank of China. China Railway Fifth Survey and Design Institute Group Co., Ltd. (CRFSD) undertook the electrification survey for the project and their design scheme passed the assessment at the first review. During construction, CCECC's subsidiary CRCC provided faster and better engineering services than local Turkish contractors in track-laying, power grid, electrification, and communication. For example, Chinese enterprises could erect hundreds of electric poles each day, while their Turkish counterparts only have a few.[18] For another, when catenary hangers were being installed in the tunnel, the Turkish employer demanded an equidistant arrangement with the maximum span, whereas Chinese designers suggested that the hangers be set up near the entrance of the tunnel. Such a suggestion was to ensure catenary reliability in response to high wind pressure brought on by the high speed of trains, as proven in high-speed railway construction in China. The employer refused to do so at first in consideration of increased cost, but later they promptly approved the Chinese designers' plan after discussing it with Spanish consulting engineers.

Secondly, Chinese enterprises had to overcome weaknesses such as inadequate adaptation to foreign working conditions and technical standards, low-level mechanization in construction, and deficiency in upgrading engineering experience into theory. When the Chinese contractors won the bid for the Ankara-Istanbul high-speed railway, they had already built the Qinhuangdao-Shenyang Passenger Dedicated Railway and just commenced construction of the 115-km-long Beijing-Tianjin Intercity Railway (launched in 2005 and completed in 2008). When the Ankara-Istanbul railway project was in progress, China also started the construction of the 1300-km-long Beijing-Shanghai High-speed Railway (launched in 2008 and completed in 2011).

of Ankara-Istanbul High-Speed Railway Project in Turkey.

[18] Zhou Xiaodong. (2015). Ankara-Istanbul High-Speed Railway project in Turkey-China's first overseas high-speed railway project. *International Project Contracting and Labour Services*, (9).

Nevertheless, the second phase of the Ankara-Istanbul Project only covered a distance of 158 km but took eight years to complete. The reasons are, on the one hand, the difficulties and complexities caused by the high bridge-tunnel ratio of the line, and on the other hand, the inadequate adaptation of Chinese enterprises to foreign working conditions and technical standards. Although the Chinese enterprises signed the contract in 2006, the construction did not begin until 2008. To be more specific, the Chinese enterprises had to spend a lot of time coordinating with their Turkish counterparts in the adjustment of plans and schemes, and the language barrier slowed down the communication between both sides. In addition, they also needed time to learn all the European standards adopted in the Project. Though 80% of the Chinese standards were similar to Europe's, the process control was more stringent. Take rails, for example. All European standards had to be met in the process, from iron ore verification to ingot casting and track production. As a matter of fact, only 5% of the products in the project were manufactured in China, and the Chinese enterprises mainly provided labor services. The first attempt at Chinese firms' application of European standards significantly impacted the progress of the project and caused a delay in completion.

At the beginning of the Ankara-Istanbul railway project, the Chinese firms were weak in the mechanization of railway construction. For example, China's shield technology was not mature enough then, and the mechanization level of tunnel construction in Turkey was much higher than that in China. Besides, Chinese enterprises failed to upgrade their abundant construction experience accumulated into systematic theories, and such experience was generally not recognized in the international market.

These deficiencies were obvious in the construction process of DT26 tunnel of the project. According to the survey results of Chinese enterprises, DT26 tunnel was to be constructed using the New Austrian Tunneling Method. However, the project's progress was delayed as a result of the destruction of the railway subgrade caused by the construction of a highway nearby. Under pressure from the Turkish employer, the consortium thus agreed to tunnel with the shield machine supplied by German enterprises. Though the tunneling speed was fast, the shield machine was stuck in the tunnel due to the supplier's inaccurate judgment of geological conditions. Eventually, this section was rerouted. In fact, the tunnel survey results of Chinese enterprises were more accurate and the construction plan more practical, but the Turkish employer refused to accept them because they did not trust the Chinese enough. When problems with tunnel construction arose, the Turkish employer was more willing to hire European tunnel experts, design institutes, and survey companies to solve the problems, but the decisions made by these European experts and companies did not fully align with the interests of the Turkish employer. Experienced as Chinese companies were in high-speed railway engineering, they could not convince the Turkish employer on this issue with either theoretical models or shield technology lagging behind that of European countries. In general, China's technical deficiency has to a certain extent led to the blind trust of the Turkish employer in European technology, causing a delay in completion as well as losses suffered by Chinese enterprises.

As China's first one in this field, the Ankara-Istanbul high-speed railway project involves only line construction, and Chinese firms are not required to provide complete equipment. Since the Turkish employer adhered to European technical standards, the project did not boost the export of China's railway equipment. During construction, the Turkish employer consulted and discussed with experts from Italy, Spain, and even South Korea on the design of high-speed trains, ordering some products for line tests. Never did they turn to China for a purchase order. Finally, the Turkish employer ordered seven Velaro high-speed trains with a contract value of EUR 285 million from German Siemens in 2013, and another 10 Velaro trains valued at EUR 400 m in 2015, while China's export of high-speed railway equipment to Turkey was minuscule.

On July 25, 2014, the Ankara-Istanbul high-speed railway opened to traffic. Recep Tayyip Erdogan, the then Prime Minister of Turkey, attended the opening ceremony and took a ride on the train. The Turkish railway department also expressed appreciation for the work of the Chinese enterprises and welcomed them to bid for other Turkish railway projects in the future.

In general, this project enabled Chinese enterprises to familiarize themselves with the European standards in the construction of high-speed railways, laying the foundation for future entry into the markets of Europe and other countries.

7.2.3 *Preliminary Global Expansion of China's Railway*

The beginning of the twenty-first century attested to China's advance in railway technology by leaps and bounds, and the achievements were especially exceptional in the field of high-speed railway technology. Starting with the Ankara-Istanbul railway project in Turkey, Chinese enterprises made their debut in global railway construction. Upon the completion of the project, China was ahead of other countries in comprehensive competitiveness. Thus, the global expansion pattern of Chinese railways' "going global" began to take shape.

Firstly, the Jakarta-Bandung High-Speed Rail Project in Indonesia marks the "going global" of China's high-speed rail standards. Jakarta-Bandung railway connects Jakarta, the capital of Indonesia, to the country's fourth largest city, Bandung, covering a total length of 150 km. It is not only the country's first high-speed railway but also the first in Southeast Asia. Since 2008, Japan had planned to export its Shinkansen high-speed railway technology to Indonesia. Indonesia, on the other hand, suffering from the low speed and capacity of existing railways, considered renovating the old lines with Japanese technology.

Since the Beijing-Shanghai High-Speed Railway opened to traffic in 2011, China has become the country with the longest mileage of high-speed railway in the world, and Chinese enterprises have made extraordinary achievements with the completion of multiple high-speed railway projects at home and abroad after the Qinhuangdao-Shenyang Passenger Dedicated Railway. In this context, Indonesia gradually developed an interest in China's high-speed railway technology. In March 2015, Joko

Widodo, the President of Indonesia, visited Japan and China respectively and took a ride on both the Shinkansen and Beijing-Tianjin high-speed trains. In March and April of the same year, President Xi Jinping met with President Joko twice, and a memorandum of understanding was signed on the construction of the Jakarta-Bandung high-speed railway, formally unveiling the competition between China and Japan over the Indonesian project.

In July 2015, the Indonesian government officially announced its plan to build the Jakarta-Bandung high-speed railway, accepting bids from Japan and China for the project. From July to August 2015, the Chinese and Japanese governments sent special envoys to Indonesia to submit feasibility reports.

By the end of September 2015, Indonesia finally agreed to China's plan for the Jakarta-Bandung high-speed railway, and the total contract sum determined at that time was USD 5.135 billion. On October 17 of the same year, China Railway International Group (CRIG) signed an agreement with four state-owned enterprises in Indonesia to establish a joint venture in Jakarta, with Indonesia holding 60% of the shares and China 40%. China is entrusted with the design and construction of the railway, supply of equipment, and operation. On March 16, 2016, the Indonesian government signed a franchise agreement with the joint venture in Jakarta, manifesting that China had officially won the Jakarta-Bandung high-speed railway contract.[19]

China Railway Group Limited and Power Construction Corporation of China are responsible for the construction works, China Railway Rolling Corporation (CRRC) for the manufacture of rolling stock, and China Railway Signal & Communication Corporation Limited (CRSC) for the supply of signal equipment. In January 2016, the commencement ceremony was held, and the construction of a 5-km pilot section also started in March. At the same time, land acquisition was underway in Indonesia. In April 2017, the EPC contract of the Jakarta-Bandung High-Speed Railway was signed in Jakarta. In July 2017, construction of the Walini Tunnel along the line commenced. By mid-2018, more than 20 breakthroughs had been achieved in critical works, signifying that project implementation was in full swing.

Indonesia had definitely benefited from the Japan-China competition. Although China bore the financial risks of the project to a certain extent, the overall advantages outweighed the disadvantages. First of all, Indonesia's choice of the Chinese plan meant that China had the competency to compete with Japan and broke its monopoly on high-speed rail technology in Asia. Moreover, the Jakarta-Bandung high-speed railway, setting an exemplary model, would affect not only the subsequent planning for the 750 km of high-speed railway in Indonesia but also high-speed railway constructions in the whole of Southeast Asia. Most importantly, the high-speed railway evidenced the "going global" of China's comprehensive high-speed railway system, technical standards, and entire industrial chains, and the

[19] Founded in December 2014, China Railway International Group, a subsidiary of China Railway Group Limited, is responsible for promoting China's high-speed railway.

breaking through of the defense in the international market dominated by Western countries.

Secondly, China has become a major global player in the export of railway equipment. In 1980, China exported railway equipment to less than 10 countries, and now its export covers more than 80 countries and regions. The product categories for export have been upgraded from railway accessories to ordinary vehicles and diesel locomotives, and then to high-speed EMUs, high-power electric locomotives, and diesel locomotives, as well as rolling stock with various body types, materials, and traction modes.

When Chinese locomotives were first exported to Myanmar in 1995, the export volume of China's railway equipment was only USD 64 million, accounting for 0.9% of the global market share. However, this figure climbed to USD 3.8 billion in 2014, accounting for 13% of the global total, with an average annual increase of 23%. In 2009, China North Locomotive & Rolling Stock Corporation Limited (CNR) exported 20 diesel locomotives to New Zealand, which was the first time Chinese locomotives were exported to developed countries. In 2012, CNR exported two six-axle broad-gauge AC drive electric locomotives to Belarus, and the Chinese locomotive made its first appearance in the European market. In the same year, China South Locomotive & Rolling Stock Corporation Limited (CSR) exported six SDA1 diesel locomotives (called Pandaroo in Australia) to Australia, registering the debut of China's independently developed locomotives in developed countries.

Since its establishment in 2015, CRRC has become the company with the most complete supply chain of railway equipment in the world. In recent years, the business growth of CRRC has largely been driven by overseas orders, and the share of overseas revenue has risen rapidly. In 2013, the overseas revenue of CSR and CNR was RMB 14 billion (equivalent to USD 2.18 billion), accounting for only 7% of the total of the two companies.[20] The overseas revenue of CRRC accounted for 12% in 2015 and 18% in 2016. Overseas contracts signed by CRRC are still on the rise after 2017, but the share of overseas revenue has declined slightly due to faster expansion of domestic business. Overall, China's export capacity of railway equipment is swiftly approaching those of Bombardier and Alstom.

Thirdly, China's overseas market for railway projects has been significantly expanded. On the one hand, Africa is still the main market, while railway projects in the Middle East and Southeast Asia are also important for the expansion. On the other hand, Chinese enterprises, which used to engage only in railway construction, have begun to partake in the operation of some completed railways in recent years.

In addition to the above-mentioned Ankara-Istanbul high-speed railway in Turkey, Chinese enterprises have also completed the following railway projects in recent years.

[20] In 2013, the overseas revenue of Bombardier, accounting for 92.6% of the company's total revenue, was USD 16.8 billion, half of which was earned from the export of transportation equipment; Alstom's overseas revenue, accounting for 89.3% of the company's total revenue, was USD18.1 billion, about a quarter of which was earned from the export of transportation equipment.

7.2.3.1 Haramain High-Speed Railway Project

Launched in 2009 and completed in 2017, the railway is a 480-km-long high-speed rail line that links two muslin holy cities of Medina in the north and Mecca in the south, passing through Rabigh and Jeddah with a maximum design speed of 360 km/h. The first phase of the project was implemented by a consortium consisting of China Railway 18th Administration Group Co., Ltd. and two Saudi companies. The Chinese firm was only charged with the construction of structures such as major railway bridges, new highway bridges, and culverts in the section from Mecca to Jeddah, with a contract value of about USD 380 million.

7.2.3.2 North-South Railway (NSR) Project in Saudi Arabia

This railway starts from Jalamid, and travels via Hail and Buraidah to Riyadh, the capital of Saudi Arabia, and then to the eastern port of Damman. With a total length of 2,400 km, it is the main channel for the transportation of Saudi mineral resources and also the longest desert railway in the world. China Railway 18th Administration Group Co., Ltd., China Railway 15th Administration Group, and China Civil Engineering Construction Corporation (CCECC) all participated in constructing a section of 400 km with a contract value of USD 524 million. The railway opened to freight transportation in June 2011 and to passengers in February 2017.

7.2.3.3 Mecca Light Rail Project

With a contract value of about USD 1.773 billion, the Mecca Light Rail Line connects Mecca and Mount Arafat and extends about 18 km. China Railway Construction Corporation Limited (CRCC) acted as the main contractor and China Railway 18th Administration Group Co., Ltd. undertook civil engineering works. It is the first overseas railway project constructed by Chinese enterprises under an EPC contract (The EPC contractor coordinates all design, procurement, construction, operation, and maintenance work and ensures that the whole project is completed as required in time). The contract was signed in February 2009 and the light rail was put into service in November 2010. At present, the railway is operated by Chinese enterprises under the concession agreement signed with the Saudi government.

7.2.3.4 Railway Rehabilitation Project in Uzbekistan

Uzbekistan has since 2006 planned to rehabilitate and renovate its domestic railways, and Chinese enterprises have taken part in a number of railway rehabilitation projects in the country. The first railway for rehabilitation, with a total length of about 830 km, starts from Keles Station on the border of Kazakhstan in the north, passes via Tashkent, Samar Khan, Malakand, Karhi and Qumkurgan, and reaches

Termez Station near the border of Afghanistan. China National Technical Import and Export Corporation (CNTIC), in collaboration with China Railway Electrification Administration Group (CREBG), rehabilitated and electrified the line section by section. The railway, spanning 505 km from Keles to Karhi, opened to traffic in August 2015. Electrification of the 325-km-railway from Karhi to Termez was officially launched in 2016 and the line was put into operation in January 2018. In addition to the above lines, China Railway Tunnel Group (CTG) also constructed the Angren-Pap Kamchiq Tunnel in the country and completed the work in February 2016. Over the past 10 years, Chinese enterprises have rehabilitated more than 1,000 km of railways in Uzbekistan and built some other key projects. These projects are of great significance for the final connection of railways in China, Kyrgyzstan, and Uzbekistan.

7.2.3.5 Railway Projects in Algeria

Chinese enterprises have completed many small-scale railway projects in Algeria, including the supply and laying of 67 km of railway tracks and turnouts in 2005, the construction of a railway marshalling station in 2006, the renovation of 94 km of railway line, turnout and ballast in 2007, and the railway line correction and construction of a 55-km-section of the double-track railway (including a tunnel of about 7 km) in 2009. European standards were applied to all the above projects completed by CCECC.

7.2.3.6 Lagos-Kano Standard Gauge Railway Project in Nigeria

In October 2006, CCECC officially signed a contract with the Nigerian government to upgrade the existing railway into a new standard gauge one. The total length of the Lagos-Kano railway is 1,315 km, from the largest city of Lagos in the south to Kano in the north, the second largest city in Nigeria. In 2006, CCECC planned to reconstruct the Abuja-Kaduna Railway, which links Abuja and Kano with a total length of 186.5 km, a design speed of 150 km/h, and an investment of about USD 850 million. However, the project was launched in 2011 and completed in 2014 due to a shortage of funds and the impact of political changes in Nigeria. It is the first modern railway built on the African continent in full compliance with Chinese technical standards.

7.2.3.7 Railway Rehabilitation Project in Angola

The Benguela railway, with a total length of 1,344 km, runs through Angola from the Atlantic port city of Lobito in the west to the border city of Luau adjacent to the Democratic Republic of the Congo, connecting the Congolese section. It is the western section of the future railway line between the Atlantic and Indian Oceans in

Africa.[21] Launched in 2007 and completed in August 2014, the project, with a total investment of USD 1.83 billion and a design speed of 90 km/h, is the longest railway built overseas in accordance with Chinese standards. It was surveyed by the China Railway Shanghai Design Institute Group and constructed by the China Railway 20th Administration Group.

7.2.3.8 Addis Ababa–Djibouti Railway Project

The Ethiopia-Djibouti railway, also known as the Addis Ababa-Djibouti Railway, connects Ethiopia's capital Addis Ababa and Djibouti City, with a total length of more than 750 km, a design speed of 120 km/h and a project investment of USD 4 billion. CCECC started construction work in 2010, and the line opened to traffic in October 2016. It is the first electrified line in East Africa built in compliance with Chinese standards. Chinese enterprises were also responsible for the design, procurement, construction, supervision, financing, and operation. Currently, CCECC is the operator of the railway.

7.2.3.9 Mombasa-Nairobi Railway Project

The Mombasa-Nairobi railway is a standard-gauge railway that connects Nairobi, the capital of Kenya, and the largest port city Mombasa with a total length of 485 km and a design speed of 120 km/h (diesel locomotives are temporarily used). The contract price for the project is USD 3.8 billion, of which 90% was financed by the Export-Import Bank of China. The existing Mombasa-Nairobi railway with a gauge of 1 meter, built in the colonial period more than 100 years ago, could no longer meet the current transport needs of Kenya. Therefore, in 2011, the Kenyan government signed a memorandum of understanding with China Road & Bridge Corporation (CRBC) on the construction of a new railway. In 2014, the execution of the works started. All the equipment for the project was from China and the construction was carried out in compliance with Chinese standards. In December 2016, tracks were laid on the line, and the railway opened to traffic in May 2017. It was officially put into operation in 2018. The Mombasa-Nairobi Railway, operated by Chinese enterprises, serves as an important support for the development of the whole East African railway network with Kenya at its center, which will connect six countries in East Africa, including Tanzania, Uganda, Rwanda, Burundi and South Sudan.

[21] Operated by the Benguela Company, the railway is also known as the "Benguela Railway".

7.2.3.10 Addis Ababa Light Rail Project

Located in Addis Ababa, the capital of Ethiopia, this project is the first urban light rail in East Africa with a total length of 34 km and a contract price of USD 475 million. Ethiopia issued a tender invitation for the project in 2008, and the consortium formed by China Railway Eryuan Engineering Group Co., Ltd. and China Railway No. 2 Engineering Group. Co., Ltd. won the bid. Starting in 2012 and officially put into operation in September 2015, the project was constructed in compliance with China's railway technical standards, and the vehicles were also imported from China. Chinese enterprises operate the railway after its completion.

After 40 years of reform and opening up, the competency of Chinese railways to "go global" has improved rapidly, especially after the entry into the era of high-speed railway, and the undertakings have developed from the initial "going global" of labor and simple equipment to that of the whole industrial chain covering scheme design, technical support, construction, equipment manufacturing, operation, and maintenance. Chinese railways have in the past years expanded their share in the world market and are now striding steadfastly into a brighter future.

7.3 Challenges and Opportunities of Chinese Railways' "Going Global"

7.3.1 New Opportunities for "Belt and Road" and Chinese Railways' "Going Global"

"Silk Road Economic Belt" and "The 21st Century Maritime Silk Road" are collectively referred to as "Belt and Road", an international initiative proposed by China in 2013. In this Initiative for cooperation, the legacy and icon of the Ancient Silk Road are crystallized into an action plan, under which Chinese Railways' "going global" has gained unprecedented significance.

In response to the Belt and Road Initiative, all countries along the route pursue the direction of cooperation in "policy coordination, facilities connectivity, unimpeded trade, financial integration, and people-to-people bonds". As an integral part of economic activity, transportation infrastructure can help improve production efficiency by accelerating the movement of labor and products to fully integrate the economic advantages in different areas. In the view of human history, the golden ages of economic development are always accompanied by large-scale construction of transportation infrastructure. Generally speaking, the development level of a country is basically consistent with that of transportation infrastructure. The more developed a country is, the more roads, railways, ports and airports per unit area or capita it possesses. The gap between developed and developing countries is often reflected in the disparity in transportation infrastructures. The Belt and Road

Initiative aims to promote the common development of all the countries along the route with the construction and improvement of transportation infrastructures.

Covering a rather large landscape, the countries along the Belt and Road hold enormous potential for the development of transport infrastructure. Located at both ends of the Belt and Road are respectively the developed Western European economic circle and the dynamic China economic circle. The former boasts a relatively complete transportation infrastructure system, while the latter has improved such a system significantly in recent years with construction competency developing at a rapid pace. However, the scenarios are not so encouraging for countries lying in between the two ends of Belt and Road. According to relevant statistics, only 10 countries have relatively completed transportation facilities, including Czech, Vietnam, Kazakhstan, Russia, Uzbekistan, Singapore, Malaysia, the United Arab Emirates, Turkey, and Poland, but the remaining countries have a considerable gap to fill, particularly the 14 land countries neighboring China. Some countries such as Laos, Afghanistan, Mongolia, and Kyrgyzstan even suffer from very backward transportation infrastructure. With the progress of multi-level interconnection networks driven by the Belt and Road Initiative, those countries will be provided access to upgrading their transportation facilities.[22] It is estimated that infrastructure improvements achieved through the Belt and Road Initiative could reduce the shipment time from East Asia to the Middle East from 15 days to 13 days and the trade costs by 10%.[23]

The Belt and Road Initiative mainly encompasses the regions in Asia, Europe, and Africa, which are geographically categorized as "World Island" and cover a land area of 84 million square km. On this "Island", the straight-line distances are more than 10,000 km both from Lisbon in Portugal, the westernmost end of Europe, to Shanghai in eastern Asia, and from northern Russia to the southern end of the African continent. The Belt and Road Initiative aims to form a regional association across Asia, Europe, and Africa and thoroughly solve the land transportation problem within the "World Island".

To date, global land and sea transportation are severely imbalanced. During the period of capitalist colonial expansion, the flow of people and cargo mainly depended on maritime transportation. After the advance of the aviation industry, people relied on airplanes for cross-country travelling, while global cargo shipment was still dominated by maritime transport.

Currently, maritime transportation accounts for more than 80% of the total volume of visible trade. In this regard, globalization is, to a large extent, that of maritime transportation. The economic centers all over the world are also basically developed on the basis of maritime (or water) transport. Among the top 25 cities by GDP in North America, 18 are port cities, and in South America, six are seaports among the top 10 cities by GDP. Europe, the birthplace of railways, has an extremely

---

[22] Zhai Kun, & Wang Jimin. (2017). *The Belt and road initiative report on five-connective index* (p. 109–112). Beijing: Economic Daily Press.

[23] World Bank. (2019). *Belt and road economics: opportunities and risks of transport corridors* (p. 48–49). Washington, DC: World Bank.

sophisticated railway system. However, among the top 15 European cities by GDP, seven are coastal cities. In contrast, economic development falls behind in countries and regions not closely linked to maritime transport. Notably, five of the 10 countries with the lowest per capita income are landlocked, though only 44 of the world's more than 220 countries are landlocked.

Despite its many irreplaceable advantages such as tracklessness, readily adjustable routes, large carrying capacity, and low freight cost, maritime transport has its weaknesses. First of all, most of the habitable areas on Earth have no seaports and can only rely on land transport, and one-fifth of the countries are landlocked without any ports. Out of the 14 landlocked countries bordering China, nine have no seaport. Secondly, maritime transport cannot meet the demands arising from diverse geographic and economic conditions.

For example, Urumqi in China's Xinjiang Uygur Autonomous Region, is only 1,000 km away from northern Pakistan, but land transport is very inconvenient. If transported by sea, the goods must be transported first by rail to eastern ports in China thousands of kilometers away, and then again by sea to the destinations in northern Pakistan nearly ten thousand kilometers apart. Maritime transport under such circumstances is not only time-consuming and laborious but also costly. Therefore, no matter how important maritime transport is, land transport stands as an indispensable link in economic globalization. Among all the land transport modes, railways have obvious advantages such as speed, coverage, safety, and all-weather operability. At present, coal, iron ore, steel, copper, aluminum, timber, and other bulk cargoes are sold all over the world through sea-rail transportation, and the flow of other cargoes also hinges on railway transportation. For example, more than half of the cargoes from East Asia to the east coast of North America are transported by railway, and nearly one-third of the general cargo is transported by rail within Europe.

Unfortunately, the railway is currently a vassal of maritime transport and ports, far from being independent. In terms of transportation routes, the connections via maritime transport at the edge of the Eurasian continent are fully developed with large numbers of seaports scattered alongside the coast from Amsterdam to Cape Town, from Basra to Mumbai, and from Singapore to Shanghai. But land transport on the continent remains underdeveloped. The railway networks are densely concentrated at the east and west ends of the Eurasian continent, but quite sparse in the hinterland. In particular, there was once no railway line in North Asia in Siberia and Central Asia around the Qinghai-Tibet Plateau. Western scholars in the nineteenth century even equated the improvement of transport conditions in this area with a revolution. At the beginning of the twentieth century, countries favoring land transport such as Switzerland, Germany, and Russia planned to connect the railways in Asia and Europe to form a land bridge across the Far East and Western Europe and established some railway cooperation organizations for this purpose. However, before the end of World War II, there were only two long-distance transcontinental railways in the world, the Siberian Railway built by Russia and the Baghdad Railway built by Germany. After World War II, with the development of railway construction in the Soviet Union and China, the New Eurasian Continental Bridge was built, stretching from the eastern coast of China to major cities along the Atlantic coast of Europe, but it is far from satisfactory.

Railway transport level in the hinterland of the Belt and Road falls below those in North America and Europe. For example, in North America, Canada, the United States and Mexico have been connected by railway networks since 1994, and passengers and goods flow freely between the three countries without the hindrance of railroad borders. Even a large number of railway businesses in the United States are directly operated and managed by Canadian companies. In the European Union, railways are also basically interconnected. As early as 1991, the framework for the integration of EU railways was established according to the *Council Directive on the Development of the Community's Railways* (91/440EEC), published by the EU Council, which includes the unification of technical standards and safety management system, the establishment of the European Union Agency for Railways and the opening-up of the railway transport market. Currently, the European Union has achieved barrier-free transit of cross-border trains, which allows trans-European international trains to travel unobstructed between major European cities. In contrast, the development of the Trans-Asian railway network trails behind. Eighteen countries, including China and Russia, signed the Trans-Asian Railway Network Intergovernmental Agreement in 2006 to build a railway as long as 810,000 km in Asia. However, before the advent of the Belt and Road Initiative, nearly half of the more than 60 countries along the "Silk Road" economic belt—including China's neighbors such as Laos, Bhutan, Nepal, India, Cambodia, Pakistan, Thailand, Bangladesh, Myanmar, Tajikistan and Afghanistan—still had no railway connecting to China.

An overreliance on maritime transport would lead to inadequate development of inland areas and an abnormal distribution of population and economic belt on the coast. The imbalanced development between maritime and land transport in countries along the Belt and Road implies that further inside the country lies the "World Island", which creates obstacles to economic development and population convergence. The only way to rectify the situation is to build a dense railway network branching through the "World Island".

> "The Belt and Road Initiative and the connectivity endeavor are compatible and reinforcing", said President Xi Jinping, "if the Belt and Road are likened to the two wings of a soaring Asia, the connectivity is like their arteries and veins".[24]

With the promotion of land connectivity between countries all over the world, railway transport will definitely receive more attention for its large carrying capacity, continuity and low cost. High-speed railways will further enhance the advantages of land transport and prompt unprecedented changes in the spatial pattern of connectivity. Although China started in the railway field later than some countries, it is, after all, the only country in the world that can build railways on a large scale, in a variety of regions, and under highly diversified climatic conditions in the past

[24] Xi Jinping. (2014, November 9). Connectivity for development and partnership on cooperation—dialogue on strengthening connectivity partnership. *People's Daily.*

20 years. With its experience and technology radically upgraded, China will undoubtedly play a predominant role in railway construction once infrastructure connectivity projects are in full swing along the Belt and Road.

7.3.2 Latest Progress in "Going Global" of Chinese Railways

During overseas expansion, Chinese railway enterprises have encountered difficulties that cannot be overcome in the short term. Nevertheless, corresponding measures will be found to solve them in the long run. Inevitably, setbacks and failures exist in the "going global" process of China's high-speed railways. But the steps forward cannot be blocked and such a move is of great significance to China and the majority of developing countries. At present, China Railway's planned or ongoing projects overseas mainly include the following.

7.3.2.1 Moscow-Kazan High-Speed Railway Project

The high-speed railway, with a total length of 770 km and a maximum design speed of 400 km/h, starts from Moscow in the west and extends to Kazan in the Republic of Tatarstan in the southeast. The railway is divided into four sections in Russia's plan, including Moscow-Vladimir, Vladimir-Nizhny Novgorod, Nizhny Novgorod-Cheboksary, and Cheboksary-Kazan. In spite of a memorandum of cooperation on this project signed between China and Russia, only Russian companies initiated the construction of the 180-km Moscow-Vladimir railway line. China Railway Eryuan Engineering Group Co., Ltd. participated in the survey and design of this project in 2015, with an accepted contract amount of about RMB 2.4 billion (equivalent to USD 374 million). At present, relevant equipment is manufactured in Russia for the Moscow-Kazan high-speed railway. In June 2016, Russia entered into an agreement with CRRC to jointly set up a factory in Russia and manufacture at least 100 EMUs for Moscow-Kazan high-speed railway, for which China Railway Signal and Communication Co., Ltd. (CRSC) will provide the train control system and signaling and communication scheme.

7.3.2.2 Hungary-Serbia Railway Project

The 350 km-long railway starts from Budapest in the north and connects Belgrade in the south, including about 166 km in Hungary and 184 km in Serbia. Instead of building a high-speed railway, the employer of this project intended to reconstruct the existing lines, which meant the transformation of the existing single-track railways into double-track ones and the construction of new double-track railways in some sections. The railway, with a maximum design speed of 200 km/h, serves as an electrified mixed passenger and freight line with a total investment of about

RMB 10 billion (equivalent to USD 1.56 billion). This project, to be implemented by a consortium formed between two subsidiaries of China Railway and Hungarian Railways, was officially launched in November 2017 and is expected to be completed in 2024.

7.3.2.3 China-Kyrgyzstan-Uzbekistan Railway Project

With a total length of 523 km, the railway is planned to pass through Kyrgyzstan from Kashgar, Xinjiang Uygur Autonomous Region of China, and reach Andijan, Uzbekistan, thus shortening the distance between five Central Asian countries and Middle East countries like Turkey by nearly 900 km. Meanwhile, this line is expected to transport 18.1 million tons of freight annually and will serve as an important passage under the Belt and Road Initiative.

As early as 1997, the governments of China, Kyrgyzstan, and Uzbekistan expressed their support for the project, and the section in China was even included in China's *Sixth Five-Year Plan*. However, this project was delayed due to political turmoil in Kyrgyzstan and Uzbekistan and did not start until 2013 when the feasibility study was conducted. During this period, China launched railway reconstruction in Uzbekistan, making full preparations to connect railways in China, Kyrgyzstan, and Uzbekistan. The relevant railways between China and Uzbekistan have been basically completed so far, while the construction in Kyrgyzstan has not yet begun. Kyrgyzstan is always concerned about the amount of investment, gauge, and return of the project. Thanks to the efforts of China and Uzbekistan, Kyrgyzstan's willingness to cooperate has increased recently. In June 2019, China and Kyrgyzstan signed a declaration establishing a comprehensive strategic partnership, which specifically mentioned speeding up the construction of China-Kyrgyzstan-Uzbekistan railway. In August of the same year, the leaders of Kyrgyzstan and Ukraine also stressed the importance of this project. In the near future, significant breakthroughs may be achieved.

7.3.2.4 China-Laos Railway Project

The China-Laos Railway, a part of the Trans-Asian Railway, begins from Yuxi, Yunnan province of China in the north and ends in Vientiane, the capital of Laos in the south, between which it crosses Mohan, the border city connecting China and Laos. The railway covers 91 km in China, and the rest 418 km line in Laos is known as the Mohan-Vientiane Railway. With a total investment of more than RMB 50 billion (equivalent to USD 7.78 billion), this electrified railway at a design speed of 160 km/h is the first one directly extending from the China railway network to connect overseas railways. It is also the first of its kind to be built by Chinese enterprises. The China-Laos Railway was jointly built by Chinese and Laotian enterprises. Six Chinese enterprises took part in the project including China Railway No. 2 Engineering Group Co., Ltd. All the equipment and technical standards adopted for this project are from China and most personnel involved in the construction were directly dispatched by the Chinese side. The construction of this railway started in 2016 and is expected to be completed in 2021.

7.3.2.5 China-Thailand Railway Project

China-Thailand railway is also a part of the Trans-Asian railway, with a total length of 867 km and a design speed of 160 to 180 km/h. The railway is planned to stretch from Nong Khai to Bangkok and Rayong in the south, connect the above-mentioned Laos-Mohan railway in the north, and reach the Gulf of Thailand in the south. China and Thailand first negotiated for the project in 2009 but did not sign the memorandum of understanding until 2014, due to Thailand's domestic problems. The Thai government did not issue approval documents until July 2017 because of environmental impact assessment, financing, and other issues. In December of the same year, the construction of the section in Nakhon Ratchasima officially started. All the design, construction technology, and equipment for the China-Thailand railway are introduced from China, and the railway is expected to be completed around 2024.

7.3.2.6 Tel Aviv Light Rail Project of Israel

The Tel Aviv light rail project is part of Israel's 176-km national light rail system containing seven lines, of which the Red Line is 23 kilometers long. In May 2015, the consortium formed by China Railway Tunnel Group Co., Ltd. and an Israeli company succeeded in winning the bid for part of the Red Line, which includes 16 cross passages, two tunnels, and six underground stations, valued at about RMB 5 billion (equivalent to USD 779 million) in total. In November 2017, China Civil Engineering Construction Corporation and Shenzhen Metro Group Co., Ltd. partnered with the Israeli contractor again and won the contract for the operation and maintenance of the Red Line project. In February 2018, China Railway Tunnel Group Co., Ltd. and China Railway Electrification Engineering Group Co., Ltd. jointly won the bid for the track design, construction and maintenance of the Red Line project. The vehicles of the project are also developed and manufactured by CRRC Changchun Railway Vehicles Co., Ltd. The construction for the Red Line of Israel's light rail project is programmed to be completed in 2021, and the overall construction will continue until 2030.

7.3.2.7 Nairobi-Malaba Railway Project of Kenya

Nairobi-Malaba railway, with a total length of 488 km and a design speed of 120 km/h, is an extension of Mombasa-Nairobi railway to northwest Kenya. It begins from Nairobi, the capital of Kenya, and ends at Malaba, the border city of the country. China Communications Construction Company Limited (CCCC) is responsible for the design, construction, and procurement of equipment for the project in compliance with Chinese standards.

CCCC started the preliminary preparation for the Nairobi-Malaba railway in 2016 and officially commenced construction in September 2018. The first phase of the project was completed in May 2019, and the whole project is expected to be completed in 2020.

7.3.2.8 Prospect of China-Pakistan Railway

The China-Pakistan railway, which has not yet entered the implementation phase, is designed to extend from Kashgar in Xinjiang Uygur Autonomous Region in China to Gwadar Port in southern Pakistan after passing through northern Pakistan. When President Xi Jinping visited Pakistan in 2015, China and Pakistan signed an agreement for the feasibility study on the construction of railways and other passages between the two countries. At the first Belt and Road Forum for International Cooperation held in Beijing in 2017, China and Pakistan signed a memorandum of understanding on the upgradation of ML-1, which is the main railway line in Pakistan at present. ML-1 starts from Karachi, the largest city in the south to Peshawar in the northwest after passing through Islamabad, Pakistan's capital. Pakistan now hopes that China will help extend the railway northward for hundreds of kilometers, which could then cross the Karakoram mountains and reach Khunjerab at the Pakistan-China border, and further connect Kashgar, making it part of the China-Pakistan Economic Corridor.

7.3.2.9 Iran's Railway Renovation Project

Iran put forward its plan for railway renovation at the end of the twentieth century and has cooperated with China, Russia, Germany, and other countries in such endeavors. In 2005, China Railway Electrification Engineering Group Co., Ltd. completed the electrification of some railways in Iran. In June 2015, China National Machinery Import & Export Corporation, Beijing Supower Technology Co., Ltd., and an Iranian company (MAPNA Group) jointly implemented the electrification project of the 926-km railway from Tehran to Mashhad, an eastern city of the country. This project was scheduled to be completed in 2020 with the goal of increasing the railway speed from 160 km/h to 200 km/h. In January 2018, China Civil Engineering Construction Corporation, a subsidiary of China Railway Construction Corporation Limited (CRCC), carried out the construction of the Kermanshah-Khosravi railway in western Iran, with a total length of 263 km, and the construction is expected to be completed in 2022.

7.3.2.10 Lagos-Kano Railway Reconstruction Project in Nigeria

The reconstruction project was launched in 2006, and only the Abuja-Kano section had been finished by 2014. Reconstruction of the Lagos-Ibadan section also commenced in March 2017 and was expected to be completed in 2020. With a total contract value of USD 1.581 billion, the section has a total length of 156.08 km and a maximum design speed of 150 km/h. In May 2018, the reconstruction of the Ibadan-Kaduna section started, and the project was planned to be completed in 2022. With a total contract value of USD 6.68 billion, this section has a total length of 616 km and a design speed of 150 km/h. All the above projects are constructed by China Civil Engineering Construction Corporation. By 2019, reconstruction had begun on all the sections in the original plan for the Lagos-Kano Railway in Nigeria.

In addition to the projects mentioned above, Chinese enterprises have also implemented railway projects in Libya, Sierra Leone, Chad, Sudan, Cambodia, and Senegal, and plan to execute new railway projects in Singapore and Malaysia. They are currently engaged in the construction of more than 100 overseas projects if all kinds of rail transit projects are taken into consideration.

The "going global" process of Chinese railways since reform and opening up exhibits the increasing maturity of China's railway technology and the continuous improvement of China's comprehensive national strength. It will not only be conducive to restructuring China's economy and the improvement of infrastructure in a multitude of developing countries, but also provide strong support to China's peaceful rise and an important impetus to narrow the North-South Divide. In the short term, China will under the Belt and Road Initiative cooperate with many other countries to build a railway network connecting Asia, Europe, and Africa, and Chinese railway enterprises will play a predominant role in this long-term construction process. From a longer-term perspective, the "going global" of Chinese railways will help developing countries narrow the technological gap with developed countries and make a dent in the Core-Periphery Model between developed and developing countries, promoting the building of a "community with a shared future for mankind".

8 Conclusion: Reform and Opening Up and Modernization Model of Chinese Railways

For most of the past four decades, Chinese railway workers have been under tremendous pressure. The problem they faced is simple to express: the increase in transport capacity failed to meet the ever-growing demands for transportation. Various complex situations, intricate contradictions and stringent constraints were like a maze challenging the resolution of this seemingly simple problem. Now, there is a sense of relief to be out of the maze at last.

At the beginning of 2019, China Railway (CR) announced a series of exciting figures, "In the past four decades, Chinese railway operating mileage increased from 52,000 kilometers to more than 131,000 kilometers, an increase of 154.4%;... The number of passengers transported by the national railways increased from 807 million to 3.317 billion, an increase of 311.0%, and the volume of freight transported rose from 1.075 billion tons to 3.19 billion tons, an increase of 196.7%". Such increases showcase that, "Chinese railways has basically been able to get rid of the serious shortage in transport capacity, and can gradually meet the growing demands". These achievements, of course, cannot be made without reform and opening up, "with this policy, railway reform continues, providing enterprises with stronger development momentum and operational vitality. Chinese railway enterprises keep on opening to the outside world and expand their overseas business scope, enhancing international influence and competitiveness of Chinese railways".[1]

In the previous chapters, we have depicted the specific situations and difficulties faced by Chinese railways in the different stages of the past four decades, the responses and measures proposed accordingly, and new ideas and methods for addressing a new problem brought about by the former one. Now, we should look back to the overall questions raised at the beginning of this book instead of specific questions in different stages, and focus on the answers provided by the Chinese Railways.

[1](2019, January 4). R*eport of Lu Dongfu at the Work Conference of China Railway. People's Railway Daily*.

X. Zhang, *China's Railway Expansion in 40 Years of Economic Reform and Opening Up*, Understanding China,
https://doi.org/10.1007/978-981-95-0333-9_8

8.1 Back to the Starting Point: What Is the Modernization of Chinese Railways?

Undoubtedly, the goal of modernization for Chinese railways is widely accepted, and railway reform and opening up are vital to achieving this goal. But what does railway modernization entail? There's no easy answer. For quite a long time, the connotation of the modernization of Chinese railways had remained unclear for two main reasons. First, before the twenty-first century, Chinese railways lagged behind those of developed countries in all aspects, falling short of modernization even by more lenient standards. In such circumstances, it seems unrealistic to propose specific indicators for railway modernization. Second, modernization is a dynamic process, evolving with different measurement indicators at different stages of economic and social development. Therefore, in the documents of the central government and MOR, expressions like "promoting railway modernization" or "improving the level of railway modernization" are commonly used to avoid clarifying what the connotations of modernization are.

In the author's opinion, the time to recognize the modernization of Chinese railways and clarify its connotation was in 2003 for the first time. At a meeting held by MOR that year, it was proposed that the modernization of Chinese railways should be realized by 2020, in which "the transport capacity of Chinese railways will be able to meet the requirements of national economic progress, its main technical equipment will reach the level of developed countries, and its service quality will meet the needs of passengers and freight owners. What's more, a management system and operation mechanism in line with the socialist market economy will be established, a high-quality workforce will be developed, and the railway staff will enjoy a moderately prosperous life in many aspects".[2] After analyzing this goal, it is evident that the expression remains vague. If China does not realize modernization in general in 2020, then the "requirements of national economic progress" and "the needs of passengers and freight owners" cannot be considered "modernized". Apart from this point, the target still emphasizes the two indicators of "transport capacity" and "technical equipment", which are very important but far from covering all aspects of railway modernization.

In 2008, the minister of MOR published an article specifying the indicators of railway modernization. It was pointed out that such modernization in developed countries manifested in five aspects, namely adequate competence, high-speed passenger transportation, heavy-haul railways for freight transportation, proper management, and convenient services. Thus, Chinese railways should achieve those targets in order to modernize itself. The development goals of Chinese railways by 2020 were set as follows, a comprehensive railway network with strong capacity, world-class technologies and equipment, advanced and appropriate operation and

[2] Editorial Department of the People's Republic of China Yearbook. (2004). *MOR establishes and implements the idea of leapfrog development of railways* (p. 664). Beijing: The People's Republic of China Yearbook.

management, high-quality services of world standards, a professional team capable of supporting the railway modernization, and a moderately prosperous life for railway staff.[3] Such descriptions were more specific than those of 2003.

Owing to the notorious corruption case in 2011 and the subsequent dissolution of MOR in 2013, the goals were removed from the official documents accordingly, and did not reappear until 2018. In that year, it was pointed out in the nationwide work conference that to build a country with a strong transportation network, the railways should be the top priority, that is, the railways should be modernized in the first place. During that period, the new goals were known as three "leadings" and "improves" for short,[4] to be specific, the mileage and quality of the railway network would reach the world-leading level, the railway technologies, equipment, and innovation would reach the world-leading level, and the safety of railway operation and management would reach the world-leading level; the railway enterprises' competence in innovating their systems and mechanisms would be improved, the status and functions of the railways in the comprehensive transportation system would be improved, and railways' contribution to the national strategies and the social and economic progress would be improved. Although the development goals of the railway network in 2020, 2025, and 2035 were also put forward in the conference report, the exact time of "modernization of the railways" remained ambiguous. According to the "two-step" approach outlined in the Report to the 19th CPC National Congress, China would achieve basic modernization by 2035, which coincides with the development goal of railways in 2035, "making Chinese railways an important symbol and component of the great modern socialist country of strength", thus it may be inferred that 2035 would be the year when Chinese railways could be modernized.

In fact, according to the appraisal of China Railway, three "leadings" have been largely accomplished. Therefore, the focus of the reform should be the three "improves", which brings us back to the introduction of this book: the central issue of the reform of Chinese railways is not its development, but how much contribution it can make to the progress of the country and the society. Over the past decades, although the railway industry has made great progress, it also became the "weak link" to the economic and social development at the end of the 1970s, and the "bottleneck" at the beginning of the 1990s. Not until 2016 can we say, "in general, the current pressure on the transportation capacity of Chinese railways has been basically lifted, the constraints have been basically removed, and it can meet the basic demands of the economic and social progress". Indeed, upon the introduction of the reform and opening up, and even in the 1990s, three "leadings" were still a pipe dream. Even though China has seen rapid growth in its economy and society, and realized "world leadings" in railway network, equipment, and operation, it can still

[3] Party Group of MOR. (2018). A review and prospect of the modernization of the Chinese railways. *Chinese Railways*, (1).

[4] Lu Dongfu. (2018, January 4). Railway leading the transportation power to make greater contribution to the sustainable and sound development of economy and society. *People's Railway Daily.*

only "basically lift" the pressure, "basically remove" the bottleneck, and meet the "basic" demands. China has a long way to go to perform more than the "basically".

However, nothing in the world can be done once and for all. Even in 2035 or 2050, Chinese railways will be confronted with various new problems and challenges, which entail new goals and solutions accordingly. The question is whether a generally effective and definite path has been found after 40 years of exploration to enable Chinese railways to maintain the "world-leading" status in key indicators and pursue "world-leading" status in all fields, thereby continuously developing itself in tandem with the escalating social demands. Or, we may ask in a more direct manner: whether a modernization model of the Chinese railways has taken fundamental shape via constant exploration and adjustments in the past four decades since the introduction of reform and opening up.

8.2 A New Question: Has a Modernization Model of the Chinese Railways Taken Shape after the Introduction of Reform and Opening Up?

The question implies great risks, and can hardly be answered, because nowadays transportation technologies are being upgraded at a rapid speed, and new modes of modern businesses are constantly emerging. Compared with industries like information or biology, the railway industry is more "traditional" in terms of technology change and corporate form, and slower in the process of transformation. Yet, it cannot do without technological innovation. Technological innovation, especially disruptive technology, is more or less unpredictable and elusive. Perhaps a question like this may be easier to answer: regarding the systems and mechanisms, are there relatively valuable experience or reliable forms which we can still learn from or apply? Moreover, as mentioned in the introduction, many obstacles were hindering the development of Chinese railways. Four decades have elapsed, are those obstacles still there? If they are, in what forms? On the basis of the previous discussion, we attempt to review the solutions or evolution of the questions: the reform and opening up, and the government and the market.

8.2.1 Reform and Opening Up

Nowadays, we are used to treating the words "reform and opening up" as a single proper noun in Chinese, just like "science and technology". However, when we were reviewing literatures, we found that the formation of the phrase had witnessed a long and complex process,[5] proving that "reform" and "opening up" are inseparable.

It is generally believed that one of the significant features of Chinese economic reform is that the reform in non-public sectors promotes the reform in public sectors. Reform in non-public sectors means that we permitted, and then encouraged the development of individual economy, private economy, and overseas investment within state-owned economy. Three advantages, including high efficiency, increasing market shares, and attractiveness to talents of non-public sectors put great pressure on the state-owned economy, forcing it to accelerate the pace of reform. In other words, the reform in non-public sectors is opening up, which advances the reform in turn.

The reform and opening up in railway industry move forward hand in hand but have different representations. When browsing through the chapters, you can see that "opening up" is not mentioned until Chapter Three, which means "opening up" has not become the important event in the history of Chinese railway progress since 1992, and played a different role in railway development from that in other industries.

However, that does not mean Chinese railways bear no pressure, especially the external pressure, because the strong dissatisfaction from the public calls for the reform in railway industry, and attracts continuous attention from the Central Government (especially during the Spring Festival Travel Rush). There indeed exists competitors with railways such as highways, aviation and other transportation modes. With the rapid expansion of expressways and aviation, market shares of railways were gradually reduced, which made the leading officers of MOR anxious. Generally speaking, when experiencing the reform trend in all industries and sectors, the railway industry with its own pressure and momentum, however, had to resort to reform itself during the 1980s without sufficient conditions for reform in non-public sectors.

In fact, if we had experienced the arduous and complex Chinese reform and then reviewed in retrospect the Contract Responsibility System launched in 1986, we may consider the proposal at that time too radical rather than conservative, because the reform acted like a pioneer in the absence of macro-environment and supporting reforms. From the perspective of macro-environment, we did not clarify the goal of

[5] It is generally believed that the Third Plenary Session of the Eleventh Central Committee of the Chinese Communist Party held in December 1978 marks the beginning of the reform and opening up, which is also well-known in Japanese academics. However, Japanese scholars believe that the reform and opening up has, in fact, witnessed a long transition process. They found that the word 'reform and opening up' had been mentioned at the first time on 18th May 1984 in *People's Daily*. Besides, the word was mentioned twice in the report of *People's Daily* in 1984, 16 times in 1985, 38 times in 1986, and up to 501 times until 1987. By the way, 'reform and opening up' was first mentioned on 28th March 1986 in *Selected Works of Deng Xiaoping*. Japanese scholars think that as a matter of fact, 1978, as the beginning year of the reform and opening up, was determined in the next years. Reference to: 「現代中国史における一九七八年の画期性について」、『中国改革開放への転換—「一九七八年」を越えて』(2011): pp. 121–122.

Chinese economic system reform and were still exploring it in the relatively open industries such as textile and household electric appliance industries.

In view of the supporting reforms, China did not start large-scale reform in investment and financing system, which is a particularly important reason why we did not accomplish the intended goal of the Contract Responsibility System. Throughout the 1980s, our country estimated that railways did not unleash its transport capacity and was lack of state financial support, so we did not approve the large-scale construction of new railways. MOR did not agree to that estimation, instead, they advocated the Contract Responsibility System and autonomy to accumulate railway construction funds. However, raising sufficient funds was extremely hard under the conditions of unestablished market-oriented investment and financing channels.

In 1988, pricing reform in the transition from planned economy to market-oriented economy led to the drastic increase in prices for raw materials and costs of railway construction. Besides, the Contract Responsibility System did not change the situation fundamentally, and MOR did not have the authority to set the prices, and the channels to raise funds were further narrowed and blockaded. Then, there came the endless loop: without well-established network, railways cannot meet the rapidly increasing demands for transport, not to mention raising construction funds. To solve the fund problems fundamentally, the railway industry should open its door to welcome diversified investments in the market. Therefore, it is also applicable for Chinese railway to advance reform through opening up.

With consensus reached on the need to reform, the approach to reform mattered. Market economies are prevalent in the world's major countries, each with disparate railway development modes. In the 1990s, China had not yet reached a consensus on what mode should be adopted in Chinese railways. Although the State Council approved *the Thirty Articles for Railway Reform* in 1994, the reform goal could only be generally expressed as "under the national macro control, the market playing the basic role in resource allocation including railway transport capacity to establish a railway management system and operation mechanism suitable for the socialist market economy". What "management system and operation mechanism" are "suitable for the socialist market economy"? We have a long way to go before we can find them. But in fact, we can be committed to the latter part of the goal, namely "accelerating railway construction, expanding transportation capacity and improving management as well as economic benefits". More to the point, with the transport capacity became insufficient, the existing lines would be exploited, leading to even more inadequate capacity, and such impasse can only be ended by "accelerating railway construction" to "expand transportation capacity".

Though no more well-known reform like the Contract Responsibility System had been practiced on a large scale since then, China has never given up the attempt of overall railway reform. Media reported that the first attempt was made between 2000 and 2003. In 2000, the Economic Planning Research Institute under MOR was appointed to draw up a reform plan for the railway system. Revised for almost 30 times, this plan was not approved by the State Council. The Development Research Center of the State Council was immediately commissioned by MOR to compile a research plan for the railway system reform, and they submitted the reform plan which was likewise rejected. Later, MOR invited the World Bank to independently complete a reform plan for the railway system which was submitted again to the State Council in 2003 whose stance was still ambiguous.[6] Louis Thomson, the World Bank's project leader, advocated for integrating network and operations—a view that clashed with the prevailing "separation of network transportation" philosophy of MOR at the time.[7]

In 2007, China's railway reform began again. The tasks of the National Development and Reform Commission in that year included "studying and formulating an overall plan for the reform of railway investment and financing system and the railway system".[8] It was supposed to be one of the preparations for the National People's Congress the next year but it turned out to be unsettled in the end.

After 1992, the most favorable factor for Chinese railways was that the central government finally changed the policy of "tapping the potential and expanding capacity" and advocated accelerating the construction of new railway lines. The economic reform goal of building a socialist market economy also made it possible to establish diversified fund raising channels. Since 1998, the central government has implemented a proactive fiscal policy and invested more financial funds in infrastructure, which benefited Chinese railways. During the reform and opening up, learning from other countries' experience, MOR should not only build more railways, but also build better and more modern ones, namely high-speed railways. Such undertaking involved funds and also the acquisition of advanced technology, and both issues required a wider opening up to the outside world on a larger scale.

As a matter of fact, by opening wider to the outside world, the railway departments not only raised the construction funds, but also obtained the high-speed railway technology. It seemed somewhat paradoxical that the opening up of railways mainly began to expand following the failure to prepare an acceptable reform plan for the railway system in 2003, and such opening up did not promote the in-depth reform of railway system. The following decade indeed witnessed the "development" of railways but no "reform" at all, according to the opinions of the outside world and the general public. Until 2012, the State Council decided to carry out a research and design a reform plan for railway system again in accordance with the requirements of the separation of government and business functions as well as the separation of government and investment entities, and encouraged private capital to

[6] Li Wei. (2012, July 22). Review of railway reform: Three difficulties and failures. *Oriental Morning Post.*

[7] Louis Thomson et al. (2000). China's railway reform: The great challenge of the railway industry. *Railway Economics Research,* (2).

[8] (2007, June 28). Circular of the General Office of the State Council on transferring and issuing the Opinions of the National Development and Reform Commission on Strengthening the Reform Of Economic Structure in 2007 (GBF [2007] No. 47).

flow into this field.[9] In 2013, MOR was dissolved at the decree of the National People's Congress.

If we say that the railway reform remained stagnant for more than a decade, it is difficult to explain that the railway development exemplified by high-speed railway technology can be achieved only by increasing capital investment and introducing foreign advanced technology, and what role the important decision on the dissolution of MOR had played in railway development. In the absence of system and mechanism adjustment, can the cutting-edge technology be smoothly linked up with the existing system? Can the system of MOR be smoothly transitioned into that of the post-MOR one? Perhaps, we need to develop a new understanding of the market economy in the railway system, especially the relationship between the government and the market.

8.2.2 Government and Market

In the early stage of reform and opening up, the integration of government and business functions was the predominant problem among many institutional ones within the railway system. In the past 40 years, the general development of railways has always been market-oriented, and the central problem to be dealt with has been the relationship between "government and market". The rapid or slow progress of the railway depends on the understanding of this relationship. The complexity of dealing with the relationship between government and enterprises lies in that it is often regarded as a symbol to evaluate whether railways have been reformed or not. The first decade of the twenty-first century, also the key period for the preparation and decision-making of China's high-speed railway, has left such an impression on researchers and commentators.

In fact, the above understanding can be found within railway departments. The basic ideas and implementation steps for the railway to go to the market were stated in 1994 in the *Thirty Articles for Railway Reform* which also put forward that the combination of the individual characteristics of railway reform and the common requirements of national reform is necessary to deepen the railway reform. The relationship between social public welfare and business operations, and unified transportation command and independent operation of enterprises shall be managed in an appropriate way with the correct market orientation. In the process of going to the market, corresponding policies should be adopted to promote reform due to the different characteristics of railway transportation, industry, infrastructure, material supply, marketing and diversified enterprises.

Railway reform shall be deepened in accordance with the following principles: unified planning, classified guidance, integrated implementation, orderly promotion

[9] (2012, 18 March). Circular of the State Council on transferring and issuing the Opinions of the National Development and Reform Commission on Strengthening the Reform of Economic Structure in 2012 (GBF [2007] No. 47).

and so on. According to the document, "the social public welfare and business operations" as well as "the unified transportation command and independent operation of enterprises" are the central issues to be solved, and a market-oriented strategy shall be further advanced in railway reform. The marketization of railways is an unstoppable trend, but how to realize it still needs to be considered. In the context of the development of a socialist market economy, it can only be pointed out in general terms that railways still have "individual characteristics" under the common requirements of national reform, this can be proved by the facts that unlike other industries, railways cover "social public welfare" under unified transportation command, and classified guidance shall be practiced in different sectors with their own characteristics in the railway system.

Can "public welfare" not be covered by market-oriented operation? Does "unified transportation command" mean that there can only be one operation entity in the national railway? These questions, which are discussed widely today, could not be answered in 1994. Both MOR and the central government held the similar view that MOR should function as both administrative and business entities.[10] In 1998, China's socialist market economic system began to take shape and there was a new understanding of the relationship between government and enterprises in Chinese railways. The document of the State Council points out that separation of government and business functions shall be implemented in the railway sector. According to the characteristics of the industry and the current situation, governance, social welfare management, and business functions shall be distinguished and separated gradually through reform.[11]

Since then, the once affiliated equipment manufacturing enterprises, colleges and universities, primary and secondary schools had been spun off from MOR and the "social welfare management function" had been weakened, but how to define the "governance function" and "business management function" has remained a problem despite the dissolution of MOR in 2013.

When it comes to functions, on the one hand, China Railway Corporation is required to focus on railway passenger and freight transportation services, and adopt the strategy of diversified business operations. On the other hand, "it is expected to be responsible for the unified dispatching and command of railway transportation, the operation and management of national railway passenger and freight transportation, the public welfare transport related to national economy and people's livelihood, and provision of services such as special transport, emergency rescue and disaster relief transport".[12] In a nutshell, China Railway Corporation shall create a well-coordinated framework that integrates the enterprise's operational activities with its social responsibility initiatives. In addition, as mentioned

[10] General Office of the State Council. *Allocation of functions, internal institutions and staffing plan of MOR* (GFB [1994] No. 16).

[11] General Office of the State Council. *Allocation of functions, internal institutions and staffing plan of MOR* (GFB [1998] No. 85).

[12] *Reply of the State Council on relevant issues concerning the establishment of China Railway Corporation* (GH [2014] No. 47).

above, railways, the main player in the implementation of the strategy to build China's strength in transportation, should be well-developed to support other strategies for the prosperity of the country.

However, whether China Railway Corporation implemented the national strategy or undertook the public welfare tasks, it obviously followed the market-oriented principle instead of the one of planned economy. Actually, during the technological innovation of high-speed railway, the large-scale R & D plans and construction projects can be implemented through the market-oriented collaboration among the enterprises, universities and research institutes, under the guidance of a strong leadership (of MOR in the past and China Railway Corporation in the future).

The pace of railway market-oriented reform had been accelerated again since 2017. In March of that year, China Railway Corporation preliminarily determined a three-step reform plan. The first step was to reform the corporate system of China Railway Investment Co., LTD. and other non-transport enterprises. The second was to reform the corporate system of 18 railway administrations across the country, and the third was to reform the corporate system of China Railway Corporation itself. In August 2017, a timetable for the basic completion of China Railway Corporation's reform within the year was put forward. On November 19, 2017, 18 railway administrations of China Railway Corporation changed their enterprise identity and transformed into Group Co., Ltd., marking an important achievement in the reform of the railway corporate system and a solid step towards the transformation of Chinese railways from a traditional transportation production-oriented enterprise to a modern transportation operation-oriented enterprise. At the same time, it was officially announced that the organizational reform of China Railway Corporation was successfully completed, and its personnel quota of government offices, internal institutions and staffing were reduced by 10.3%, 26.6% and 8.1% respectively.[13]

At the Central Economic Working Conference at the end of 2018, the central government clearly proposed to accelerate the shareholding reform of China Railway Corporation. On December 5 of the same year, China Railway Corporation was approved to change its name into China State Railway Group Co., Ltd. (China Railway), according to the State Administration for Market Regulation. After the restructuring, China Railway as a solely state-owned enterprise was officially established, which can be described as an important measure in deepening the reform of the railway system. On June 18th, 2019, with the approval of the State Council, China Railway was unveiled in Beijing.

[13] *In-depth interpretation of railway reform*, from https://www.sohu.com/a/206157092_763498

8.3 Who Represents Chinese Railways

We instinctively referred to Ministry of Railways (MOR) when we talked about Chinese railways, because MOR was responsible for all matters related to the railway, including railway construction, locomotive and vehicle manufacturing, signal equipment, dispatching command, operation and maintenance before 2000. But after 2000, especially after the dissolution of MOR in 2013, who represents Chinese railways has been in question.

It is generally known that China Railway Corporation and the National Railway Administration could represent Chinese railways because they performed the functions and accepted assets of MOR. In theory, the National Railway Administration, as the government department shall plot out the development of Chinese railways. Or at least, it should function as the regulatory authority of China Railway Corporation. However, in practice, all sectors generally take China Railway Corporation as the original MOR for the following reasons: China Railway Corporation inherited the office building of MOR, and the first general manager was the former Minister of MOR; China Railway Corporation is directly led by the State Council, while the National Railway Administration at the vice-ministerial level is affiliated with the Ministry of Transport. In possession of a great magnitude of assets, China Railway Corporation probably also consciously or unconsciously has the same self-positioning. Its documents and reports always give people an illusion that MOR still exists. However, from the perspective of its level and scale, China Railway Corporation is only an enterprise. Over the past few years, the central government has constantly urged it to accelerate the pace of market-oriented reform in order to become a real enterprise. Further, in terms of business scope, China Railway Corporation is only a transportation enterprise, which is far from covering all aspects of the railway system.

China Railway Corporation is indeed a large-scale transportation enterprise. In 2018, the gross revenue of railway transportation reached a record high 772 billion yuan. The General Manager of the Corporation commented that "the business results are better than expected".[14] However, this figure does not have significant advantages compared with that of the two railway construction enterprises spun off from the former MOR in the early twenty-first century, among which China Railway Group Limited (CREC) obtained a gross revenue of 740.4 billion yuan in 2018,[15] while China Railway Construction Corporation Limited (CRCC) earned a gross revenue of 730.123 billion yuan.[16] CRRC Corporation Limited, the rolling stock

---

[14](2019, January 4). *Report delivered by Lu Dongfu at the Work Conference of China Railway Corporation. People's Railway Daily.*

[15]*2018 Annual Report of China Railway Group Limited.* China Railway Group Limited and China Railway Construction Corporation Limited, formerly known as enterprises were subordinate to the former MOR for survey, design and construction, were separated from MOR in 2000. Of course, the two enterprises earns a considerable operating income beyond railway business.

[16]*2018 Annual Report of China Railway Engineering Group Limited.*

manufacturer, also produced a gross revenue of 219.083 billion yuan in 2018 and grew into the world's largest high-speed train manufacturer, with a turnover of more than 200 billion yuan for four consecutive years. Its products were exported to 102 countries and regions across six continents, and its annual export exceeded 8 billion US dollars, accounting for 83% of the global market share, spearheading the global rail transit industry.[17] This dominance compelled traditional train manufacturing giants Alstom of France and Siemens of Germany to attempt to merge their train manufacturing departments.[18] Even the comparatively smaller China Railway Signal & Communication Group Corporation Limited (CRSC), restructured from the former China Railway Signal & Communication Corporation, made a gross revenue of nearly 40 billion yuan in 2018.[19]

In fact, after the separation of sideline businesses from core businesses in railway transportation in 2000, MOR can directly manage passenger and freight transportation as its core business. Therefore, strictly speaking, MOR at that time could not represent Chinese railways.

Nevertheless, we can still use Chinese railways to refer to the industry, just as Japan's, Germany's, and France's railways, although these countries never had a government department like MOR. What really matters to the public is the railway's image. After 40 years of reform, opening up and construction, the public's perception of Chinese railways has changed completely. 40 years ago, transportation modes were limited, people had to choose the railway as their main option for travel. Although there are more choices of travel modes 40 years later, most people prefer to travel by high-speed railway. 40 years ago, railway stations were always associated with words like *crowded* and *chaotic*. 40 years later, a new railway station means the rise of a new business district and even a new city, which also brings new pressure and impetus to the railway. The railway is still plagued by a series of new problems: huge debt, low operating income of high-speed railways in central and western China, various risks in the process of going global, etc. More than 100 years ago, we struggled to learn how to build railways and take back the ownership of railways. 70 years ago, there was not even a railway in the vast western region. 40 years ago, we tried to resolve how to gradually master the technology of the high-speed railway with a speed of 250 km/h. Thinking of these previous difficulties, we are full of confidence in solving the current problems and move ahead.

---

[17] *2018 Annual Report of CRRC*. The predecessor of CRRC consists of many locomotive manufacturing and maintenance enterprises subordinate to the former MOR. In 2000, the former CRCC separated from MOR and established CNR and CSR. In 2015, CNR and CSR merged into CRRC.

[18] EU's rejection of the merger of Siemens and Alstom: Excessive Worry About the Competition from Chinese Enterprises, from http://www.cankaoxiaoxi.com/finance/20190208/2371154.shtml

[19] *2018 Annual Report of China Railway Signal & Communication Corporation Limited.*

Bibliography

Archives and History Center of China Railway Corporation
Archives of Southwest Jiaotong University
Archives of Chengdu Railway Bureau
People's Daily 1976–2018
People Rail 1978–2018
Railway Affairs 1991–2018
Railway Transport and Economy 1979–2018
Railway Economics Research 1995–2018

Archives and History Center of the Ministry of Railways. (1999). *China railway yearbook (1999)*. Beijing: China Railway Publishing House

Beijing-Shanghai High-Speed Railway Co., Ltd. (2011). *Introduction to Beijing-Shanghai high-speed railway*. Beijing: China Railway Publishing House

Cai Qinghua. (2000). *Technological innovation project of Chinese railway*. Beijing: China Railway Publishing House

Chen Puru. (1985). Speech at the working conference of national traffic safety. *Railway Transport and Economy,* (1), 1–9

China Academy of Railway Sciences. (1985). *Research on railway development strategy* (internal materials)

China Railway Society, & China Academy of Railway Sciences. (2008). *The great speeding-up of Chinese railway.* Beijing: China Railway Publishing House

China Railway Corporation. (2015). *Rapid development of Chinese high-speed railway.* Beijing: China Railway Publishing House.

Department of Science and Technology, Ministry of Railways of the People's Republic of China. (1992). Plan in advance: Outline of the "Ten Year Programme of Railway Scientific and Technical Development" as well as the "8th Five-Year Plan" of national economy development. *Chinese Railways*, (1), 1–6

Editorial Board of Fifty Years of Development of China Railway and Locomotive. (1999). *Fifty years of development of China railway and locomotive (1949–1999).* Beijing: China Railway Publishing House

Editorial Board of China's High-Speed Railway Facing the 21st Century. (1999). *China's high-speed railway facing the 21st century*. Beijing: China Railway Publishing House

Editorial Board of History of Chinese Railway Construction. (1999). *History of Chinese railway construction.* Beijing: China Railway Publishing House

X. Zhang, *China's Railway Expansion in 40 Years of Economic Reform and Opening Up*, Understanding China, https://doi.org/10.1007/978-981-95-0333-9

Editorial Board of History of Chinese Railway Construction. (2003). *History of Chinese railway construction.* Beijing: China Railway Publishing House

Editorial Board of History of Chinese Railway Construction. (2009). *History of Chinese railway construction.* Beijing: China Railway Publishing House

Electric Railway Committee of China Railway Society. (2014). *Chinese electric railway construction (1985–2012).* Beijing: China Railway Publishing House

Fang Ju. (2006). *History of Chinese railway (1881–2000).* Beijing: Beijing Jiaotong University Press

Fang Yuanming. (2010). *Strategic research on "going global" of Chinese railway industry*. Hefei: Anhui University Press

Frank Dobbin. (2008). *Forging industrial policy: The United States, Britain, and France in the railway age* (Zhang Wangcheng, & Zhang Haidong, Trans.). Shanghai: Shanghai People's Publishing House

Fu Zhihuan. (1999). *A review of five glorious decades of the Republic: Railways volume*. Beijing: Economic Press China

Fu Zhihuan. (2004). *Exploration and practice of railway reform and development in China.* Beijing: China Railway Publishing House

Fu Zhihuan. (2002). The development of Chinese railways across the century. *Chinese Railways*, (11), 11–13+10

Fu Zhong. (1982). Discussion on the speed of railway construction in China. *Journal of Northern Jiaotong University*, (2), 69–78+182

Gao Lipeng (2008). The four-year implementation of "Mid-and long-term railway network plan". *Railway Freight Transport*, (2), 26

Gao Tao, & Huang Chenguang. (1996). *History of Chinese railway (1876–1995).* Beijing: China Railway Publishing House

Gao Yan. (2017). The integrated development of railway and modern comprehensive transportation system. *Theory Studying and Exploration*, (3), 41–43

Gaotie Jianwen. (2017). *The speed of a major power: The rise of Chinese high-speed railway.* Changsha: Hunan Science and Technology Press

Guolin, Fu Zongliang, & Peng Jiaxian. (1991). Priority shall be given to the construction of Beijing-Tianjin section of Beijing-Shanghai high-speed railway. *Chinese Railways*, (11), 21–25

Han Jiangping. (2007). *China railway yearbook (2007).* Archives and History Center of Ministry of Railways

Han Jiangping. (2008). *China railway yearbook (2008).* Archives and History Center of Ministry of Railways

Han Jiangping. (2016). *China railway yearbook (2015).* China Railway Corporation

Hao Lei. (1988). *Introduction to Chinese railway construction.* Beijing: China Railway Publishing House

Hu Guangrong. (1979). Discussion on the investment proportion of railway construction in China. *Railway Transport and Economy*, (1), 2–11

Hua Maokun. (2002). *The way to speed up Chinese railway.* Beijing: China Railway Publishing House

Huang Ming. (2005). The direction of investment and financing reform for Chinese railways. *China Transportation Review*, (10), 10–12

Huang Sichuan. (1996). *The important manuscripts on the reform and development of Chinese railways in 1995.* Beijing: China Railway Publishing House

Huang Sichuan. (1997). *The important manuscripts on the reform and development of Chinese railways in 1996.* Beijing: China Railway Publishing House

Huang Sichuan. (1998). *The important manuscripts on the reform and development of Chinese railways in 1997.* Beijing: China Railway Publishing House

Huang Sichuan. (1999). *The important manuscripts on the reform and development of Chinese railways in 1998.* Beijing: China Railway Publishing House

Jia Guangzhi. (2015). *Research on related issues of Chinese railways' "going global" strategy*. Beijing: Scientific and Technological Information Research Institute of China Academy of Railway Sciences

Jia Xinmin. (2005). *Railway dictionary* (3rd Ed.). Beijing: China Railway Publishing House

Jin Zhiguo et al. (2006). *Century railway to heaven: The Qinghai-Tibet railway tour*. Beijing: Sino-Culture Press

Ju Jiaxing. (2002). Speed increase, a grand achievement at the turn of the century: A review of the course of railway speed increase. *Chinese Railways*, (11), 14–23

Kou Linlin. (2013). Exploration on the China's railway financing model in the new era. *Construction Economy*, (5), 98–100

Lei Fengxing. (2013). *Speeding ahead: China's rapid transit railways*. Beijing: China Intercontinental Press

Leng Meng. (2016). *National name card: The story behind high-speed railway*. Beijing: The Eastern Publishing Co., Ltd

Li Chunbo. (1983). Doing a good job in checking and compiling and tapping the potential of transportation. *Railway Transport and Economy*, (1), 12–15

Li Jixiang et al. (1990). *Contemporary China: Railway undertakings* (Volume I and II), Beijing: China Social Science Press

Li Renhan. (1999). Comparison between maglev train and wheel-rail train. Strategic Study of CAE, (1), 71

Li Xuesong. (1999). *Game theory and economic transformation: On China's railway reform*. Beijing: Social Sciences Academic Press

Li Yining, Lin Yifu, & Zheng Yongnian et al. (2015). *Understanding the Belt and Road*. Beijing: Citic Press

Liu Jianzhang. (2001). *I'm ninety years (1910–2000)*. Beijing: China Railway Publishing House

Liu Jitang. (1979). Past, present and future of railway: Discussion on several problems in development. *Harbin Railway Science and Technology*, (6), 1–17

Liu Ruilin. (1980). Several problems in the development of Chinese railway transportation. *Railway Transport and Economy*, (1), 1–8

Louis Thomson, et al. (2000). Chinese railway reform: Great challenges for the railway industry. *Railway Economics Research*, (2), 3–10

Lu Chunfang et al. (2013). *China's high-speed railway*. Beijing: China Railway Publishing House

Lu Feng. (2019). Break through the fog: Reveal the origin of Chinese high-speed railway technology advances. *Journal of Management World*, (9), 164–194+200

Luo Qingzhong, Jia Guangzhi, & Chang Jing. (2013). *Foreign railway reform*. Beijing: China Railway Publishing House

Luo Weiyi. (2010). *Tazara railway operation and management*. Beijing: China Railway Publishing House

Lv Yinghua. (1994). *China's railway advancing in the reform*. Beijing: China Railway Publishing House

Lv Zhengcao. (2004a). *Lv Zhengcao memoir*, Beijing: People's Liberation Army Publishing House

Lv Zhengcao. (2004b). *The autobiography of General Lv Zhengcao*, Shenyang: Liaoning People's Publishing House

Miao Qiulin. (1994). *China's railway transport*. Beijing: China Railway Publishing House

Ministry of Railways of the People's Republic of China. (2000a). *Major technology policies of the railway*. Beijing: China Railway Publishing House

Ministry of Railways of the People's Republic of China. (2000b). *Marching into the northwestern China*. Beijing: China Railway Publishing House

Ministry of Railways of the People's Republic of China. (2003a). *Building great corridors*. Beijing: China Railway Publishing House

Ministry of Railways of the People's Republic of China. (2003b). *Aiming at the southwestern China*. Beijing: China Railway Publishing House

Ministry of Railways of the People's Republic of China. (2003c). *The Chinese railway construction across the century (1998–2002)*. Beijing: China Railway Publishing House

Ministry of Railways of the People's Republic of China. (2003d). *The great decision of the century*. Beijing: China Railway Publishing House

Ministry of Railways of the People's Republic of China. (2013). *Major railway technology and policies of the People's Public of China*. Beijing: China Railway Publishing House

Ou Guoli. (2001). *Transport market change and China's railway market reform*. Beijing: China Railway Publishing House

Peng Kaizhou. (2000a). *The important manuscripts on the reform and development of Chinese railways in 1999*. Beijing: China Railway Publishing House

Peng Kaizhou. (2000b). *China railway yearbook (2000)*. Archives and History Center of the Ministry of Railways

Peng Kaizhou. (2001a). *The important manuscripts on the reform and development of Chinese railways in 2000*. Beijing: China Railway Publishing House

Peng Kaizhou. (2001b). *China railway yearbook (2001)*. Archives and History Center of the Ministry of Railways

Peng Kaizhou. (2002). *The important manuscripts on the reform and development of Chinese railways in 2001*. Beijing: China Railway Publishing House

Peng Kaizhou. (2011). Transform the development mode of railways, further promote diversified operation and continuously improve the quality and efficiency of railway development. *Chinese Railways*, (5), 1–8+30

Planning Statistics Departments of China Railway, & Statistics Committee of China Railway Society. (2014). *Statistics chronicle of Chinese railway (1876–2010)*. Statistics Committee of China Railway Society

Qian Ming. (2011). *China railway yearbook (2010)*. Archives and History Center of the Ministry of Railways

Qian Ming. (2012). *China railway yearbook (2011)*. Archives and History Center of the Ministry of Railways

Qian Ming. (2013). *China railway yearbook (2012)*. Archives and History Center of the Ministry of Railways

Qian Ming. (2014). *China railway yearbook (2013)*. Archives and History Center of the Ministry of Railways

Qian Ming. (2015). *China railway yearbook (2014)*. China Railway Corporation

Qu Hong. (1997). Multiple sources of funds for railway development: Chinese railway financing strategies. *Chinese Railways*, (9), 10–12

Ren Xin, & Zhou Yu. (2016). Chinese railway investment and financing: Status analysis, problem analysis, and reform policy orientation. *Macroeconomics*, (5), 16–22

Research Group of Five-Connective Index on the Belt and Road Initiative of Peking University. (2017). *The Belt and Road Initiative: Report on five-connective index*. Beijing: China Economic Times

Safety Committee of China Railway Society, & Editorial Board of Records of Chinese Railway Security. (2012). *Records of Chinese Railway Security (1876–2011)*. Shanghai: Shanghai Jiaotong University Press.

Sheng Guangzu. (2012). Creating a new situation of railway scientific development & better serving the economic and social development and the people. *Railway Economics Research*, (1), 1–9

Statistics Center of Ministry of Railways. (2008). *Collection of the historical statistics data of Chinese railways (1949–2006)* (internal resources). Statistics Center of Ministry of Railways

Statistics Center of the Ministry of Railways. (2010). *Concise Statistics data of the railways in 2008*. Beijing: China Railway Publishing House

Sun Lianjie. (1995). *Deng Xiaoping and China's railway*. Beijing: Central Party School Press

Sun Yongfu. (1999). Promoting institutional reform: Transforming government functions and accelerating the separation of government and enterprises in railway. *Railway Economics Research*, (1)

Sun Yongfu. (2004). *Collection of railway construction management*. Beijing: China Railway Publishing House

Sun Yongfu. (2007a). *Research on permafrost engineering of the Qinghai-Tibet railway construction*. Beijing: China Railway Publishing House

Sun Yongfu. (2007b). *Research on environmental protection of the Qinghai-Tibet railway construction*. Beijing: China Railway Publishing House

Sun Yongfu. (2007c). *Research on health security of the Qinghai-Tibet railway construction*. Beijing: China Railway Publishing House

Sun Yongfu. (2008). *Research on the major role of Beijing-Kowloon railway in economic and social development*. Beijing: China Economic Press

Sun Yongfu et al. (2017). Research on the development strategy of Chinese railway "going global". Strategic Study of CAE, (5), 1–8

"The Belt and Road" Information Center of the State Information Center. (2017). *Report on "the Belt and Road" information*. Beijing: The Commercial Press

The Editorial Board of Records of Shanghai Railway. (1999). *Records of Shanghai railway*. Shanghai: Shanghai Academy of Social Sciences Press

The Editorial Board of Chinese Railways at the Turn of the Century. (2002) *Chinese railways at the turn of the century*. Beijing: China Railway Publishing House

The Editorial Center of Chinese Railway History. (1996). *Memorabilia of Chinese railway (1876–1995)*. Beijing: China Railway Publishing House

The Editorial Group of Wan Li's Opinions on the Railway Reform and Construction. (1997). *Wan Li's opinions on the railway reform and construction*. Beijing: China Democracy Legal System Publishing House

The Office of the Editorial Board of The History of the Chinese Railways of Archives and History Center of the Ministry of Railways. (2007). *An overview of the Chinese railways (1876–2005)* (internal resources). The Office of the Editorial Board of *The History of the Chinese Railways* of Archives and History Center of the Ministry of Railways

The Political and Publicity Department of the Ministry of Railways. (2004). *Chinese railways and the leapfrog development*. Beijing: China Railway Publishing House

The Secretariat of China Railway Society. (2002). *Reformation & scientific technical innovation of China railway after joining WTO*. Beijing: China Railway Publishing House

Tian Yongxiu. (2016). Chengdu-Chongqing railway opened to traffic and public recognition. *Journal of Southwest Jiaotong University (SOCIAL SCIENCE)*, (6), 7–13

Tomoki Kamo, Iida Masafumi, & Ken Jinbo. (2011). *The shift toward China's reform and opening-up: Beyond 1978*. Tokyo: Keio University Press.

Wang Huiying. (2015). *Research on China's railway development and reform*. Beijing: Tsinghua University Press

Wang Linggui. (2015). *On One Belt One Road by worldwide major think tanks*. Beijing: Social Sciences Academic Press

Wang Puqu, & Liu Shangxi. (2015). *The Belt and Road knowledge quiz*. Beijing: People's Publishing House

Wang Xiong. (2016). *China speed: Development of China's high-speed rail*. Beijing: Foreign Languages Press

Wang Xinyuan. (2002). *Chronicle of the international multimodal transportation of Chinese railways (1950–1999)*. Beijing: China Railway Publishing

Wang Yiwei. (2016). *The world is connected: The logic of the Belt and Road Initiative*. Beijing: The Commercial Press

Wang Zhanquan, Liu Tao, & Guo Peihong. (2018). Research on planning thought and concept innovation of comprehensive transportation system based on building a powerful transportation country. *Journal of Transport Management Institute Ministry of Transport*, (4), 31–34

Wang Zhiguo. (2004). *China railway yearbook (2004)*. Archives and History Center of the Ministry of Railways

Wang Zhiguo. (2005). *China railway yearbook (2005)*. Archives and History Center of the Ministry of Railways

Wang Zhiguo. (2005). *The important manuscripts on the leaping development of China's railways in 2004*. Beijing: China Railway Publishing House

Wang Zhiguo. (2006). *The important manuscripts on the leaping development of China's railways in 2005*. Beijing: China Railway Publishing House

Wang Zhizhong, & Wei Liying. (2001). *Research on China's railway reform and development (1978–1998)*. Beijing: Contemporary China Publishing House

Wang Zhuting, & You Juxiang. (1984). On the construction of Chinese railway network. *Development of Railway Science and Technology*, (7), 7–12

Wei Wenying. (1987). *Railway ideological and political work*. Shenyang: Liaoning Education Press

Wu Changyuan. (1994). *The important manuscripts on the reform and development of Chinese railways in 1993*. Beijing: China Railway Publishing House

Wu Changyuan. (1995). *The important manuscripts on the reform and development of Chinese railways in 1994*. Beijing: China Railway Publishing House

Wu Jianhong et al. (2001). *WTO and Chinese railways*. Beijing: China Railway Publishing House

Wu Limin. (2006). *China railway yearbook (2006)*. Archives and History Center of the Ministry of Railways

Wu Limin. (2009). *China railway yearbook (2009)*. Archives and History Center of the Ministry of Railways

Xi Jinping. (2017). *Working together to advance the Belt and Road initiative: Speech on the Opening Ceremony of Belt and Road Forum for international cooperation*. Beijing: People's Publishing House

Xie Gaochao. (2002). *China railway yearbook* (2002). Archives and History Center of the Ministry of Railways

Xie Gaochao. (2003). *The important manuscripts on the reform and development of Chinese railways* in 2002. Beijing: China Railway Publishing House

Xie Gaochao. (2003). *China railway yearbook* (2003). Archives and History Center of the Ministry of Railways

Xie Gaochao. (2004). *The important manuscripts on the reform and development of Chinese railways* in 2003. Beijing: China Railway Publishing House

Xu Fei. (2017). *The Belt and Road and the global strategy of China's high-speed rail*. Shanghai: Gezhi Publishing House, Shanghai People's Publishing House

Xu Limin. (2012). Research on the development of separating passenger and freight transportation in China. *Railway Transport and Economy*, (11), 20–27

Xu Zenglin. (1999). *Fifty years of the railroads of the People's Republic of China (1949–1999)*. Beijing: China Railway Publishing House

Yan Pengfei. (2004). Reform on China's railroads investment system: A late choice for China's railroads. *Fortune World*, (10), 14–19

Yearbook of China Integrated Transport Press. (1986). *Yearbook of China transportation & communications (1986–2015)*. Beijing: Yearbook of China Integrated Transport Press

Yu Chuan. (1999). *Finance and accounting of Chinese railways: A review and prospect of the past five decades*. Beijing: China Railway Publishing House

Yu Qianwen. (1999). *Railway plan management in contemporary China*. Beijing: China Railway Publishing House

Zhang Xueyong. (2018). External influencing and internal reasoning: The course of China's railway reform and opening up from the perspective of changes over 100 years. *Journal of Southwest Jiaotong University (SOCIAL SCIENCE)*, (6), 1–8

Zhang Youmin, & Tang Jie. (1992). *Encyclopedia of China's reforms: The reform of transportation*. Dalian: Dalian Publishing House

Zhang Zhizhong. (2007). *A history of locomotives in China* (Volume I & II). Jinan: Shandong Education Press
Zhao Guoqing. (1994a). *Challenges and solutions*. Beijing: China Railway Publishing House
Zhao Guoqing. (1994b). *Reference and evaluation*. Beijing: China Railway Publishing House
Zhao Haipei, & Wu Jianhong. (1995). *Theory and practice of China's railway going to the market*. Beijing: China Railway Publishing House
Zhao Jian. (2016). *The reform and reorganization of China's railway and research on high-speed railway*. Beijing: Economic Press China
Zhao Lei. (2015). *One Belt One Road: China's civilizational rise*. Beijing: CITIC Press Group
Zhou Aiming. (2016). *The Qinghai-Tibet railway in China*. Beijing: Sino-culture Press
Zhou Xiaodong. (2015). Ankara-Istanbul high-speed train project in Turkey: The first Chinese high-speed railway project going global. *International Project Contracting & Labor Service*, (9), 50–52
Zhou Xinjun. (2012). Chinese railway reform: Particularity, dilemma, and basic ideas. *Economic Affairs*, (4), 70–75
Zhuang Zheng. (1999). *The construction of Chinese railways*. Beijing: China Railway Publishing House

MIX
Papier aus verantwortungsvollen Quellen
Paper from responsible sources
FSC® C105338

If you have any concerns about our products,
you can contact us on
ProductSafety@springernature.com

In case Publisher is established outside the EU,
the EU authorized representative is:
Springer Nature Customer Service Center GmbH
Europaplatz 3, 69115 Heidelberg, Germany

Printed by Libri Plureos GmbH
in Hamburg, Germany